TOPICS IN SOLID STATE
AND QUANTUM ELECTRONICS

UNIVERSITY OF CALIFORNIA
ENGINEERING AND PHYSICAL
SCIENCES EXTENSION SERIES

Howard Seifert, Editor • Space Technology

Robert L. Pecsok, Editor • Principles and Practice of Gas Chromatography

Howard Seifert and Kenneth Brown, Editors • Ballistic Missile and Space
    Vehicle Systems

George R. Pitman, Jr., Editor • Inertial Guidance

Kenneth Brown and Lawrence D. Ely, Editors • Space Logistics Engineering

Robert W. Vance and W. M. Duke, Editors • Applied Cryogenic Engineering

Donald P. LeGalley, Editor • Space Science

Robert W. Vance, Editor • Cryogenic Technology

Donald P. LeGalley and Alan Rosen, Editors • Space Physics

Edwin F. Beckenbach, Editor • Applied Combinatorial Mathematics

Alan S. Goldman and T. B. Slattery • Maintainability: A Major Element
    of System Effectiveness

C. T. Leondes and Robert W. Vance, Editors • Lunar Missions and
    Exploration

J. E. Hove and W. C. Riley, Editors • Modern Ceramics: Some Principles
    and Concepts

J. E. Hove and W. C. Riley, Editors • Ceramics for Advanced Technologies

Joseph A. Pask • An Atomistic Approach to the Nature and Properties of
    Materials

John F. Brahtz, Editor • Ocean Engineering

John F. Brahtz • Coastal Zone Management

Edwin F. Beckenbach and Charles B. Tompkins, Editors • Concepts of
    Communications: Interpersonal, Intrapersonal, and Mathematical

W. D. Hershberger, Editor • Topics in Solid State and Quantum Mechanics

TOPICS IN SOLID STATE
AND QUANTUM ELECTRONICS

Papers from a statewide lecture series, University of California,
March 1970

The Authors

BENJAMIN LAX

HERBERT KROEMER

W. J. EVANS

RICHARD M. WHITE

J. E. MERCEREAU

W. V. SMITH

AMNON YARIV

MONTE ROSS

B. KAZAN

A. S. GROVE

MARTIN CAULTON

PETER J. WOJTOWICZ

The Editor

W. D. HERSHBERGER

University of California
Los Angeles

JOHN WILEY AND SONS, INC.,
NEW YORK · LONDON · SYDNEY · TORONTO

Library of Congress Cataloging in Publication Data:
Main entry under title:

Topics in solid state and quantum electronics.

(University of California engineering and physical sciences extension series)
"Papers from a statewide lecture series, University of California, March 1970.
1. Semiconductors—Addresses, essays, lectures. 2. Quantum electronics—Addresses, essays, lectures. I. Lax, Benjamin. II. Hershberger, William Delmar, 1903–   ed. III. Series: California. University. University of California engineering and physical sciences extension series.

TK7871.85.T66   621.3815   75-169163
ISBN 0-471-37350-8

Printed in the United States of America.

10 9 8 7 6 5 4 3 2 1

THE AUTHORS⎯⎯⎯⎯⎯⎯⎯⎯⎯⎯⎯⎯⎯⎯⎯

MARTIN CAULTON, David Sarnoff Research Center, RCA Corporation, Princeton, New Jersey

W. J. EVANS, Bell Telephone Laboratories, Incorporated, Murray Hill, New Jersey

A. S. GROVE, Intel Corporation, Mountain View, California

B. KAZAN, International Business Machines Corporation, Thomas J. Watson Research Center, Yorktown Heights, New York

HERBERT KROEMER, Electrical Engineering Department, University of Colorado, Boulder, Colorado

BENJAMIN LAX, Francis Bitter National Magnet Laboratory, Massachusetts Institute of Technology, Cambridge, Massachusetts

J. E. MERCEREAU, Department of Physics, California Institute of Technology, Pasadena, California

MONTE ROSS, McDonnell Douglas Astronautics, St. Louis, Missouri

W. V. SMITH, IBM Research Laboratory, Switzerland

R. M. WHITE, College of Engineering, Department of Electrical Engineering and Computer Sciences, University of California, Berkeley, California

PETER J. WOJTOWICZ, David Sarnoff Research Center, RCA Corporation, Princeton, New Jersey

AMNON YARIV, Electrical Engineering Department, California Institute of Technology, Pasadena, California

PREFACE

The areas of solid state electronics and quantum electronics are characterized by rapid change and the existence of a large and productive research effort. The disciplines basic to these topics are electromagnetics, quantum mechanics, and statistical mechanics. The development of new devices and systems has necessitated the formulation and solution of new theoretical problems, while theory in turn has yielded information on ultimate performance and limitations and indicated promising directions for experimental work. As a result of this activity and progress, the scientist and engineer in these and neighboring areas is required to keep both his knowledge and his approach up to date. The purpose of the present book is to serve his specific needs.

Topics in solid state and quantum electronics in which progress is most rapid include avalanche diodes, high-frequency ultrasonic oscillators and amplifiers using the piezo-electric effect, oscillators employing the Gunn effect, and solid state lasers. The last two of these, in particular, require for their understanding a knowledge of the quantum mechanics of the solid state. A treatment of superconductivity, as applied to electronics, is included as a matter of interest and necessity; engineering applications are currently being exploited. A treatment of solid state lasers would obviously be incomplete without some comparisons with gaseous lasers. The fruitful work with lasers in harmonic generation and parametric effects are treated in a separate chapter. Laser systems are discussed, as well as the application of integrated circuits, in one frequency range, to computers; and in a second frequency range, to the development of microwave devices and circuits. Since advances in materials are a key to the future, this topic is included, along with the application of materials to image pick up and display devices. There is no attempt at completeness of coverage; rather the choice of topics was determined by (a) current research activity and progress, and (b) the promise of new applications.

The book is the result of efforts by the School of Engineering and Applied Science of the University of California, Los Angeles, to serve engineers, scientists, and recent graduates whose courses of study did not include the material they need in their present work. The topics were chosen and speakers of distinction were selected with the aid of an advisory committee consisting of Paul D. Arthur, Rubin Braunstein, J. C. Dillon, Donald B. Hummel, N. Bruce Kramer, D. B. Langmuir, W. G. Oldham, Bernice Park, and T. F. Tao. Mrs. Park and J. C. Dillon, in particular, of University Extension facilitated operations by making needed arrangements. Chapter 3 was written by Dr. W. J. Evans of the Bell Telephone Laboratories while Dr. D. L. Scharfetter gave the oral presentation in the course.

The speakers prepared lectures for delivery to audiences in four locations in California and in addition prepared their respective chapters for publication. Thanks are due to each speaker-author whose efforts served to make this series a success.

This statewide series of lectures was presented in the Spring of 1970.

W. D. HERSHBERGER

Los Angeles, California
December 1971

CONTENTS

LIST OF TABLES CONTAINING USEFUL NUMERICAL INFORMATION

TOPICS IN SOLID STATE
AND QUANTUM ELECTRONICS

Chapter 1

SOLID STATE PHYSICS—AN OVERVIEW

Benjamin Lax*

Francis Bitter National Magnet Laboratory†
Massachusetts Institute of Technology
Cambridge, Massachusetts

1.1 INTRODUCTION

The study of solids consists of the theoretical and experimental investigation of the mechanical, electrical, magnetic, optical, and thermal properties. Often one thinks of solid state physics as the science of different classes of materials and their physical properties. Classification may be made in terms of the study of metals, semiconductors, dielectrics, superconductors, and magnetic materials. The classification can also be made in terms of the broad class of phenomena that are observed in solids. For example, it is of primary importance to study the crystal structure, its elastic and mechanical properties, lattice vibrations or phonons, thermal conductivity, band structure, electrical conductivity, superconductivity, dielectric properties, magnetism and magnetic resonance of different varieties, optical and magnetooptical properties, and so on. In fact, we could almost classify solids in terms of any one of the major categories just listed. Perhaps one of the most interesting classifications is in terms of the resistance of materials that span 40 orders of magnitude from molecular crystals of very high resistivity to superconductors of vanishingly small resistivity. In between we have the insulators and dielectrics, semiconductors, semimetals, and the real metals like copper and silver.

*Also Physics Department, Massachusetts Institute of Technology
†Supported by the U.S. Air Force Office of Scientific Research.

Another classification that is often used in describing solids is in terms of the bonding of the atoms to one another in a solid. The three most familiar are the metallic, covalent, and ionic bondings. In the metallic case, the conduction electrons are free to run around throughout the whole crystal and serve as the glue that binds the ions in the crystal. At the other extreme are the covalent crystals in which the electrons are shared by the atoms. These may be tightly bound as in diamond or loosely bound as in grey tin. Similarly, the ionic crystals can be very tightly or loosely bonded, depending on the atoms that participate in the ionic bonding in which the metallic atom, like sodium, gives up its electron to chlorine to form an attractive bond between the positive and negative ion by coulomb interaction. Obviously, the smaller the size of the ions the greater the coulomb attraction and the harder the material to compress or to melt. This is illustrated quite clearly in the scale of hardness for some oxides, sulfides, selenides, and telluride compounds.

1.2 CRYSTAL STRUCTURE

One of the fundamental problems of solid state physics is that of predicting the crystal structure which atoms assume in forming a crystalline solid. There is no theory for this. However, from experimental information obtained mostly from X-rays we can determine the exact nature of the microscopic structure of the unit cell of a crystal that is the building block of solids. Once we know the crystal structure and the atoms we can, in principle, calculate the electronic states, the vibrational spectrum of the lattice, and the magnetic structure. In practice, it is sometimes difficult to predict the qualitative aspects and more difficult to predict the quantitative nature of the solids. Nevertheless, the crystal structure plays an important role in determining the behavior of the solid. The crystal structure is classified into 14 basic lattices called Bravais lattices. Many of the actual crystals are more complicated than the basic lattices shown in the diagrams of Fig. 1. There may be many atoms within the structure whose location may have important effects on such properties as optical nonlinearity, ferroelectric, and piezoelectric properties. The theoretical analysis of electron states, magnetic structure, and lattice phonon spectra starts with the structure of the crystal lattice. The energy diagram associated with each of these fundamental spectra reflects the crystal symmetry faithfully. The electron energy spectra are usually determined by optical and transport measurements, where-

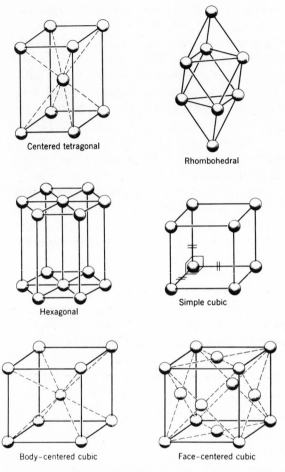

Centered tetragonal

Rhombohedral

Hexagonal

Simple cubic

Body-centered cubic

Face-centered cubic

Fig. 1.   Six of the 14 basic Bravais lattices of crystal structures.   (From Wein-reich.[1])

as the magnetic and lattice spectra are known primarily from neutron scattering.

1.3 LATTICE VIBRATIONS AND THERMAL PROPERTIES

Many important phenomena of basic and applied nature depend on the properties of lattice vibrations. The model, which is a classical one, still serves a useful purpose in explaining the nature of the lattice spectrum. Atoms, in an ordered array, are assumed to interact with a linear restoring force between nearest and next nearest neighbors, etc. When the many-body equations of motion, in accordance with Newton, are written for the crystal, the solution describes the appropriate spectrum. This is usually plotted as frequency versus wave number, which is shown in Fig. 2. This diagram, which has been obtained from neutron scattering, shows many important features of lattice vibrations. The lower curves show both longitudinal and transverse branches of the acoustic mode. These are intimately related to the elastic vibrations of solids which are used for generating different frequencies of sound in transducers. The upper branches are called the optical modes because ions of the lattice vibrate 180° out of phase at frequencies that can only be seen by optical techniques in the infrared region. Today Raman scattering by lasers in which the frequency of the scattered light is reduced by that of the optical mode frequency gives us a great deal of information about lattice phonons or quanta at selected regions of the spectra. Such a phonon shifted peak from Raman scattering is shown in Fig. 3.

Another powerful tool for investigating the detailed nature of the lattice spectrum has been by neutron diffraction. Thermal neutrons from reactors have the appropriate de Broglie wavelength to be diffracted by crystals to create a monochromatic source. This is then inelastically scattered from the crystal under study and a third crystal analyzes the resultant energy. Thus such a three-crystal spectrom-

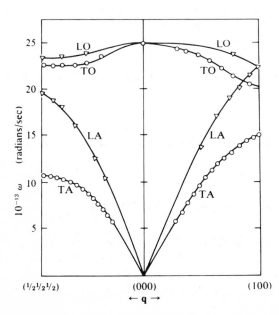

Fig. 2. The phonon spectrum of the diamond structure along two principal crystal directions as obtained from inelastic neutron scattering as a function of phonon wave number q. (From Blakemore.[2])

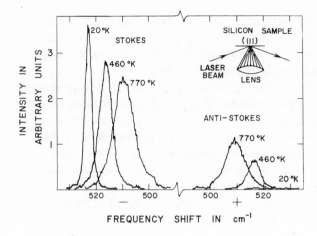

Fig. 3. Raman scattering of optical phonons from the surface of silicon by an argon laser. (After Hart, Aggarwal, and Lax.[3])

eter can produce the detailed spectra shown.

The importance of phonons is that they play a dominant role in the thermal properties of solids. In fact, heat and phonons in this context are synonymous. Heat capacity, the ability to absorb heat, depends on the phonon spectrum. This was one of the earliest applications of lattice theory by Einstein and then Debye. Einstein assumed only the existence of the optical branch and obtained a good account for the high-temperature range of the specific heat. Debye assumed a simple model of the acoustical mode alone and did somewhat better in a variety of crystals by defining a quantity known as the Debye temperature. Today, with modern computers combined with neutron data on phonons, we can account for the specific heat quite well, at both the low-temperature and high-temperature ranges.

The interaction of phonons with electrons and electron spin systems plays a significant role in determining the figure of merit of practical devices. For example, the line width of ferromagnetic resonance in microwave ferrite devices is such a parameter. The study of this phenomenon by physicists has led to the improvement of circulators, isolators, and phase shifters, and also to the development of new devices such as the ferrite power limiter. These now provide essential components for solid state duplexers.

The nonlinear interactions of phonons, which are the higher-order terms in the potential of the atom of the lattice, are responsible for the scattering of phonons by multiphonon processes. The collisions of phonons that have been experimentally demonstrated are the processes which are responsible for the process of heat transfer in solids. In a particular process called the "Umklapp" process a phonon of large momentum is generated such that the phonons transfer momentum to the crystal as a whole. These nonlinear phonon interactions are the ones that are responsible for the expansion of a solid. As these increase in amplitude with temperature to a sufficient degree, the expansion achieves the critical value which results in melting.

Another thermal phenomenon that is of great practical importance is the heat transferred by electrons and holes in semimetals and semi-conductors. The existence of the two types of carriers has led to the development of electronic energy converters and refrigerators. In these devices the competition of heat transfer by phonons and elec-trons is, in fact, not desirable and limits the efficiency of such de-vices. Hence it is desirable to use heavier atomic compounds to minimize the phonon contribution, and to optimize the electronic heat conduction and the efficiency of the device.

1.4 ELECTRONIC STATES

The question of determining the electronic states of solids either by theoretical or experimental techniques and (usually by both) is a central problem of solid state physics. Energy states of electrons are of great relevance to the performance of transistors, diodes, bulk de-vices, optical and infrared detectors, and to all practical applications involving electrons and holes. Energy states of electrons in solids have their origin in the atomic states of the original component atoms. These states in the isolated atom are discrete. However, when the atoms are brought together to form a solid, each of these states broad-ens to form bands analogous to the formulation of pass-band filters from single resonant circuits, and many of them become coupled. The formation of bands makes it difficult to analyze the energy states by the usual techniques of atomic spectroscopy. Hence a whole new field of solid state spectroscopy and related transport measurements has evolved to study in detail the band structure of solids.

In working out the band structure, we find that the energy is a function of momentum in a complicated way, which reflects the sym-metry of the crystal structure in a well-defined manner. There are a number of mathematical techniques which consider the atomic wave functions and the symmetry of the crystal to develop the form of the energy bands. The parameters of these treatments are then deter-mined from various experimental data. Some of these combined theoretical-experimental techniques have given us a very good model of the bands in a variety of solids which are of great interest today. The results of such analysis and experiments are shown in Fig. 4

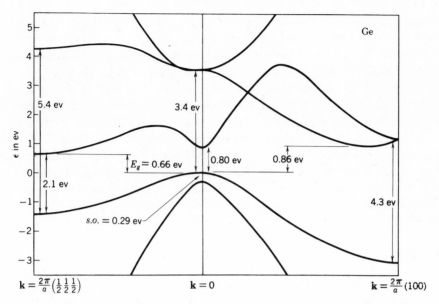

Fig. 4. The band structure of germanium. The band parameters near the energy gap correspond to those at $T = 300\,°K$. Energy bands have been calculated to fit data from optical experiments. (From Kittel.[4])

which gives the band structure of germanium along two crystal directions.

1.5 CYCLOTRON RESONANCE

One of the most basic techniques that has given us a detailed picture of energy bands is cyclotron resonance. This phenomenon was initially performed by utilizing high resolution, highly sensitive microwave techniques. The crystal to be studied was usually immersed in liquid helium inside a microwave cavity. Resonance absorption was observed by varying the magnetic field of modest value obtained from an electromagnet. In this way the effective mass tensors and the anisotropic character of the cyclotron resonance were obtained in n-type and p-type germanium and silicon samples by orienting the magnetic field along different crystalline axes. Later on, to extend the usefulness of this important method of cyclotron resonance, millimeter spectrometers were built, still using conventional methods. Such materials as indium antimonide and diamond provided only partial results. However, modified techniques provided considerable success in highly pure metals. From these results new information about the Fermi surfaces of such materials was obtained.

In order to extend the usefulness of cyclotron resonance to new materials of limited purity, it was necessary to go to much higher frequencies in the far infrared and submillimeter regions. This necessitated the development of high field pulsed and water-cooled magnets. Recently, the combination of the latter, and molecular gas lasers in the submillimeter region have provided the most useful cyclotron resonance spectrometers in existence to date. Many new semiconductors and other good crystals are now being successfully investigated. Figure 5 shows a typical spectrum obtained from such a submillimeter spectrometer at 337 μ in p-type indium antimonide.

1.6 OPTICAL AND MAGNETO-OPTICAL METHODS

Old standard methods for determining the energy gap of a semiconductor were to transmit infrared radiation through a thin sample as a function of the wavelength and then observe the absorption edge at a wavelength corresponding to the energy gap. In recent years, more sophisticated reflection techniques have not only provided quantitative information about the fundamental energy gap but have also exhibited reflection peaks corresponding to strong absorption between deeper valence bands to higher conduction bands than those associated with the fundamental edge. This, combined with theoretical work, has provided one of the most powerful tools for studying the energy scale of the band structure.

Similarly, using optical methods and high-magnetic fields at low temperatures, additional quantitative information about the energy

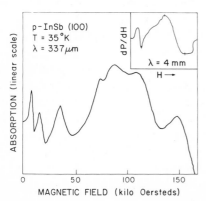

Fig. 5. Submillimeter cyclotron resonance of holes in InSb, showing the fine structure of light holes on the left and heavy holes on the right, due to quantum effects. Inset shows unresolved resonance at 4 mm. (After Button, Lax, and Bradley.[5])

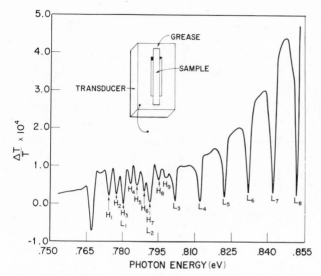

Fig. 6. The magneto-optical spectrum of the indirect interband transition in germanium, as obtained by piezomodulation of the transmission. The symbols *H* and *L* indicate transitions between heavy and light holes to electrons, respectively. (After Aggarwal, Zutek, and Lax.[6])

band extrema was obtained. The quantum effects of this modern magneto-optical technique showed up in the form of absorption or reflection peaks in semiconductors and semimetals, respectively. More-refined modulation techniques permitted similar observations for transitions between deeper valence bands and higher conduction bands. From this type of data were obtained effective masses or curvature of the energy bands, anomalous spin values of the electron states and often, such states were submerged below the Fermi energy or submerged below the normal valence and conduction states. Thus this method complemented the cyclotron resonance and transport techniques which could only be used to study those electrons near the Fermi energy or at the band extrema. Figure 6 shows the magneto-optical spectra in germanium at low temperatures.

1.7 LASERS

Solid state physics and the detailed study of optical properties made an early and important contribution to the invention and development of lasers. The optical properties of ruby, which consist of a chromium ion in sapphire, and similarly, the optical properties of

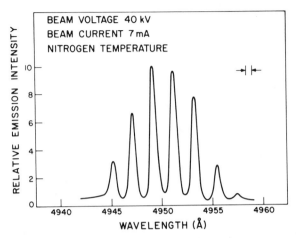

Fig. 7. Spectral output of an electron pumped CdS laser above threshold, showing the structure of the Fabry-Perot modes of the optical cavity of the crystal.
(After Hurwitz.[7])

niobium ion in glass and other host crystals, gave rise to the well-
known, high-powered solid state lasers. These, of course, have been
used to study a whole new class of optical phenomena in solids which
are known as nonlinear optics. From this has evolved a series of
parametric devices that now permit tunable radiation in the visible and
infrared regions. Analogous phenomena, using high-power CO_2 lasers
with semiconductors, also permit tunable radiation in the far infrared
and submillimeter regions. In general, these require the use of high-
magnetic fields and the detailed knowledge obtained from the magneto-
optical studies of energy bands.

    Another class of lasers using semiconductors has been developed.
A number of these utilize semiconductor diodes for electrical excitation, although a number of other semiconductor lasers have been excited by electron beams and optical radiation. Now almost the entire
spectral range from ultraviolet to the far infrared has been spanned by
using these varieties of techniques of excitation of the many binary and
ternary compounds to produce semiconductor lasers. The spectral
output of a semiconductor laser of CdS crystal is shown in Fig. 7.

1.8 PARAMAGNETISM

    A traditional study of paramagnetic ions of the transition and rare
earth materials in insulating host crystals has provided a good deal of

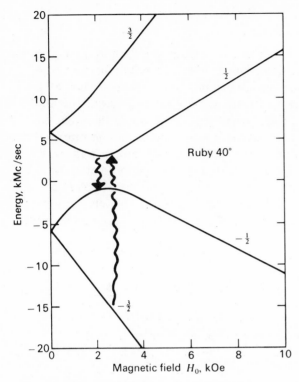

Fig. 8. The energy levels of a paramagnetic maser of ruby. The vertical arrow upward represents the pump frequency and downward the maser output from the inverted upper level.

knowledge about the basic magnetism of these atomic ions. Simple measurements of magnetization in high magnetic fields and low temperatures give information about the magnetic moment. This knowledge, combined with the quantum theory of atoms, gave the clue to the fundamental atomic building blocks of magnetism. More refined information was later obtained when paramagnetic resonance at microwave frequencies was used to study these same atoms in such host crystals. The most important practical result of this type of work was the invention and discovery of the solid state maser, shown in Fig. 8, which illustrates the paramagnetic transitions for a ruby maser. It was this work supplemented by the related optical study of these paramagnetic ions that gave rise to the early development of the ruby laser.

1.9 FERROMAGNETISM

In a paramagnet the concentration of magnetic ions is dilute; therefore, the magnetic interaction between them is smaller because the
distance is large. However, in a ferromagnetic material such as iron,
nickel, or cobalt the distance between the atoms in these materials is
small and, therefore, the interaction is much stronger than that calculated between two simple magnetic dipoles. This strong interaction,
which tends to line up the magnetic dipoles of each atom, was first
discovered by Pierre Weiss and is called the Weiss molecular field.
Today we know from quantum mechanics that this interaction is an exchange interaction, called the Heisenberg exchange, which can either
line up the spin parallel or antiparallel, depending on the distance between magnetic atoms.

The exchange interaction in magnetic metals of the transition
series can qualitatively be explained by the Heisenberg exchange mechanism. However, ferromagnetism in ferrites is more difficult because the magnetic ions are separated by oxygen atoms. Nevertheless, through a more complicated quantum mechanical exchange interaction, the spins can either line up parallel or antiparallel. This
is called superexchange. Therefore, "ferromagnetism" in ferrites and
garnets occurs when the magnetic atoms in one site exceed those in
another crystallographic site, resulting in a net magnetic moment for
the crystal as a whole. Such a crystal macroscopically behaves as a
ferromagnet, although microscopically it is quite different, and is
said to be ferrimagnetic. If the magnetic ions on the two sites are
equivalent and oppositely oriented at low temperatures, the net magnetism of the sample as a whole can cancel to zero. This is called
antiferromagnetism.

Experimentally, we can identify ferromagnetic and antiferromagnetic materials by a number of different experimental techniques. The
oldest among these is the measurement of the magnetization or susceptibility as a function of temperature. This appears quite different
for the three materials. In the case of the ferromagnet, above certain
temperatures (called the Curie temperature) the material becomes
paramagnetic. For the ferrimagnetic materials the magnetization may
become zero, and then reverse itself before it reaches the Néel temperature (the equivalent of the Curie temperature); then it becomes
paramagnetic. This situation is illustrated in Fig. 9. The antiferromagnetic material is one of the most unusual materials of all. Its
magnetic phase diagram, shown in Fig. 10, displays three different
spin configurations when the plot of the magnetic field versus temper-

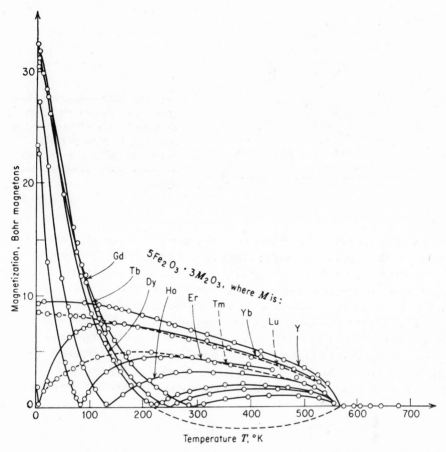

Fig. 9. Magnetization curves of the ferrimagnetic rare earth garnets as a function of temperature. [After Bertaut and Pauthenet.[8])

ature is made. One region is antiferromagnetic. A second is called the "spin-flop" region. In a third region, at sufficiently high magnetic fields, the spins tend to line up with the magnetic field; at low temperatures they are essentially ferromagnetically aligned. At higher temperatures, above the triple point, the spins are paramagnetically aligned.

Each of these magnetic materials exhibits magnetic domains in which small regions of the material line up along different equivalent crystallographic directions, such that the net magnetization, in the absence of an external magnetic field, is zero. This is a well-known phenomenon in certain types of iron magnets used in electrical machinery. When the magnetic field is applied, depending on the geom-

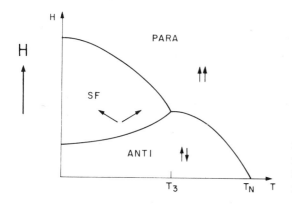

Fig. 10. Magnetic phase diagram of an antiferromagnet, showing the spin flop and paramagnetic regions at higher fields, in addition to the antiferromagnetic region at lower fields.

etry and the material, the domains disappear and at certain values of the magnetic field the magnetic domains are parallel to the magnetic field and the material is fully magnetized. Because of imperfections, impurities and strain in the material, this magnetization process involves losses as the magnetization in the forward and reverse directions are cycled as a function of the magnetic field. The process exhibits a hysteresis loop which is a measure of these irreversible losses. When these hysteresis loops have a "square configuration" with the suitable quantitative properties, as in ferrites and thin films, they are extremely useful and important as memory elements in computers.

1.10 SPIN WAVES

Since the spin dipoles of magnetic atoms in ferromagnets and ferrimagnets are aligned parallel or antiparallel to one another in a magnetic field, each of these spins precess, usually in phase with its neighbor, in the presence of an external field. This mode is called uniform precession. However, by introducing a disturbance, it is possible to alter the phase of precession between neighboring spins in a regular pattern such that the variation of the phase has a sinusoidal pattern. This type of a disturbance is called the spin wave. Indeed, the spin configuration of different types of magnetic materials can be probed experimentally by inelastic neutron diffraction. This technique measures the energy or frequency of the spin waves as a function of the wave number of the spin wave in a complicated magnetic material which has two or more spins per unit cell. The spin wave modes exhibit both acoustical and optical branches in the spin wave spectrum

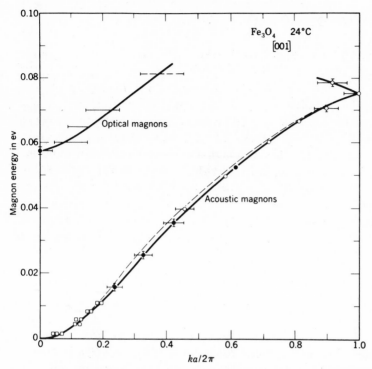

Fig. 11. Dispersion relations for the acoustical and optical branches of spin waves (magnons) in Fe_3O_4, as obtained from neutron scattering by Brockhouse and Watanabe. (IAEA Symposium, Chalk River, Ontario, 1962; from Kittel.[4])

obtained from neutron diffraction, as shown in Fig. 11. The properties of the spin waves are very important in a number of experiments, both in ferrites and in thin films of ferromagnetic materials. The properties of these spin wave phenomena of low wave number have been studied by microwave resonance techniques. Spin waves play an important role in the resonance losses in microwave devices. Some of these devices are the nonreciprocal isolators, circulators, and phase shifters that use ferrites or garnets. Others such as the magneto-elastic interaction of spin waves with the acoustic phonons in ferrites or garnets are now used as delay lines in radar applications. Antiferromagnetic materials and other magnetic oxides are beginning to play an important role in the development of similar circuit devices in the submillimeter and far infrared region, which will be used in conjunction with lasers for future communication systems.

1.11 SUPERCONDUCTIVITY

Until about 10 years ago superconductivity was one of the subjects of pure research that physicists studied as a scientific curiosity. Today, however, superconductivity has become a very practical phenomenon with a number of important applications. Certain materials exhibit zero conductivity below a critical temperature, varying between 1 and about 20 °K. They also show an interesting magnetic behavior. Below a certain critical value the magnetic field is excluded from the superconductor. Above that critical value the magnetic field destroys superconductivity and the material becomes a normal conductor. The critical field is temperature-dependent, and one can draw a phase diagram of critical field versus temperature which obeys a parabolic relation. Another interesting feature is the critical current in a superconducting wire. This, too, has a critical value, corresponding to the critical magnetic field created by the current in the surface of the wire.

Superconductivity was not understood until recently when Bardeen, Cooper, and Schrieffer developed the quantum theory that explains the phenomenon. The phenomenon is based upon the formation of two bound electrons which are coupled to each other by acoustic phonons to form the bound pair. This pair travels through the lattice without being scattered, thus accounting for the superconductivity. In addition, the theory shows that superconductors have small energy gaps similar to that of semiconductors. This energy gap is associated with the formation of the superconducting state of the bound pairs of electrons. The best way to demonstrate the existence of an energy gap is by means of electron tunneling between superconducting oxide metal sandwiches. The thin film of superconductors separated by a thin film of oxide at low temperatures shows anomalous conducting behavior associated with the tunneling of electrons across the oxide field. By differential techniques, one or two sets of peaks in the current voltage relation as a function of temperature, give an accurate measurement of the energy gap. From a series of such measurements the energy gap versus temperature is shown to agree very well with this theory. This is illustrated in Fig. 12.

1.12 APPLICATIONS OF SUPERCONDUCTIVITY

Two important applications of superconductivity depend on what is known as Josephson tunneling and type II superconductivity, respective-

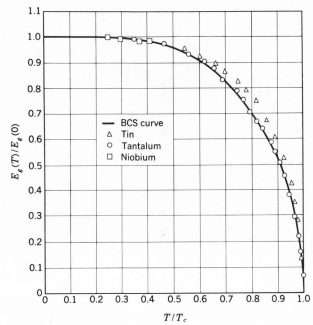

Fig. 12. Energy gap of superconductors versus temperature in terms of reduced parameters. (After Townsend and Sutton.[9])

ly. The Josephson tunneling is tunneling between two superconductors
separated by a very thin film of oxide of the order of 10 Å such that
the electrons can tunnel as bound pairs without breaking up. This
phenomenon was shown to exhibit a resonance characteristic in which
the resonant energy is equal to the applied d-c voltage imposed on the
Josephson junction. This pehnomenon has now given rise to a very
sensitive microwave and far infrared detector. Another property of a
Josephson junction is that it is sensitive to the phase of the superconducting current associated with the pairs of electrons passing into it.
Using this property one can build a superconducting loop containing
two Josephson junctions on either side of the loop between the two
junctions. When a variable magnetic field passes through the loop perpendicular to its cross section, the current oscillations through the
bridge measure the fundamental fluxoid passing through the loop. This
fundamental fluxoid, which is of the order of 10^{-7} G/cm^2 confirms the
existence of bound pairs. Thus, by appropriate techniques, one can now
build a Josephson magnetometer of this type to measure fields of the
order of 10^{-9} G. In fact, we have measured the magnetocardiography
of the human heart by such a magnetometer by means of a magnetically
shielded room.

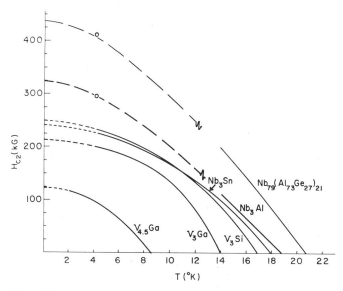

Fig. 13. Upper critical field of superconducting alloys as a function of temperature. (After Foner *et al.*[10])

The phenomenon of type II superconductivity occurs in hard magnetic superconductors, usually involving binary to ternary alloys, e.g., niobium tin, niobium aluminum, and niobium germanium-aluminum. These materials have the property of breaking up into regions or domains of normal and superconducting filaments below the upper critical temperature, when the magnetic field is applied at lower but modest critical value. The ratio of the filaments varies up to an upper critical field at which the superconductivity is completely destroyed. This upper critical field is very high and can range anywhere from 100 kG to several hundred kG in different materials. Recent measurements have shown that niobium aluminum and niobium germanium alloys have critical fields of 300 and 400 kG respectively, at liquid helium temperatures. Their characteristics are illustrated in Fig. 13. Materials such as niobium tin have been fabricated into wires that have been used to build high field superconducting magnets of the order of 100 kG. These magnets are of great practical importance as research tools for nuclear magnetic resonance, solid state and plasma physics, bubble chambers, and ultimately for the achievement of fusion devices in the future. High field superconductors are also being considered for the levitation of high-speed trains and for use as high-power transmission lines.

REFERENCES

1. G. Weinreich, *Solids, Elementary Theory for Advanced Students*, John Wiley & Sons, New York, 1965.
2. J. S. Blakemore, *Solid State Physics*, W. B. Saunders, Philadelphia, Pa., 1969.
3. T. R. Hart, R. L. Aggarwal, and B. Lax, *Phys. Rev.*, B1, 638 (1970).
4. C. Kittel, *Introduction to Solid State Physics*, 3rd ed., John Wiley & Sons, New York, 1966.
5. K. J. Button, B. Lax, and C. C. Bradley, *Phys. Rev. Lett.*, 21, 350 (1968).
6. R. L. Aggarwal, M. D. Zutek, and B. Lax, *Phys. Rev.*, 180, 800 (1969).
7. C. Hurwitz, *Appl. Phys. Lett.*, 8, 121 (1966).
8. F. Bertaut and R. Pauthenet, *Proc. Inst. Elec. Engrs. (London)*, 104B, 261 (1956).
9. P. Townsend and J. Sutton, *Phys. Rev.*, 128, 591 (1962).
10. S. Foner, E. J. McNiff, B. T. Mathias, T. H. Geballer, R. H. Willens, and E. Corenzwit, *Phys. Lett.*, 31A, 349 (1970).
11. W. R. Beam, *Electronics of Solids*, McGraw-Hill, New York, 1965.

Chapter 2

GUNN EFFECT—BULK INSTABILITIES

Herbert Kroemer

Electrical Engineering Department
University of Colorado
Boulder, Colorado

2.1 INTRODUCTION

At the 1963 Solid State Device Research Conference in East Lansing, Michigan, J. B. Gunn presented a paper in which he described his recent discovery of microwave current oscillations in a simple piece of gallium arsenide, when he applied a d-c bias above a few thousand volts per centimeter (Fig. 1).*[1,2] This discovery brought to a successful culmination a search for a negative differential bulk conductivity in semiconductors, but at the time neither Gunn himself nor anybody else appeared to have recognized this. Nor was it realized that what would later turn out to be the theory of the effect had already been published, as a speculation, in two papers by Ridley and Watkins,[3] and by Hilsum,[4] with a third paper by Ridley[5] to appear in a matter of months. To the extent that Gunn's paper attracted any attention at all, it was because of the high frequencies involved, of the order

$$f = \frac{v}{L} \tag{1}$$

---

*The literature on the Gunn effect is vast. We are quoting here only selected papers that have a direct bearing on specific points discussed in this article. The reader interested in a complete bibliography is referred to the excellent "GSR" bibliography, which is very complete up to the end of 1968, and which lists papers, with titles, by year: T. K. Gaylord, P. L. Shaw, and T. A. Rabson; "Gunn Effect Bibliography," *IEEE Trans. Electron Devices*, 15, 777–788 (October 1968); "Gunn Effect Bibliography Supplement," 16, 490–494 (May 1969).

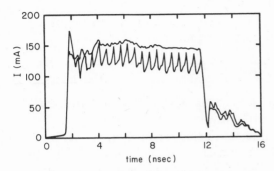

Fig. 1. The current through a GaAs crystal, just below the Gunn effect threshold, and just above. Note that the oscillations take place "downward" from the prethreshold current. (After Gunn.[2])

where $v \cong 10^7$ cm/sec is the drift velocity of the electrons and L is the electrical length of the device, typically below 10^{-2} cm. But apart from these high frequencies, Gunn's discovery was widely regarded as "another one of those freak oscillations," which had frequently been observed during the preceding few years and which were invariably caused by some minority carrier transit time effect or piezoelectric effects. For once, things turned out differently.

Today, this Gunn effect forms the basis of an entire new class of microwave devices, particularly for microwave generation. In the laboratory, frequencies as high as 150 GHz,[6] and pulse power as high as 6 kW (at around 1.8 GHz[7]) have been achieved. Devices with more moderate performance levels such as around 100 mW at X-band, are today commercially available, with noise levels comparable to those of microwave tubes if not better, and all of this at only a few volts of d-c bias. But this is far from what appears to be the ultimate performance limits of these devices; in the next few years commercial Gunn effect devices can be expected to change the entire field of solid state microwave electronics. To provide an understanding for this development is the purpose of this article.

While mysterious at the time of its discovery, the Gunn effect is now well understood. In particular, it is clear today that the effect is a negative conductance effect which is a consequence of the fact that the electrons in GaAs exhibit a *negative differential mobility*.[8, 9] This means that the drift velocity of the electrons will increase with increasing electric field only up to a certain point. Beyond that point an increasing electric field will actually lead to a decreasing electron drift velocity, as shown in Fig. 2.[10] This behavior is not exclusive to GaAs but it is most pronounced there. In GaAs the maximum drift velocity is about 2.2×10^7 cm/sec, occurring at an electric field of about 3.2 kV/cm. Beyond that point, the *differential mobility*,

$$\mu = \frac{dv}{dF} \tag{2}$$

becomes negative.* The same is, of course, true for the differential bulk conductivity.

We shall discuss the origin of this behavior in Section 2.2, its consequences in the remainder of this article. In Section 2.3 we shall treat the static behavior and the small-signal response of a crystal with a negative differential mobility. In Section 2.4 we shall discuss the oscillatory space charge instabilities of such a crystal, i.e., the Gunn effect proper. In Section 2.5 we shall discuss the effect of a resonance circuit on these oscillations.

2.2 HOT ELECTRON PROPERTIES OF GALLIUM ARSENIDE

2.2.1 Basic Energy Band Structure Concepts

The negative mobility behavior of the electrons in GaAs, shown in Fig. 2, is a direct consequence of the energy band structure of GaAs, specifically of its conduction band structure.

Energy band structures of solids are commonly expressed mathematically in terms of the functional dependence, $\epsilon(\vec{k})$, of the energy ϵ of the electrons, on their *wave vector* $\vec{k}$. For those readers who are not familiar with the concept of a wave vector, some explanation of this term appears in order. †

Ever since the discovery of electron diffraction effects it has been a well-established fact that, on a scale of atomic dimensions, electrons must be described in terms of their wave properties. The wavelength of an electron wave in *free space* is known to be given by

---

*Since electrons are negatively charged, they move in the direction opposite to the true electric field $\vec{E}$. In order to avoid numerous confusing minus signs we use the quantity

$$\vec{F} = -\vec{E}$$

in most of our calculations and refer to it as the electric field.

†See, e.g., C. Kittel, *Introduction to Solid Physics*, 3rd ed., John Wiley & Sons, New York, 1966.

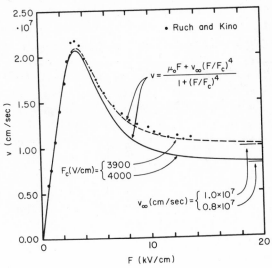

Fig. 2.   The velocity versus field characteristics of hot electrons in GaAs.   The curves do not represent theoretical values, but an analytic approximation to the experimental data.   (Experimental points after Ruch and Kino.[10])

$$\lambda = \frac{h}{m_0 v} \qquad (3)$$

where h is Planck's constant, m_0 the mass of the electron, and v the absolute value of its velocity.   Since the velocity is a vector, the inverse relationship between λ and v makes it awkward to express the dynamics of the electron waves in terms of λ; one introduces a wave vector $\vec{k}$ which has the direction of the propagation of the electron wave, the magnitude of which (the so-called wave number) is given by

$$\left|\vec{k}\right| = k = \frac{2\pi}{\lambda} = \frac{m_0 v}{\hbar} \qquad (4)$$

where

$$\hbar = \frac{h}{2\pi} \qquad (5)$$

is another form of Planck's constant.   Expressed in terms of k, the kinetic energy of a vacuum electron can then be written, using Eqs. 3 and 4,

$$\frac{1}{2} m_0 v^2 = \epsilon(k) = \frac{\hbar^2 k^2}{2m_0} \qquad (6)$$

However, Eq. 3, and therefore Eq. 6, hold only for electrons in free space.   Inside a solid the electron waves get strongly modulated by the periodic array of atoms, and the simple relationships of Eqs. 3 and 6 between v and λ, and between ϵ and k^* give way to more complicated relationships which are characteristic of the particular solid.

In fact, the dynamics of the electrons inside a semiconductor is fully determined if for each of its bands the $\epsilon(\vec{k})$ relationship is known. It can be shown that the electron velocities are then simply given by

$$v_x = \frac{1}{\hbar}\frac{\partial \epsilon}{\partial k_x}, \quad v_y = \frac{1}{\hbar}\frac{\partial \epsilon}{\partial k_y}, \quad v_z = \frac{1}{\hbar}\frac{\partial \epsilon}{\partial k_z} \tag{7a}$$

or symbolically,

$$\vec{v} = \frac{1}{\hbar}\nabla_k \epsilon \tag{7b}$$

Note that Eq. 7a reduces to Eq. 3 when Eq. 6 is valid. But Eq. 7 holds generally, regardless of the detailed form of $\epsilon(\vec{k})$.

Under the influence of an electric field the free electrons in a solid are accelerated. In free space this acceleration would be governed by Newton's law,

$$\frac{d\vec{v}}{dt} = \frac{q\vec{F}}{m_0} \tag{8}$$

where q is the electronic charge and $\vec{F}$ the electric field. Written in terms of $\vec{k}$, Eq. 8 becomes

$$\hbar\frac{d\vec{k}}{dt} = q\vec{F} \tag{9}$$

Inside a solid, where Eq. 3 no longer holds, these two forms of Newton's law are no longer equivalent. Now one of the most fundamental results of solid state physics is that the k-form Eq. 9 remains valid, not the v-form Eq. 8. Thus $\vec{k}$, rather than $\vec{v}$, changes with a constant rate, under the influence of a constant field.

It is possible, however, to express Newton's law in a modified v-form, by differentiating Eq. 7a and inserting Eq. 9. For the x component one obtains

---

*In solid state physics, the wave vector $\vec{k}$ always refers to the wavelength and propagation direction of the *carrier* wave, not to that of the modulation. This is true despite the fact that, in contrast to radio waves, the modulation wavelength of electron waves in solids is usually shorter than the carrier wavelength.

$$\frac{dv_x}{dt} = \frac{1}{\hbar}\left[\frac{\partial^2\epsilon}{\partial k_x^2}\frac{dk_x}{dt} + \frac{\partial^2\epsilon}{\partial k_x\partial k_y}\frac{dk_y}{dt} + \frac{\partial^2\epsilon}{\partial k_x\partial k_z}\frac{dk_z}{dt}\right]$$

$$= \frac{q}{\hbar^2}\left[\frac{\partial^2\epsilon}{\partial k_x^2} F_x + \frac{\partial^2\epsilon}{\partial k_x\partial k_y} F_y + \frac{\partial^2\epsilon}{\partial k_x\partial k_z} F_z\right] \tag{10}$$

and similar expressions result for the y and z components. They can be combined into the vector form

$$\frac{d\vec{v}}{dt} = \frac{q\vec{F}}{m^*} \tag{11}$$

where m^* is not a scalar, but a tensor, the inverse of which has the tensor components

$$\left[\frac{1}{m^*}\right]_{ij} = \frac{1}{\hbar^2}\frac{\partial^2\epsilon}{\partial k_i\partial k_j} \quad i,\, j = x,\, y,\, z \tag{12}$$

Equations 11 and 12 indicate that a free electron in a solid is accelerated *as if* it were an electron in free space, not with its true mass m but with an *effective mass* m^*, which in general is a tensor.

In the free-space case, when Eq. 6 is valid, m^* of course reduces to the scalar m_0.

2.2.2 The Energy Band Structure of GaAs

In Fig. 3 we are showing a cross section through the $\epsilon(\vec{k})$ diagram for the conduction band of GaAs, up to about 0.4 eV, and for $\vec{k}$-vectors in the crystallographic [100] and [$\bar{1}$00] directions.[11] Just as in the case of free-space electrons the lowest electron energy occurs for $k = 0$, at the bottom of what is called the *central valley* of this band structure. But in contrast to free-space electrons additional energy

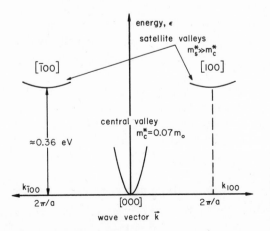

Fig. 3. A cross section through the conduction band structure of GaAs, along the $\langle 100\rangle$ directions. (After Ehrenreich.[11])

minima exist, which are so-called *satellite valleys*, at some higher energies, along the different $\langle 100 \rangle$ and (not shown) $\langle 111 \rangle$ directions of $\vec{k}$-space.

This band structure differs drastically from that of the most common semiconductors, silicon and germanium. In both of these semiconductors the lowest energies do not occur at $k = 0$ but in silicon the $\langle 100 \rangle$ satellite valleys, and in germanium the $\langle 111 \rangle$ satellite valleys have the lowest energy. Since the valence band maxima in almost all practically used semiconductors happen to occur at $k = 0$, germanium and silicon are referred to as *indirect-gap* semiconductors, and GaAs and several other substances with similar band structures as *direct-gap* semiconductors. The distinction is important. For example, it accounts for the fact that germanium and silicon are much better materials than GaAs for (bipolar) transistors, but that they are useless for injection lasers, while GaAs is excellent for this purpose. As we shall see presently, the direct-gap band structure is also an essential ingredient for the Gunn effect.

For sufficiently small values of the wave number k one can expand $\epsilon(\vec{k})$ into a Taylor series about $k = 0$, containing no linear terms, and ending with the quadratic terms:

$$\epsilon(\vec{k}) = \alpha_{xx}k_x^2 + \alpha_{yy}k_y^2 + \alpha_{zz}k_z^2$$
$$+ \beta_{xy}k_xk_y + \beta_{yz}k_yk_z + \beta_{zx}k_zk_x \tag{13}$$

Since GaAs is cubic, the three α's must be the same and the three β's must vanish. By making use of Eq. 12 this can be rewritten

$$\epsilon(\vec{k}) = \frac{\hbar^2 k^2}{2m_c^*} \tag{14}$$

where the effective mass m_c^* has again reduced to a scalar, as in the case of free-space electrons.

It is now typical for many direct-gap semiconductors that this effective mass of the central valley is considerably lower than the free electron mass. In GaAs, $m_c^* \cong 0.07 m_0$. These electrons are therefore much more readily accelerated than free-space electrons, which leads to high electron mobilities, despite the existence of strong electron scattering inside a semiconductor. In GaAs, the room temperature, low-field mobility μ_0 is about 8000 cm^2/V-sec, about twice the value of germanium, four times that of silicon, and very much higher than that typical of metals. We shall see shortly that this high mobility is essential to the Gunn effect.

The effective masses in the satellite valleys are very much larger. We have indicated this in Fig. 3 by showing a very much weaker dependence of ϵ on $\vec{k}$, near the bottom of the valleys. Exact values for these effective masses are not known, and little is known about the anisotropy of each individual valley. *

A detailed discussion of the effective masses of the satellite valleys would go far beyond the scope of this article—and beyond our needs. All that can be said here is that the combined set of satellite valleys[†] behave similarly to the way in which a single isotropic valley with a scalar $m_s^* \cong m_0$ would behave. As it turns out, this limited amount of knowledge is entirely sufficient for our purpose.

The location of the center of the satellite valleys in k-space is almost certainly exactly at the so-called surface of the Brillouin zone. For the $\langle 100 \rangle$ valleys this is at $k = 2\pi/a$, for the $\langle 111 \rangle$ valleys at $k = \pi\sqrt{3}/a$, where a is the lattice constant of GaAs. However, according to Eq. 7 the velocity of an electron does not depend on the absolute value of k, but only on the slope of the $\epsilon(\vec{k})$ relationship, i.e., on the location of $\vec{k}$ relative to the center of the valley. The exact location of the valleys, although known, has therefore little influence on the Gunn effect, and on the level of *our* discussion it does not enter at all.

What *is* important is the *energy* of the lowest set of satellite valleys. It varies greatly from semiconductor to semiconductor. In GaAs the $\langle 100 \rangle$ valleys are generally believed to be the lowest set of valleys, and they are located about 0.36 eV above the central valley.[11] As we shall see in the next section, it is a very favorable value, in the sense that it is very large compared to the thermal energy kT ($\cong 0.026$ eV at room temperature), but still small compared to the energy gap between valence and conduction band ($\cong 1.35$ eV). There exists some question as to how close in energy the $\langle 111 \rangle$ valleys are to the $\langle 100 \rangle$ valleys and whether they are not actually even lower than the latter. While a very close set of $\langle 111 \rangle$ valleys, or an inverted order, would have some influence in a *quantitative* computation of the GaAs $v(F)$ characteristic, it would in no way alter the qualitative explanation of this characteristic, or even its general shape. We shall therefore ignore the $\langle 111 \rangle$ valleys from now on.

---

*Since the satellite valleys are not located at $k = 0$, they need not be isotropic, not even in a cubic semiconductor.

[†]Note that there is one $\langle 100 \rangle$ valley for *each* of the different $\langle 100 \rangle$ directions, and the equivalent holds for the $\langle 111 \rangle$ valleys.

2.2.3 The GaAs $v(F)$ Characteristic

Since the satellite valley energy in GaAs is large compared to kT, almost all the electrons are located near the bottom of the central valley, at least at room temperature and for low electric fields. Because of their low effective mass, they have the well-known, high-electron mobility of GaAs. But this high mobility also makes it rather easy for a sufficiently strong electric field to accelerate the electrons, i.e., to heat them up, to rather high energies of the order of the satellite valley energy. As this happens, the electrons can scatter into the satellite valleys, where they have a much higher effective mass and a much lower mobility, and where they contribute much less to the current.[3,4] The current is further reduced by the strong intervalley scattering between the different satellite valleys.

If the intervalley transfer from the central valley to the satellite valleys were to take place only very gradually with increasing field, it would merely lead to a slow decrease of the conductivity, but the overall current would keep rising with increasing field, albeit nonlinearly. But as it happens, the transfer sets in rather abruptly with increasing field, and the current *decrease*, due to electron transfer into the low-mobility satellite valleys, is stronger than the current *increase* due to the velocity increase of those electrons that remain in the high-mobility central valley. As a result, the overall current drops and the crystal exhibits a bulk negative differential conductivity.

This, in short, is the basic origin of the negative differential mobility of the hot electrons in GaAs. The elaboration of this mechanism into a quantitative theory is a tedious matter which goes beyond the scope of this article. The interested reader is referred to the relevant literature; it need merely be said that the theoretical results now agree rather closely with the best experimental data, those of Ruch and Kino,[10] already shown in Fig. 2. While the latter are neither the only nor even the first experimental data on the velocity-field characteristic of GaAs, they are so far the only data that represent a *direct* velocity measurement (essentially a Haynes-Shockley experiment), rather than inferring the drift velocity from some other measurement. Moreover, among all the existing data they agree most closely with those conclusions concerning the $v(F)$ characteristics, which one can draw from observations of the Gunn effect such as the location and sharpness of the velocity peak, as well as the steepness of the velocity falloff beyond that peak.

Unfortunately, Ruch and Kino's data cease at a field of about 14 kV/cm, with the velocity still decreasing at that point. For higher fields one has to rely on other evidence. Potential probing experiments on uniformly propagating mature Gunn domains suggest a nearly constant drift velocity of about $8 \cdot 10^6$ cm/sec over the entire range of domain fields for which such potential probing data exist, i.e., from about 60 kV/cm all the way to the avalanche breakdown field, somewhere between 250 and 300 kV/cm.

Various data taken on GaAs avalanche diodes such as oscillation frequency data and avalanche impedance measurements suggest an even lower drift velocity in the avalanche field range, around 300 kV/cm, of about 6×10^6 cm/sec or possibly even less.[12] This value is not necessarily in contradiction to the domain probing value. Quite possibly, the two sets of data indicate that $v(F)$ drops gradually from a value between 8 and 10×10^6 cm/sec around 60 kV/cm, to about 6×10^6 cm/sec at 300 kV/cm.

The various data quoted so far leave the field range from 15 to 60 kV/cm unaccounted for. Some theoretical papers have suggested a velocity minimum in this range. Since such a minimum would have a noticeable effect on the space charge dynamics of the Gunn oscillation, particularly the LSA mode (see Section 2.5), this question is not without importance. At the time of this writing there exists no experimental evidence for, and little evidence against, such a minimum. Perhaps the only thing that can be said is the following: The potential probing experiments on uniformly traveling, mature Gunn domains and the application of Butcher's equal-areas rule (see Section 2.4.2) suggest that, if there exists such a minimum, it must be quite shallow.

Throughout the remainder of this article we shall for simplicity assume that there is neither a velocity minimum, nor a gradual falloff of the $v(F)$ characteristic toward the highest fields. Instead we shall assume a true "flat-valley" characteristic. In this case one can approximate the $v(F)$ characteristic over its entire field range well by a law of the form

$$v(F) = \frac{\mu_0 F + v_\infty (F/F_c)^K}{1 + (F/F_c)^K} \tag{15}$$

where μ_0 is the low-field mobility of GaAs ($\cong 8000$ cm^2/V-sec), v_∞ the asymptotic velocity, F_c some characteristic field, and K some "excitation exponent." A law of this form would arise physically if one would assume that the excitation of electrons from the low-mass, low-energy central valley into the high-mass, high-energy satellite valleys obeys a simple power law

$$\frac{n_2}{n_1} = \left(\frac{F}{F_c}\right)^K \tag{16}$$

a possibility first suggested by Kroemer[13] as a purely phenomenological description without any further physical basis. Equation 15 then would arise if one would further assume that the light electrons have a field-independent mobility μ_0 and the heavy electrons a field-independent velocity v_∞. The latter assumption cannot possibly be correct for low fields, but then there are no heavy electrons at low fields, so that this low-field error does not matter very much. As Hilsum[4] has pointed out, the assumption of a field-independent mobility for the light electrons is likely to be quite good for GaAs around room temperature.

As Fig. 2 shows, a rather excellent fit of Eq. 15 to Ruch and Kino's data can be obtained with

$$\mu_0 = 8000 \text{ cm}^2/\text{V-sec}, \quad K = 4$$

$$F_c = 3900 \text{ V/cm}, \quad v_\infty = 1 \cdot 10^7 \text{cm/sec} \qquad (17)$$

However, this v_∞ value appears to be slightly too high to agree with the high-field $v(F)$ data. In order to agree with both sets of data, a smaller value of v_∞ and a noninteger value of K would have to be used. But since the main use of Eq. 15 is in the computer simulation of Gunn effect phenomena, and since noninteger exponents greatly increase the computation time, several workers in this field have rather accepted a slight discrepancy with Ruch and Kino's data and have instead worked with the data:

$$\mu_0 = 8000 \text{ cm}^2/\text{V-sec}, \quad K = 4$$

$$F_c = 4000 \text{ V/cm}, \quad v_\infty = 8 \cdot 10^6 \text{ cm/sec} \qquad (18)$$

This curve is also shown in Fig. 2.

2.3 THE D-C AND SMALL-SIGNAL A-C PROPERTIES OF A NEGATIVE MOBILITY CRYSTAL

2.3.1 Introduction

While we are interested in the Gunn effect primarily as an oscillatory phenomenon, it is nevertheless very useful to consider first the static properties of a medium of negative conductivity, i.e., the properties it would have in the absence of any oscillations, regardless of whether such a nonoscillating situation is experimentally realizable. There are two reasons for such a consideration:

1. It greatly enhances the understanding of the origin of the various forms of oscillations that *do* take place.

2. It represents the actual behavior of devices in which the total

ionized donor density, per unit cross-sectional area of the device, stays below some critical value.[8,9]

Such "subcritical" devices will in general not oscillate, except under special conditions. The value of the critical doping is not sharply defined; it depends weakly on many parameters such as the bias field strength, the circuit impedance, the nature of the contacts, the presence of inhomogeneities in the crystal, etc. But in most cases the critical value falls in the range 0.5 to 1.0×10^{12} cm^{-2}.[9,14] Since the donor density per unit *area* is simply the product of the density per unit *volume*, N, times the length of the device L, this fact is usually expressed as a critical $N \cdot L$ product,

$$(N \cdot L)_{crit} = 0.5-1.0 \times 10^{12} \text{ cm}^{-2} \tag{19}$$

Apart from its purpose to aid in the understanding of the Gunn effect oscillations proper, the present section then presents the elements of the theory of subcritical devices, i.e., devices with an $N \cdot L$ product less than the critical $N \cdot L$ product. While not of the same importance as above-critical devices, such devices are still of considerable interest in their own right, because they exhibit a voltage-controlled negative conductance at frequencies around the transit-time frequency,

$$f_T = \frac{1}{\tau_T} \tag{20}$$

where τ_T is the transit time of the electrons through the device.[9,15] Sometimes, the negative conductance phenomenon repeats itself at some of the integer multiples of f_T such as $2f_T$ and $3f_T$, but these higher-frequency negative conductances are inevitably much weaker than the one at $f = f_T$ if they occur at all.

The negative conductance that occurs at the transit-time frequency is of interest for its potential as a two-terminal microwave amplifier,[15] and it has indeed been studied extensively for this purpose. The reader interested in the *experimental* facts is referred to the literature; in this book we wish to concentrate on the theoretical fundamentals of the phenomenon. The reason for our bypassing of the experimental situation is that the currently existing data appear still very heavily plagued by material problems. This follows from the fact that a subcritical device is made of a semiconductor which must of necessity be of a higher resistivity than an above-critical device of the same thickness. Almost all of the experimental data were obtained at a time when these high resistivities could be obtained only by heavy compensation with large concentrations of deep impurity levels. This leads to extensive

trapping effects. Since crystals with the same net density of free
electrons may differ drastically in the total number of impurity and
trapping centers, the existing amplifier data have a very poor reproducibility. Both because of the trapping and because of the poor reproducibility, the present experimental situation can, therefore, hardly be considered representative of the ultimate potential of subcritical
amplifying devices. Recently, GaAs technology has made very rapid
progress in eliminating these difficulties, but no data on subcritical
devices from such new material are available yet.

Even with our restriction on the theoretical fundamentals we shall
not be able to present a complete and rigorous theory of the small-
signal a-c conductance of a subcritically doped GaAs crystal. The
reason is simple: Such a theory does not exist, and it would of necessity have to be a purely numerical, i.e., computer-based, theory.
Like most aspects of the Gunn effect, the problem is far too complicated to be amenable to a rigorous closed-form solution, and whatever
closed-form treatments exist are treatments that make one far-reach-
ing simplifying assumption or another. Each of these existing treatments has a certain validity range—this is what distinguishes the different treatments—but none of them can claim any general validity.
We shall briefly discuss the more important of these treatments in
Section 2.3.4. But for our own purposes we shall follow a different
treatment which, rather than attempt to give as accurate as possible
a quantitative account, stresses the underlying physical *concepts*,
particularly those concepts that are important to the understanding of
normal Gunn oscillations but which are more naturally and more easily
introduced by a discussion of subcritical devices.

2.3.2 The Static Behavior of a Negative-Mobility Crystal

In this section we consider the static properties of a negative-mo-
bility crystal such as a GaAs crystal, with "ohmic contacts." The
term "ohmic contacts" is ill-defined; in order to be specific, we shall
temporarily assume that the contacts consist of a heavily doped section of the same semiconductor as the main section of the crystal. We
shall generalize this situation shortly.

Intuitively, one might expect that such a crystal would exhibit a
static current-voltage characteristic with a region of voltage-controlled, negative differential conductance, similar to that of the tunnel
diode. In fact, however, this is not the case. It had already been
shown by Shockley by 1954,[16] that the cathode contact in such a structure would inject excess electrons into the interior of the crystal. The
density of injected electrons increases with increasing bias voltage,

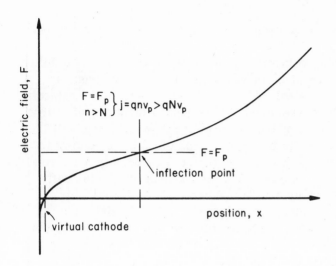

Fig. 4. The static electric field distribution in a crystal with negative differential mobility.

and it always increases more rapidly than the drift velocity decreases, leading to an increase rather than a decrease of the current itself.

This situation is illustrated in Fig. 4 which shows the electric field as a function of distance in such a structure. This electric field, somewhere within the crystal, must increase from the low values at the cathode contact to the high values corresponding to the applied bias. In the process, it must pass through a plane where it equals the peak velocity field F_p of the velocity field characteristic. By definition, within this plane, the velocity is equal to the peak velocity, v_p, and the current density is given by

$$j = qnv_p \tag{21}$$

Since this plane is in a region of an increasing field, it must also be in a region of negative space charge, i.e.,

$$n > N \tag{22}$$

Consequently,

$$j > qNv_p \tag{23}$$

where N is the doping density. Furthermore, this current must be constant throughout the entire length of the device. However, since the velocity everywhere else is lower than in the threshold plane, one must conclude that the electron density must everywhere be higher than in this plane. Since a space charge already existed in this plane, it

follows that there must be an even larger space charge density throughout the entire crystal.

   With increasing bias voltage, the entire field curve will get shifted
upward. As a result, the plane in which the electric field passes
through the threshold value will shift to the left, closer to the cathode.
While this happens, the curve also gets steeper, indicating, of course,
that the excess of the electron density over the doping density will increase. Because the velocity in the threshold plane is, by definition,
constant and equal to the peak velocity, the current density must also
increase with an increase in bias. This is, of course, intrinsically
what is contained in Shockley's 1954 theorem.

   The above "proof" of Shockley's theorem is hardly rigorous. In
particular, it leaves open the question after the validity of Shockley's
theorem for a nonuniform doping distribution and for geometries that
are not essentially one-dimensional. This question was answered only
recently by Kroemer.[17] He has shown that Shockley's theorem holds
regardless of the device geometry and of the impurity distribution,
provided that a few assumptions are satisfied. Of these assumptions
the only one of any consequence is the one of cathode boundary conditions such that the electric field at the cathode, if not zero, does at
least not decrease with increasing current density:

$$\frac{dF_c}{dj} \geq 0 \tag{24}$$

   While such boundary conditions might appear natural to assume for
ohmic contacts, they are by no means the only possible ones. Kroemer[18] has analyzed the case of a very shallow ($\ll kT$) Schottky barrier
at the cathode, in which case Eq. 24 does not hold, and which indeed
led to a negative static differential conductance. Also, if the current-
voltage characteristic of an oscillating Gunn effect diode is traced with
an instrument that is not capable of following the oscillations, which
merely senses the *average* current through the device, this *average*
I-V characteristic usually has a negative slope. This is a simple
consequence of the fact, already indicated in Fig. 1, that the Gunn effect oscillations occur "downward" from the threshold current. Shockley's theorem does not refer to these *average* currents but only to the
currents through a truly nonoscillating device.

2.3.2 The Transient Response of a Subcritical Device

The Problem. We saw in Section 2.3.2 that the absence of a
static negative differential conductance is caused by the voltage-de-

pendent, space charge readjustment which more than compensates for any decrease of the average electron drift velocity. This result raises two problems:

1. Since the space charge readjustment requires time, Shockley's theorem says nothing about the sign of the small-signal a-c conductance of the device at high frequencies. As we have already stated in Section 2.3.1, this a-c conductance becomes negative around the transit-time frequency, and one of our tasks will have to be the theoretical explanation of this result.

2. Furthermore, if the high-frequency conductance can become negative, the mere existence of a static solution of the space charge problem says nothing about the stability of that solution. The static solution, even though existing, might actually be unstable. The onset of Gunn oscillations for $N \cdot L$ products above the critical $N \cdot L$ product is, in fact, precisely the onset of such an instability, and the derivation of this fact will be another of our tasks.

These two problems are, of course, related: If the time dependence of all small-signal a-c quantities is assumed to be proportional to $\exp(i\omega t)$, and if the resulting complex admittance, $Y(\omega)$, is known as an analytical function of the frequency ω, then the device will be unstable under constant voltage condition if $Y(\omega)$ has any zeros, for complex ω-values with a *negative* imaginary part.

This coupling of the two questions of the frequency dependence of the admittance and of the onset of instabilities is mathematically straightforward. In the literature the analysis is usually carried out in the frequency domain, i.e., for a purely harmonic signal. While this is mathematically the simplest approach, it rather obscures some of the physics, in particular, some aspects of the physics that are important for the detailed understanding of the full spontaneous Gunn effect oscillations. Therefore we use here a time-domain treatment, an analysis of the transient response of the device under the influence of a small voltage step, at $t = 0$. This voltage step can be written as a Fourier integral

$$U(t) - U_0 = \frac{\delta U}{2} + \int_0^\infty A_U(\omega) \cdot \sin(\omega t)\, d\omega \tag{25}$$

with

$$A_U(\omega) = \frac{\delta U}{2\pi} \cdot \frac{1}{\omega} > 0 \tag{26}$$

Since the sign of all the A's is positive, the sign of the conductance at any given frequency is simply given by the sign of the sine Fourier coefficient of the *current* for that frequency.

Let us now assume that all the physical quantities such as electron densities, fields, and drift velocities can be considered as consisting of some time-independent d-c value, plus some small "disturbance," such as

$$n = n_0 + \delta n(t) \tag{27a}$$

$$F = F_0 + \delta F(t) \tag{27b}$$

$$v = v_0 + \delta v(t) \tag{27c}$$

etc. Because of Poisson's equation, we have

$$\epsilon \frac{d}{dx} \delta F = q \delta n \tag{28}$$

Also, since $v = v(F)$,

$$\delta v = \frac{dv}{dF} \delta F \tag{29}$$

The Injected Accumulation Layer. During the abrupt voltage increase, there flows, of course, an infinite displacement current spike. The total charge δQ carried in this current spike is simply given by the geometric capacitance and the height of the voltage step:

$$\delta Q = C \cdot \delta U = \frac{\epsilon}{L} \cdot \delta U \tag{30}$$

Immediately after the application of a voltage step the electron density inside the device is still the same as immediately preceding, except immediately ahead of the cathode, near $x = 0$, where the above charge δQ has been injected from the cathode which supports the increased electric field.

In order to understand the behavior of a subcritical device at intermediate frequencies, it is necessary to understand the time evolution of the electron density as a consequence of the voltage step δU and the injected charge δQ. For conceptual purposes, this evolution can be divided into two contributions:

1. Density changes resulting from the evolution of that space charge delta function of initial strength δQ, which was injected at the cathode, at $t = 0$.

2. Density changes which result because the change in electric field causes a change in the local drift velocity and therefore the current is no longer divergence-free:

$$\frac{\partial n}{\partial t} = -\nabla \cdot (n\vec{v}) \tag{31}$$

For a *quantitative* description both of these contributions must be included. But from a conceptual point of view, an understanding of the first of these contributions is sufficient in order to understand the overall device behavior. The second contribution represents merely a complication which, whatever its numerical magnitude, merely distracts from the physical essentials of the device behavior. Since our own interest here is in these essentials, we shall neglect the second contribution altogether. Since the finite divergence of the current density is a consequence of the position-dependent electron density and of the associated position-dependent field and drift velocity, we can achieve the desired simplification of the problem best by simply neglecting the d-c space charge and considering only the space charge due to the originally injected charge δQ itself, including the time evolution of the latter.

Under our assumptions the injected space charge would propagate through the device with a velocity approximately equal to $v_0 = v(F_0)$. The velocity would be *exactly* equal to this value if the height of the voltage step, δU, and therefore the initial strength of the space charge layer, δQ, were negligibly small. In this case the field difference across the traveling space charge layer would also be negligible, and the fields on *both* sides of the space charge layer would be the same and equal to the bias field, F_0. We shall neglect here any effects of a finite δU and δQ on the space charge layer propagation velocity and assume this velocity to be constant:

$$x' = v_0 t \tag{32}$$

However, another consequence of the finite field difference, which is not only not negligible but actually of crucial importance is the fact that the strength of the space charge layer will actually grow as it propagates. To see this, assume that at a given time t the space charge has the strength $Q(t)$. Let

$$F(x, t) = \begin{cases} F_0 + F^- \\ F_0 + F^+ \end{cases} \text{ for } \begin{cases} x < v_0 t \\ x > v_0 t \end{cases} \tag{33}$$

be the field strengths on the two sides of the space charge layer. Then

$$\Delta F = F^+ - F^- = \frac{Q}{\epsilon} \tag{34}$$

But this also leads to a current density difference

$$\Delta j = j^{+} - j^{-} = qN \frac{dv}{dF} \cdot \quad \Delta F = \frac{qN}{\epsilon} \frac{dv}{dF} \cdot Q \tag{35}$$

and quite obviously,

$$\frac{dQ}{dt} = -(j^{+} - j^{-}) = -\frac{qN}{\epsilon} \frac{dv}{dF} \cdot Q \tag{36}$$

with the solution

$$Q(t) = \delta Q \cdot e^{t/\tau_D} = \frac{\epsilon \delta U}{L} e^{t/\tau_D} \tag{37}$$

where

$$\tau_D = -\frac{\epsilon}{qN \, dv/dF} = \frac{\epsilon}{qN(-\mu)} \tag{38}$$

is the dielectric relaxation time corresponding to the absolute value of the negative differential mobility,

$$\mu = \frac{dv}{dF} < 0 \tag{39}$$

of the semiconductor. Thus we see that the space charge layer grows exponentially with time and, of course, with distance. The overall growth between cathode and anode is simply the growth during one transit time τ_T:

$$\tau_T = \frac{L}{v_0} \tag{40}$$

$$Q(\tau_T) = Q \cdot \exp\left(\frac{\tau_T}{\tau_D}\right) = \frac{\epsilon \delta U}{L} \cdot \exp\left(\frac{\tau_T}{\tau_D}\right)$$

$$= \frac{\epsilon \delta U}{L} \cdot \exp\left(\frac{N \cdot L}{M}\right) \tag{41}$$

where

$$M = \frac{\epsilon v_0}{q |\mu|} \tag{42}$$

We are encountering the $N \cdot L$ product of the device, divided by the quantity M defined in Eq. (42). Thus we see that, in devices with $N \cdot L \gg M$, some "primary accumulation layer," injected at the cathode, will grow by a factor that is large compared to e, while growth by less than a factor e indicates a device with $N \cdot L < M$. We shall shortly be able to relate this observation to the cause of the onset of Gunn oscillations in above-critical devices.

Clearly, the traveling accumulation layer (AL) does not spread out and disperse, as it would in a medium of positive mobility, but it is held together by the negative mobility which effectively converts the Coulomb repulsion into an apparent attraction. In the absence of diffusion the layer would travel as a delta function, but diffusion effects act as dispersive forces, and the actual profile of the layer is determined by the balance between the negative mobility attraction and the diffusive dispersion.

This behavior is illustrated in Figs. 5 and 6, which represent computer simulations of the propagation and growth of the ALs in two different devices, under the influence, not of a small voltage step, but as a result of the sudden application of an above-threshold bias field. Both devices are 100 μm (= 10^{-2} cm) long. In the case of Fig. 5,[19] the donor density N is zero, and it is shown how an AL propagates into the device, which was initially empty of electrons, under a bias field of 4.7 kV/cm. Clearly visible are the very small amount of growth of the AL and the diffusion spread.

Figure 6 shows the behavior of a device with a donor density N = 10^{14} cm^{-3},[20] under a bias field of 5.0 kV/cm. Note the difference in vertical scale and the difference in the AL growth rate. This second device is just above-critical; we shall use it again in later examples.

Note also that in both examples the AL speed is not truly constant but varies over a range of about 2:1.

One of the most important aspects of a traveling accumulation layer is the time variation of the *incremental* field F^- between the layer

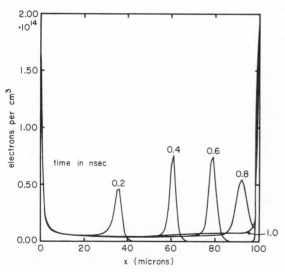

Fig. 5. Propagation of an electron accumulation layer into a crystal with zero doping, upon the application of a bias field of 4.7 kV/cm. (After Kroemer.[19])

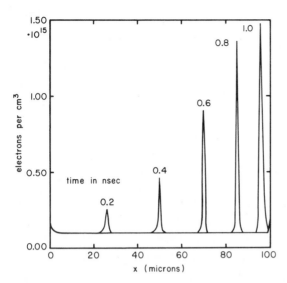

Fig. 6. Propagation and growth of an electron accumulation layer into a crystal with a donor density $N = 10^{14}$ cm^{-3}, under a bias field of 5 kV/cm. (After Kroemer.[20])

and the cathode. It can readily be calculated from Eqs. 3.14 and 3.41, and from the obvious additional condition,

$$F^- \cdot x' + F^+(L - x') = \delta U \tag{43}$$

The elimination of F^+ leads to

$$F^- = \frac{\delta U}{L}\left[1 - \frac{\tau_T - t}{\tau_T} \cdot e^{t/\tau_D}\right] \tag{44}$$

or, expressed as a function of accumulation layer position, x',

$$F^- = \frac{\delta U}{L}\left[1 - \frac{L - x'}{L} \cdot e^{x'/\lambda}\right] \tag{45}$$

where

$$\lambda = v_0 \tau_D = \frac{\epsilon v_0}{qN|\mu|} = \frac{M}{L} \tag{46}$$

is a "dielectric relaxation length," i.e., the distance along which the injected accumulation layer grows by a factor e.

The behavior of F^- for different ratios of L/λ, or NL/M, is shown in Fig. 7. Two special cases are of particular interest:

1. $NL/M = 0$. If the accumulation layer would not grow at all the field would rise linearly.

2. $NL/M = 1$. In this case the initial rate of field change is zero.

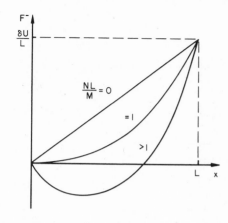

Fig. 7. The incremental electric field between cathode and accumulation layer, as a function of the accumulation layer position, for various values of the $N \cdot L$ product.

For larger values of NL/M the field will decrease initially.

For later purposes it is useful to illustrate the behavior in the two ranges of NL/M which are separated by the case $NL/M = 1$. This is done schematically in Fig. 8, in which both the electric field and the electrostatic potential are shown as a function of position for various instances during the travel of the injected accumulation layer. For clarity the strength of the accumulation layer has been greatly exaggerated. The different initial behavior of the electric field near the cathode is clearly apparent. We shall see shortly that these two dif-

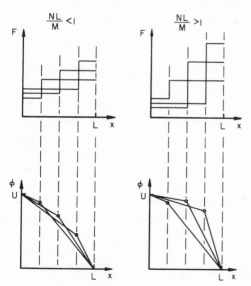

Fig. 8. The evolution of the electric field, and of the potential, as a function of position for the two different ranges of the $N \cdot L$ product.

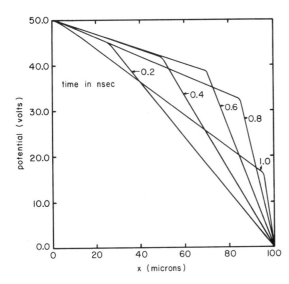

Fig. 9. The evolution of the potential in the case of the propagating accumulation layer of Fig. 6.

ferent types of behavior are related to the question of stability or instability, i.e., to the onset of spontaneous Gunn effect oscillations. In Fig. 9 we show, as an example, the potential distribution for our computer-simulation example of Fig. 6. It is clear that this falls into the category $NL > M$.

The Reinjection Feedback Concept. We have so far only discussed the evolution of the *originally* injected AL and the field variation at the cathode which takes place as a result of the movement and growth of that layer. We must now discuss the consequences of this field variation. Clearly, this field variation has the same consequences as an equal field variation, caused by an appropriate *gradual* bias variation. Such a gradual variation can, of course, be considered as an infinite number of infinitesimally small voltage steps; our results for one single discrete step can again be applied. In other words, the movement of the *discrete* primary accumulation layer causes a *continuous reinjection* of *distributed* charge at the (well-behaved) cathode. We have here a feedback phenomenon which we shall henceforth refer to as the *reinjection feedback loop*.[19] Together with the concept of the $N \cdot L$ product, the accumulation layer concept, and several concepts to be discussed later, the reinjection feedback concept is one of the crucial concepts in the understanding of the space charge dynamics of Gunn effect devices.

It is clear from the analogy of the gradual cathode field variation with an infinite number of infinitesimal voltage steps that the *rate* with

which the charge gets reinjected at the cathode is proportional to the
rate with which the field between the cathode and the primary AL rises
with time. Now the *re*injected charge does, of course, reduce the
field variations that actually occur at the cathode (ideally to zero); we
shall shortly have to discuss the effects of the *reinjected* charge on the
field at the cathode. But for the moment let us ignore these effects
and consider only the field variation and the resulting reinjection, due
to the movement and growth of the primary AL itself. Under these
circumstances, according to Fig. 7, the reinjection rate is strongest
just before the primary AL reaches the anode, the more so the larger
the $N \cdot L$ product of the device. Consequently, the reinjection charge
forms another AL in its own right, except that this reinjected layer
will not be a delta function layer but it will be somewhat more spread
out in both space and time, and correspondingly lowered in height.
However, the larger the $N \cdot L$ product, the more concentrated will the
reinjected AL be.

Ultimately, this second accumulation will also approach the anode
and, if we now apply the reinjection argument to this second-generation AL, it will give rise to a third-generation AL, and the argument
can obviously be repeated *ad infinitum*. Thus the initial voltage step
causes a strong ringing of the device at frequencies somewhat above
the transit-time frequency f_T. This means, of course, that the absolute value of the device admittance exhibits a resonance near the ringing frequency. This resonance corresponds to a negative real part of
the admittance, i.e., a transit-time negative conductance. However,
before discussing this fact we wish to pursue our reinjection feedback
loop concept a little further.

We have so far said nothing about the strength of the reinjected
accumulation layer relative to the primary accumulation layer. As long
as

$$N \cdot L < M \tag{47}$$

this second AL will certainly be weaker than the primary AL, and the
subsequent ALs will be weaker yet. This follows directly from Fig. 7
which shows that in this case the total field swing at the cathode,
caused by the primary AL, is no larger than the field swing due to the
bias voltage step but is more spread out in time. In fact, there will be
a finite reinjection rate already at $t = 0$, and the overall reinjection
rate will never drop below this initial finite value. Part of the reinjected charge will therefore simply go into a time-independent charge
background which will not contribute any current ringing at all in later
cycles. This charge background is, of course, the space charge re-

adjustment required by Shockley's theorem for the increased voltage. Only that part of the reinjected charge that does *not* go into this space charge readjustment can contribute to current ringing, and only *it* should be considered as the second-generation AL. This second-generation AL is, of course, weaker than the primary AL. The same argument can again be applied to the reinjected charge caused by the second-generation AL, showing that the third-generation AL is weaker yet, etc. The current ringing of the device is therefore a damped ringing; the loop gain of the reinjected feedback loop is less than unity.

Figure 10 shows an example of this behavior, for our device with zero doping, of Fig. 5. The current shown is the ringing caused by a sudden increase of the bias field, from the 4.7 kV/cm case of Fig. 5 to 4.9 kV/cm. The current density J_0 indicates the current density before the bias field increase. It is seen that, as soon as the device has recovered from the displacement current spike at $t = 0$, the current starts rising again, reaching a maximum at about 0.7 nsec. This maximum corresponds to the center of the second-generation, reinjected AL. There is another, much weaker maximum, at 1.4 nsec, corresponding to the third-generation, reinjected AL, etc. The third maximum is weaker yet, and the fourth is just barely discernible.

When $N \cdot L$ is greater than M, the field at the cathode dips initial-

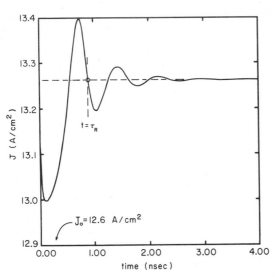

Fig. 10. Current ringing for the zero–doping device of Fig. 5 under the influence of a bias field increase from 4.7 kV/cm to 4.9 kV/cm. The current density J_0 indicates the current density before the bias field increase. (After Kroemer.[19])

ly, and the total field swing becomes larger than $\delta U/L$. The initial field dip, of course, causes a temporary depression of the electron flow from the cathode. If our uniform field approximation were strictly applicable, this would indicate the formation of a depletion rather than an accumulation layer. However, in this particular case it is to be remembered that this approximation is just that, namely, an approximation. Instead, the negative reinjection merely goes at the expense of the d-c space charge, without leading to an actual depletion layer, i.e., a layer with an electron density below the donor density.

If the subsequent field increase took place during an infinitesimally short time, the reinjected AL would now clearly be stronger than the primary AL. Once this reinjected AL disappears into the anode, it causes an even stronger AL to be reinjected. In other words, the device would exhibit growing rather than decaying oscillations; it would be unstable. Actually, of course, the field rise is still a gradual one and the reinjected AL is again spread out in time and space. As we saw earlier, this is equivalent to a damping effect, and, at least for $N \cdot L$ products that exceed the value M by only a relatively small amount, the reinjection oscillations will remain damped oscillations. But, with increasing $N \cdot L$ product the field rise not only gets larger in magnitude but also takes place over a shorter time. Ultimately, therefore, a situation will be reached where the reinjection ALs get successively stronger, and the current oscillations will increase rather than damp out; the loop gain of the reinjection feedback loop will become greater than unity.

Such a crystal will be electrically unstable under fixed bias, even in the absence of an external bias step, and oscillations of the transit-time frequency will develop spontaneously. They may arise either from noise in the bias circuit, which acts indistinguishably, of course, from an intentional voltage step δU; or they may be due to internal statistical fluctuations. Such fluctuations can be considered as microscopic accumulation and depletion layers. As they travel toward the anode, they cause the reinjection of new accumulation and depletion layers indistinguishable from those produced by an intentional voltage step; again transit-time oscillations build up.

These oscillations are the oscillations of the Gunn effect proper. Thus we see that the onset of these oscillations can be understood as the onset of a feedback instability. It takes place when the loop gain in the reinjection feedback loop exceeds unity, as a result of the growth of the space charge layers in a medium of negative mobility. We also see that this onset is indeed determined by the value of the $N \cdot L$ product, and that the critical $N \cdot L$ product must be a value somewhat larger than M. Exactly how much larger has not been determined quantitatively.

We shall turn to the detailed behavior of the Gunn oscillations in Section 2.4; for the remainder of the present section we shall consider further the problems of subcritically doped devices.

2.3.4 Transit-Time Negative Conductance

The Principle. We saw in Section 2.3.3 that, in subcritically doped devices, a sudden voltage step leads to a damped ringing of the current, at a frequency slightly above the transit-time frequency, and we pointed out that this ringing implies some resonance in the admittance at the ringing frequency. This resonance is associated with a negative conductance at that frequency. In order to show this, it would, in view of Eqs. 25 and 26 be merely necessary to show that the sine part of the current Fourier transform,

$$A_J(\omega) = \int_0^\infty J(t) \cdot \sin(\omega t)\, dt \tag{48}$$

is negative at the ringing frequency. But, in order to evaluate Eq. 48, a quantitative knowledge of $J(t)$ is required. Because of the continuous nature of the reinjection process, this in itself is a formidable problem, even under the simplifying assumptions we have made so far; it is beyond the scope of this article. We shall, therefore, be content to illustrate our point with an example, the zero-doping example of Figs. 5 and 10. If one assumes a ringing time τ_R as indicated in Fig. 10, it is clear by visual inspection of this curve that, for $\omega = \omega_R = 2\pi/\tau_R$, the integral in Eq. 48 will indeed be negative. The behavior at other frequencies is not so obvious; a numerical calculation leads to the solid curve in Fig. 11.[19] The broken curve will be discussed later.

The reader interested in a more detailed treatment of transit-time negative conductance must be referred to the literature. However, because of widely diverse and even conflicting assumptions made in different treatments, an uninitiated reader might find the different and sometimes conflicting results of the various treatments highly confusing. We are, therefore, presenting here a survey of some of the most important treatments found in the literature.

Various Theories of Transit-Time Negative Conductance. In this chapter we have discussed the behavior of subcritically doped Gunn effect devices in terms that emphasize the various concepts involved, particularly those concepts which will be important in the understanding of above-critically doped Gunn effect devices but which are best introduced by a discussion of subcritical devices. In this context the

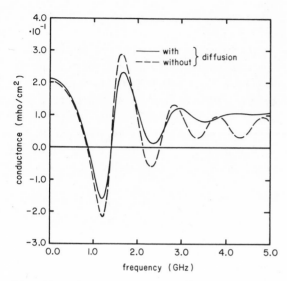

Fig. 11. The terminal conductance derived from the current transient of Fig. 10. The broken line indicates the results of a closed–form analytical approximation (Eq. 50), neglecting diffusion. (After Kroemer.[19])

question of a quantitative agreement was of relatively minor concern to us. In the literature, the emphasis is more frequently on an at least semiquantitative description. Of course, this is an equally legitimate point of view, and we are giving here a brief survey of the most important approximations found in the literature.

We say "approximations" because, as we have already pointed out, a rigorous and quantitative theory of the negative conductance of subcritically doped devices is very complicated. In fact, it is not possible in closed form but requires numerical techniques. Special cases of such numerical treatments have been given by McCumber and Chynoweth,[9] and Kroemer.[19] Like the treatment in Section 2.3.3, these numerical treatments were treatments in the time domain. Most other treatments in the literature are in the frequency domain.

Closed-form treatments require one simplification or another, and the various closed-form treatments existing in the literature differ primarily in the choice of the simplifications which are made. Most of the mathematical complications arise from the existence of a d-c space charge, and from the resulting nonuniformity of the d-c field and of various field-dependent quantities. Consequently, the most obvious possible simplification consists of neglecting the d-c space charge and the associated nonuniformities, and to treat the problem as one of growing space charge waves (Ridley[21]) in an otherwise uniform medium of negative conductivity, with suitable boundary conditions for the a-c field, the a-c current density, and the a-c charge density. If the theory is to reproduce properly the positive low-frequency conductance in accordance with Shockley's theorem, the boundary conditions must be chosen in such a way that they simulate the increase in electron density if the bias is increased above the d-c bias, right down to zero frequency. Hakki[22] has done this in a purely formal, *ad hoc* fashion; by properly selecting an adjustable parameter in his boundary condition, he was able to obtain very good agreement with experimental data and with the more rigorous numerical calculations.

However, this good agreement holds only for $N \cdot L$ products quite close to the critical $N \cdot L$ product. Hakki himself points out that, with decreasing doping, his approximation soon becomes a very poor one because the total d-c space charge *increases* with decreasing doping, not only relative to the decreasing donor density but absolutely. These difficulties arise in all theories that neglect the d-c space charge. For example, these theories invariably predict a lower doping or $N \cdot L$ product limit below which no negative conductance should occur at all. Kroemer[19] has shown that this conclusion is erroneous, at least in the absence of trapping and major inhomogeneities. Present-day, high-resistivity GaAs crystals usually do exhibit both strong trapping and strong inhomogeneities, and negative conductances have indeed not been observed, presumably for these reasons.

Closed-form theories that do not neglect the d-c space charge inevitably neglect diffusion effects instead. The oldest of these theories is the one by McCumber and Chynoweth.[9] These authors set up the proper differential equation, with boundary conditions, for the space charge waves, in the presence of the d-c space charge. The coefficients of this equation contain the local drift velocity and differential mobility. But, in order to make this equation tractable, the authors replace these coefficients by suitable constant averages. On closer inspection this is *not* the same as neglecting the d-c space charge, but it is equivalent to some drastic simplifications of the $v(F)$ characteristic. Strictly speaking, their assumptions could be interpreted as the assumption of a constant drift velocity and a constant negative mobility. Since the mobility is the derivative of the velocity, this interpretation is obviously inconsistent, but in the first order this inconsistency can be ignored, and the McCumber-Chynoweth theory may be regarded as the theory for a $v(F)$ characteristic with a constant negative differential mobility, as shown in Fig. 12a. Since such a $v(F)$ characteristic does not exhibit a positive mobility section, the finite-

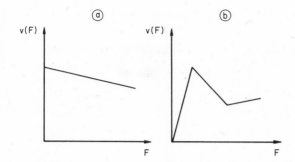

Fig. 12. Two straight–line approximations to the $v(F)$ characteristics.

thickness, positive conductivity section near the cathode is absent, too. As a result,* this theory does not exhibit a positive asymptotic conductance, but instead an oscillatory conductance, even at the highest frequencies. McCumber and Chynoweth realize that this result is in conflict, both with the experimental facts and with the results of their own numerical calculations, but they erroneously ascribe the discrepancy to the neglect of diffusion in their closed-form treatment.

Employing the result of their impedance calculations in a stability analysis, McCumber and Cynoweth give as an approximate instability criterion

$$N \cdot L > 2.09 \frac{\epsilon}{q} \frac{\bar{v}}{|\bar{\mu}|} \tag{49}$$

where $\bar{v}$ and $\bar{\mu}$ are some averages of the drift velocity and of the differential mobility of the electrons. The McCumber-Chynoweth theory does not specify the averaging process by which $\bar{v}$ and $\bar{\mu}$ are determined; in order to obtain agreement with the experimental range for the critical $N \cdot L$ product (Eq. 19), it is necessary to use a $\bar{\mu}$ value of the order 300 cm²/V-sec, somewhat implausibly small compared to the maximum negative mobility value of Ruch and Kino's data. The discrepancy is probably inherent in the McCumber-Chynoweth treatment with its oversimplification of the $v(F)$ characteristic.

This oversimplification in the McCumber-Chynoweth theory of the $v(F)$ characteristic has been removed by Mahrous and Robson[23] and by McWhorter and Foyt,[24] by assuming a three-section piecewise linear $v(F)$ characteristic of the type shown in Fig. 12b. With such a $v(F)$ characteristic the true $v(F)$ characteristic can be approximated sufficiently well so that the residual discrepancies in device behavior become quite inconsequential, particularly compared to the neglect of

*This consequence is not obvious; see Ref. 19.

diffusion effects. But this improved accuracy comes at a price. Because of the need for appropriate matching conditions between the three different field ranges, the resulting expressions for the device impedance, while technically in closed form, are so complicated that it is very difficult to gain any general insights from the formal structure of these expressions themselves. Instead, one is almost forced to a numerical computer evaluation of these closed-form expressions, thus losing the main advantage of a closed-form analytical treatment over a purely numerical treatment.

This latter advantage is retained in the limiting case studied by Kroemer,[19] the case of zero doping and no trapping, i.e., essentially, the case opposite to Hakki's.[22] In Kroemer's limit none of the complications of the other theories occur, and the real part of the device *impedance* simply becomes

$$R(\omega) = \frac{1}{\omega J_0} \int_0^{F_a} v(F) \sin\left[\frac{\epsilon \omega}{J_0}(F_a - F)\right] dF \tag{50}$$

where J_0 is the d-c current density and F_a the field at the anode end. The two in turn are related to each other and to the d-c bias through

$$J_0 = \frac{\epsilon}{L} \int_0^{F_a} v(F) \, dF \tag{51}$$

$$U = \frac{\epsilon}{J_0} \int_0^{F_a} v(F) F \, dF \tag{52}$$

These expressions hold for an arbitrary $v(F)$ characteristic. For the $v(F)$ characteristic of GaAs, there always exists a range of negative R, around the transit-time frequency, provided only that the bias current and voltage, and therefore F_a, are high enough. While Eqs. 50 through 52 are rigorous (except for the neglect of diffusion) only for zero doping, their very simplicity makes them an attractive approximation for finite $N \cdot L$ products, as long as the latter remain small compared to the critical $N \cdot L$ product.

Kroemer[19] has compared the results of this closed-form theoretical treatment with those of the computer simulation of a zero-doping device, presented in Fig. 11. Apart from the calculational procedure the conditions were chosen identical, except for the fact that the computer simulation included diffusion, while the closed-form treatment does not. This closed-form result is also included in Fig. 11, and the comparison serves to point out the role of diffusion in such devices. The results are given in the dotted curve in Fig. 11.

For frequencies below the transit-time frequency the results are in excellent agreement, indicating the negligibility of diffusion effects at low frequencies. At the transit-time frequency the diffusion leads to a slight weakening of the negative conductance. But at higher frequency the agreement rapidly deteriorates. Without diffusion, the amplitude of the conductance oscillations about the asymptotic value decreases roughly inversely proportionately to the frequency. With diffusion the oscillations die out very much faster, having essentially disappeared at four times the transit-time frequency.

We wish to conclude the present section by remarking that none of the theories briefly discussed in this section represents, by itself, a comprehensive treatment of the entire field of subcritically doped amplifying devices, and that the reader interested in a complete coverage will find it necessary to draw on all of these theories.

The Two-Port, Thin-Layer Amplifier. Although the transit-time negative conductance has been studied as a microwave amplifier, by Thim *et al.*[15] and by others, it shares with other two-terminal, negative resistance amplifiers, such as the Esaki tunnel diode, the drawback of a lack of directionality.

Robson, Kino, and Fay[25] overcame this drawback, at least partially, by constructing a fairly long but still subcritically doped device with two cathodes and two anodes (Fig. 13a). By impressing a microwave signal, not between the cathode and the anode but between the main cathode and the auxiliary cathode, the field is varied ahead of the

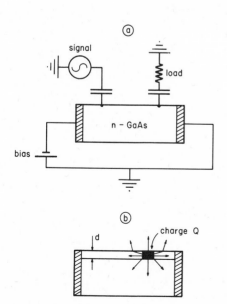

Fig. 13. (a) Two-port amplifier structure. (After Robson, Kino, and Fay[25];) (b) electric flux leakage in a thin-layer structure.

main cathode, just as in the two-terminal case of the field variation
due to the reinjection feedback loop. The field variation again launches
a space charge wave. This space charge wave gets amplified on the
way towards the anode just as a simple AL would, and an amplified
signal is picked up between the two anodes. Such a device is a truly
unidirectional, two-port, solid state, traveling wave amplifier; it will
operate over a wide band of frequencies, particularly at frequencies
above the transit-time frequency.

The signal gain G in such a structure is at most equal to the factor G_n by which the space charge waves grow. Our calculation of the
growth of a simple AL, in Section 2.3.3, is directly applicable to this
problem, and from Eq. 41 we obtain

$$G < G_n = \exp\left[\frac{\tau_T}{\tau_D}\right] = \exp\left[\frac{N \cdot L}{M}\right] \tag{53}$$

But we saw in Section 2.3.3 that such a structure will become unstable and will break into Gunn effect oscillation between the anode
and cathode if $N \cdot L$ becomes at most a few times M. Therefore, an
unmodified structure as in Fig. 13a will have only a low utilizable
gain.

In order to achieve high two-port gain, it is necessary to postpone
the onset of the Gunn instability until $N \cdot L$ products have been reached
which exceed M by a factor that is large compared to unity. This requires somehow weakening the reinjection feedback loop. Now, this
loop really consists of two parts: (a) the growth of any space charges
on their path from the cathode to the anode, and (b) the electrostatic
coupling between the growing space charge and the cathode, causing a
field variation at the cathode. The gain in the first part determines,
of course, the gain of the two-port amplifier; if this gain is to be increased, the electrostatic feedback coupling must be weakened.

This objective can be achieved easily[26-28] by employing a structure
as indicated in Fig. 13b, where the n-type GaAs consists of a thin
layer epitaxially deposited on semiinsulating GaAs, a good dielectric.
If the thickness d of the active layer is small compared to the device
length L, most of the electric flux emerging from any charge traveling
in the active layer will leak out of this layer into the substrate and, to
some extent, into the space above the active layer.* The result is a
considerable weakening of the field variations directly ahead of the

---

*The flux leakage can be greatly enhanced by covering the free surface with
a dielectric of very high dielectric constant, with or without metal backing.[29]

cathode, inside the active layer.   But this means that the reinjection
feedback loop is weakened.   In the limit of sufficiently small d the
$N \cdot L$ product criterion (Eq.) for Gunn effect instability gives way to
an $N \cdot d$ product criterion

$$N \cdot d \gtrless 1 - 2 \times 10^{11} \ \text{cm}^{-2} \qquad (54)$$

Devices not satisfying this criterion will be stable and can be used as
high-gain amplifiers, regardless of the value of L.   The design and
constructure of such amplifiers is an area of active current research.[30]

2.4 LARGE-AMPLITUDE, CONSTANT VOLTAGE, SPACE CHARGE INSTABILITIES

2.4.1 Accumulation and Depletion Layers

The Pure Accumulation Mode. In Section 2.3.3 we had shown
that the Gunn effect in the narrower sense, i.e., the onset of spontaneous current oscillations under quasiconstant bias conditions, can
be understood as a feedback instability, caused by a reinjection feedback gain in excess of unity.   This reinjection feedback gain, in turn,
was a result of the space charge growth of the traveling injected space
charge layers in a medium of negative differential conductivity.   In the
above section, our discussion of the reinjected feedback loop has remained basically a small-signal discussion.   This is entirely proper
for an externally drive *stable* system but not for a spontaneously oscillating one.   In the latter the oscillations will automatically build up to
the point that they finally become limited by the nonlinearities in the
system, and the final form of the oscillation cannot be discussed without making the nonlinearities an essential element of the discussion.

As an introduction, we extend our discussion of the injected primary AL into the nonlinear range.[13] It is at this point that our choice of
the time-domain discussion of subcritical devices shows its real merit
over the more common frequency domain discussion which is inherently limited to linear phenomena.

We saw in Section 2.3.3 that a primary AL injected at the cathode
would grow by the factor $\exp(NL/M)$ on its way to the anode.   For
$NL > M$ this growth was associated with a drop in the electric field F_c,
between the cathode and the traveling AL.   It was also stated in Section 2.3.3 that, for $N \cdot L$ products sufficiently larger than M, the current ringing would not be damped but growing.   The combination of the

two facts implies that the minimum value of F_c would drop lower with every oscillation cycle, and for sufficiently large $N \cdot L$ products this minimum field *must* ultimately drop below the threshold field F_p of the $v(F)$ characteristic. The larger the $N \cdot L$ product, the sooner this must take place.

At this point the linearized treatment of Section 2.3.3, characterized by the exponential space charge growth law Eq. 41, must be discarded. However, at this point the AL will still grow because the drift velocity on the cathode side of the AL is at its maximum possible value, and higher than on the anode side. This continued growth of the AL will not stop, as long as the cathode side of the AL stays above the value F_{ca} in Fig. 14. Here F_{ca} is the field value at which the electron *velocity* has dropped to the same value as the velocity on the anode side of the AL, where the field F_a is higher than on the cathode side.

An AL that has reached the stage at which the electron velocity is the same on both sides, and which therefore no longer grows, will be called a *mature accumulation layer*. Clearly, such a mature AL represents a limiting situation that, in general, can be achieved only asymptotically. We shall consider this limit in more detail later on; for the time being we merely assume that F_c will drop below F_p.

The consequences of this are far reaching. Such a drop will place the region between the cathode and the AL in a field range of positive differential mobility. Any space charge reinjected from the cathode will, therefore, decay rather than grow. Now we saw earlier that, for above-critical devices, the growth of a weak space charge layer which

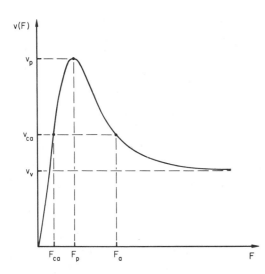

Fig. 14. An accumulation layer will be "mature," i.e., it will stop growing, when the field F_c on the cathode side has dropped to the value F_{ca} across from the field F_a on the anode side, i.e., when the electron velocity on both sides has reached the common value v_{ca}.

lies entirely in the negative mobility region would be larger than by a factor of e, possibly much larger. If the positive mobility section of the GaAs $v(F)$ characteristic had the same (absolute) slope as the negative mobility section, any space charges reinjected at the cathode would therefore decay by at least a factor of $1/e$ (over the full device length), once the field F_c upstream of the primary AL has dropped below F_p. Actually, however, the positive mobility section of the $v(F)$ characteristic is much steeper (at least by a factor 3) than the negative mobility section; consequently, the attenuation of the reinjected charge is very much stronger than the *growth* of a weak charge in the negative mobility region.

Under these circumstances it is an excellent approximation, for above-critical devices, to neglect the reinjection altogether and to assume a perfectly uniform field between the cathode and AL as long as that field stays below F_p. This is what we shall assume here. *

Ultimately, as the AL approaches the anode, the field F_c will rise again toward F_p, and the reinjection process will inject a new (initially weak) AL at the cathode. We shall discuss this process further.

Consider Fig. 15*a*, which shows the potential diagram for a localized AL, at position x', with *uniform* fields F_c and F_a on either side of itself. Obviously,

$$F_c x' + F_a (L - x') = U = \overline{F} \cdot L \tag{55}$$

where U is the bias potential and $\overline{F}$ the average or bias field. Of course, we assume

$$\overline{F} > F_p \tag{56}$$

The total charge Q in the AL obeys

$$Q = \epsilon (F_a - F_c) \tag{57}$$

and the rate of change of this charge is related to the drift velocity difference, according to

$$\frac{dQ}{dt} = qN[v(F_a) - v(F_c)] \le qN(v_p - v_v) \tag{58}$$

If we now assume that the AL moves with a finite velocity c, then we can readily see that F_c cannot stay below F_p indefinitely. For otherwise, because of Eqs. 55 and 56, F_a would have to go toward infinity. This would require an infinite space charge Q, which is of course in-

---

*This behavior is already implicit in our discussion, in Section 2.3.3 where we stated that a field *decrease* occurs at the expense of the dc space charge.

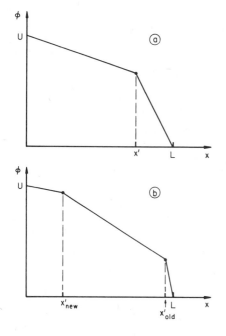

Fig. 15. (a) Potential diagram for pure AL; (b) renucleation of new AL, as the old AL approaches the anode, and the field F_c as the cathode rises to F_p.

compatible with the finite charging rate Eq. 58. But if F_a cannot become infinite, then F_c must ultimately increase beyond F_p.

As the field on the cathode side of the AL exceeds the threshold field, the reinjection feedback process becomes active again, and a new AL is injected by the cathode. Thus, for a short time, a situation exists as shown in Fig. 15b, with two ALs present simultaneously. However, since the new AL is in general formed only shortly before the old AL reaches the anode, the old AL usually disappears shortly afterward, and we again obtain a situation as we had in Fig. 15a. Clearly, there is a periodic nucleation and disappearance of accumulation layers.

Associated with the periodic variation of the potential pattern is, of course, a periodic variation of the current. At any given position in the crystal the total current is a sum of conduction and displacement currents. Both contributions are position-dependent, but their sum is independent of position and equal to the circuit current. A general theory of this current, as a function of time, bias, and doping level, is moderately difficult and goes beyond the scope of our presentation. However, it is quite easy to see that the "current peak-to-valley ratio," i.e., the ratio of maximum to minimum device current, must approach the velocity PVR, $v_p : v_v$, in the limit of $N \cdot L$ products which are large compared to M.

In this limit the displacement currents are clearly small compared to the conduction currents, except during the AL renucleation
process, on the anode side of the AL (our above considerations show
that during renucleation the displacement currents near the anode are
always comparable to the conduction currents). But the conduction
current can, of course, not exceed qNv_p, nor can it drop below qNv_v.
Furthermore, both limiting currents are actually approached in the
limit of large $N \cdot L$'s. For $j_p = qNv_p$ this is obvious; for $j_v = qNv_v$ this
follows from the fact that for such devices F_a will approach values
far out along the flat portion of the $v(F)$ characteristic, *before* the
displacement currents on the anode side can no longer be neglected.

Thus we see that the pure accumulation mode in the limit of
large $N \cdot L$ products is a mode with a large current amplitude and,
since the current wave shape turns out to be quite favorable, a mode
with potentially high power efficiency. In fact, from the device utility
point of view, it would be one of the most desirable of all Gunn effect
modes. Unfortunately, it is also one of the most difficult modes to
obtain in practice. The reasons for this will be described later.

Before closing this subsection we wish to make one qualifying re-
mark. We have, in our treatment, assumed that the AL always
moves with a finite velocity. Under certain special circumstances the
propagation speed of the accumulation layer can slow down to zero as
it gets very close to the anode. If this happens before the new AL has
been nucleated, the reinjection feedback loop becomes inoperative and
one obtains again Shockley's nonoscillatory solution, despite the fact
that the $N \cdot L$ product is above-critical. If the field near the anode gets
sufficiently high, this may lead to a stationary avalanche breakdown
of the device.

If sustained, this can lead to thermal destruction, and this behavior is indeed one of the failure mechanisms in Gunn effect devices
that are poorly designed or made from poor-quality GaAs. Since the
phenomenon does not occur in well-designed devices made from high-
quality GaAs, and since its detailed physics is complicated and only
imcompletely worked out, we shall just mention the phenomenon here.

Depletion Layer Nucleation at Imperfect Cathode Contacts. The
treatment of the pure accumulation mode in Section 2.4.1 made the
(implicit) assumption that was already inherent in most of Section 2.3,
namely, that the cathode contact is "well-behaved":

1. The electric field at the entrance into the active region of the
Gunn effect device stays below F_p.
2. The transition from the below-threshold entrance fields to the

fields inside the device is supported by a negative space charge of the mobile electrons themselves.

    3. The field at the cathode interface F_c, if it varies at all, does not decrease with increasing current.

In such structures the periodic rise of the electric field ahead of the cathode does indeed cause a pure AL to be injected by the cathode, and our description would apply.

    Actual Gunn effect device cathode contacts frequently do not appear to behave in this simple fashion but appear to nucleate a positively charged electron *depletion layer* (DL), as the field ahead of the cathode increases beyond the Gunn effect threshold field. If this DL is not followed by an AL, a purely static potential distribution as in curve *a* of Fig. 16 arises, exhibiting a "cathode fall" with typically 10 to 20 kv/cm internal fields. As Gunn and Kennedy have pointed out,[31, 32] many actual devices exhibit such a behavior, at least over a narrow voltage range. Usually, however, the DL is followed in time by an AL, also nucleated at the cathode-to-active layer interface, leading to a *dipole domain* (DD) as shown in curve *b* of Fig. 16. We shall study the properties of this second important mode, the dipole mode, in more detail in Section 2.4.2, for the time being we shall concentrate on the processes that lead to the formation of the DL and the dipole, in the first place.

    Kroemer[18] has presented an extensive theoretical discussion of

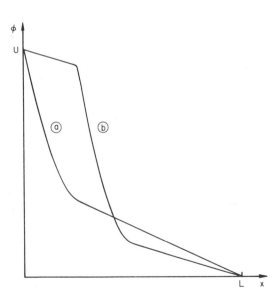

Fig. 16. Potential evolution in devices with "imperfect cathode" boundary conditions. At first an electron *depletion layer* is formed (curve *a*), the opposite of an accumulation layer. This is followed by an AL, forming a *dipole domain* (curve *b*).

various cathode models such as shallow Schottky barriers, embedded spurious p layers or high-resistivity layers, and others. This list could easily be extended further. All of these models lead to nucleation of a primary DL and most of them to the subsequent nucleation of an AL. Unfortunately, however, there exist at this time no experimental data to permit a choice between the variety of possible theoretical models. However, as Kroemer did point out, almost regardless of the exact physical model, it should be possible to express this model in terms of a phenomenological "control characteristic," i.e., a current density versus electric field characteristic $j_c(F)$ *of the cathode-to-active-layer interface itself.* A well-behaved cathode is a cathode for which the control characteristic stays above the "neutral bulk characteristic," i.e., above the curve $qN \cdot v(F)$ (Fig. 17, curve a).

In this case the cathode is able, at all values of the interface field F_c, to supply more electrons than the active layer is able to carry away, regardless of the field at the interface, leading always to AL formation at the interface.

In the opposite case, curve c in Fig. 17, the cathode cannot supply the number of electrons that the active layer would be capable of carrying away, again regardless of the interface field. The result would be a depletion of electrons in the active layer and a potential distribution as in curve a of Fig. 16. Such a situation is static rather

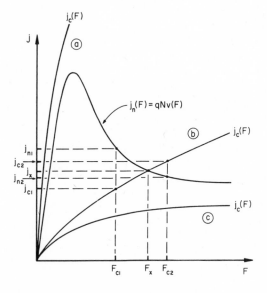

Fig. 17. The *neutral bulk characteristic,* $j_n(F)$ and three cases of *control characteristics.* (a) "well-behaved cathode"; (b) "imperfect cathode" with *crossover point*; (c) without. (After Kroemer.[18])

than oscillatory.* Consequently, it is of not interest in a discussion of oscillatory space charge instabilities; we shall omit it here.

Of the greatest interest is the intermediate curve b in Fig. 17, the one that exhibits a crossover point with the neutral characteristic, at $F = F_x$. This is the control characteristic which we shall assume for the remainder of the present section.

Assume now that in such a device the bias field is gradually raised, beginning from a value sufficiently low so that the field at the cathode interface is initially below the crossover field, say at $F = F_{c1}$ in Fig. 17. The electron current actually delivered by the cathode, j_{c1}, is then initially less than the current j_{n1} which the active layer would be capable of carrying away, for the same field. This leads to a depletion of electrons ahead of the interface, i.e., to a positive space charge layer. This will enhance the field at the cathode interface, the so-called control plane, and bring it closer to the crossover field. As long as the bias field stays sufficiently low, the field at the control plane will stay below the crossover field, despite this field enhancement, and a stationary cathode fall will occur. This is precisely what was observed by Gunn and by Kennedy. As the bias voltage is increased, the field in the control plane will ultimately exceed the crossover field going to, say $F = F_{c2}$ in Fig. 17. Now the current j_{c2} actually delivered by the cathode has increased to a value that is larger than the current j_{n2} which the active layer is capable of carrying away. Consequently, an AL finally forms, just as for well-behaved cathode boundary conditions.

As this AL grows, the electric field between the AL and the cathode drops again, just as in the pure accumulation mode. But while in the latter the AL layer buildup did not stop until the field had dropped significantly below the *threshold* field, to the point F_{ca} in Fig. 14, in the case of imperfect cathode boundary conditions the AL buildup will stop as soon as the field drops below the crossover field. At this point another—stationary—DL forms at the cathode plane. The positive space charge in this DL causes the field between the traveling AL and the stationary new DL to drop even lower; ultimately a potential distribution *more or less* similar to the one shown in Fig. 18 arises, consisting of a stationary DL near the cathode, and an AL-DL pair, a so-called dipole domain, ahead of the stationary DL.

Now, the most important fact about this configuration is that the dipole domain can move, just as a pure AL, in contrast to the station-

---

*Essentially, because the reinjection feedback loop is inoperative.

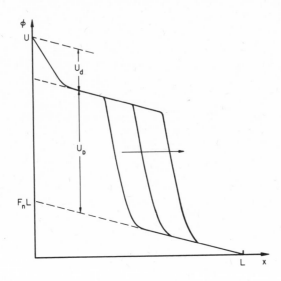

Fig. 18. Potential distribution for the *mature dipole mode*. (After Kroemer.[18])

ary DL ahead of the cathode. The detailed shape of the overall potential distribution, both initially and as a function of time, can vary over
an extremely wide range, depending on the numerical values for several relevant parameters such as the $N \cdot L$ product, the bias voltage,
the control characteristic, particularly the value of the crossover
field. We have shown, in Fig. 18 what is probably the simplest of all
possible configurations: a stationary cathode fall takes up a fixed portion U_d of the overall bias voltage U, and a second fixed portion U_D is
taken up by a uniformly traveling "mature" dipole domain, i.e., a domain that exhibits no longer any changes with time other than a uniform translational propagation from cathode to anode. The remainder
of the overall bias voltage is given by the resistive drop, $F_n \cdot L$, where
F_n is the common electric field within those electrically neutral regions
of the semiconductor which are outside of either the cathode fall region, or the dipole domain. Obviously, $F_n < F_p$.

This *mature dipole mode* actually exists; we shall study it—and
the conditions for its existence—in more detail in Section 2.4.2. However, we wish to point out here that, in many actual devices, this
division of a fixed bias voltage into three time-independent components
does not occur. For example, the cathode fall depletion layer voltage
U_d may grow at the expense of the dipole domain voltage U_D (Fig. 19),[33]
and Kroemer[18] has shown that the dipole domain may even collapse before it reaches the anode. But none of these variations affect the primary (qualitative) point of our present discussion, namely, that the
traveling dipole domain mode represents an oscillatory phenomenon,

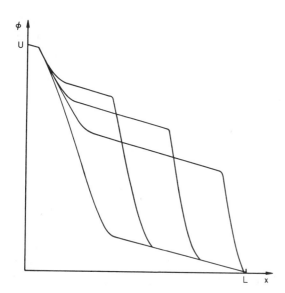

Fig. 19. A case of a shrinking dipole domain and a growing cathode fall, obtained by computer simulation[33] of a device with $N = 10^{15}$ cm^{-3}, $L = 2 \times 10^{-3}$ cm, $F_x = 7$ kV/cm, and $\overline{F} = 3.5$ kV/cm. In this particular device the *effective* cathode was not located at $x = 0$ but about 10^{-4} cm inside the device.

just as the pure accumulation mode. What happens is the following. As the dipole domain disappears (either by collapsing inside the crystal or by disappearing into the anode), the potential drop across the cathode fall increases, and thereby the electric field at the control plane increases again. If the overall bias voltage was sufficiently large to cause the control field to exceed the crossover field in the first place, then the field at the control plane would in general, exceed the crossover field again, as the dipole domain disappears. This, of course, leads to the renucleation of another dipole domain, and the entire process will repeat itself periodically.

This renucleation process is simplest in our example of the "mature" dipole domain of Fig. 18. As the dipole reaches the anode, the depletion layer disappears first; once the depletion layer has disappeared the situation near the anode becomes essentially the same as during the later phases of the pure accumulation mode, in Section 2.4.1. The only difference, then, is the field enhancement at the cathode, due to the imperfect cathode boundary conditions.

The qualifying remarks made on page 57 apply to the dipole modes as well.

2.4.2 The Mature Dipole Mode

Uniformly Propagating Dipole Domains. As discussed in Section 2.4.1, imperfect cathode boundary conditions can lead to the formation of traveling dipole domains. We pointed out that, in this case, some voltage division occurs between the cathode fall, the domain itself, and the purely resistive voltage drop. In the present section we wish to study in more detail the propagation properties of these dipole domains, in the limit that the cathode fall voltage can be neglected, i.e., that the domain voltage is given by

$$U_D = U - F_n L \quad F_n < F_p \tag{59}$$

In this case the voltage division problem does not occur, and it probably is obvious that under these circumstances the traveling domain will, in a sufficiently long crystal, ultimately assume some limiting shape which simply propagates with some uniform speed c through the crystal. In this uniformly propagating limit all local physical properties (except the electrostatic potential) such as electric fields, electron densities, and velocities depend on position and time, no longer separately, but only through the combination

$$u = x - ct \tag{60}$$

i.e., the distance measured in a frame of reference which moves along with the domain at the same speed. All time dependences in this moving frame of reference then vanish.

This situation is actually simpler than for the pure accumulation mode. We pointed out in Section 2.4.1 that for an AL the space charge growth will stop when the electron drift velocity on the low-field (= cathode) side of the AL has dropped to the same value as on the high-field (= anode) side. But the situation was complicated by the fact that the electric field on the anode side depended on the AL-to-anode distance, and thereby kept changing. If a DL is present downstream of the AL, this time dependence of the field on the high-field side of the AL is not required. The DL itself assumes, relative to the AL, the role of a "traveling anode," and if the distance between the AL and the DL reaches a stationary value, the traveling fields will assume stationary values, too.

At first glance, it might appear possible that the fields inside the dipole domain could build up indefinitely, rather than saturate, if the space charge densities in *both* the AL and the DL could grow indefinitely, and if the two space charge layers could move arbitrarily close together as this growth takes place. However, particularly on the DL side the space charge density is limited to qN, the space charge density of the ionized donors in the semiconductor. In order to support a finite electric field difference, the DL must therefore have a finite thickness, implying a finite voltage difference *across the DL itself.*

It is an elementary exercise in electrostatics to show that the maximum domain field obtainable under these circumstances is

$$F_{D,M} = \sqrt{\frac{2qNU_D}{\epsilon}} + F_n \qquad (61)$$

This upper limit is obtained if one neglects all electron diffusion effects, leading to a fully depleted DL of finite width, supporting the entire domain voltage U_D, and to an infinitely thin AL with infinite space charge density inside, with zero voltage difference across the AL. The inclusion of diffusion effects lowers the fields below $F_{D,M}$.

Clearly, then, for a given bias the domain shape must ultimately saturate, and a uniformly propagating mature dipole domain must ultimately result, at least in a sufficiently long crystal. Sufficiently long in this context means that the domain transit time τ_T is sufficiently longer than that dielectric relaxation time which is appropriate to the negative mobility section of the $v(F)$ characteristic. But this is, of course, a requirement that can, as in Section 2.3, again be expressed in terms of an $N \cdot L$ product. Numerical calculations show that, for devices with an $N \cdot L$ product of about five times the critical $N \cdot L$ product, the domain matures within about the first 20% of its travel throughout the crystal.

Such relatively large $N \cdot L$ products have been employed primarily in research devices which were built for the specific purpose of studying the physics, especially the space charge dynamics, of the Gunn effect, by potential probing.[*] Because of the experimental difficulty of making and using very fine potential probes, most such physics research structures are made longer than about 100 μ, often much longer. Because of the technological difficulty of preparing *high-quality* GaAs with doping levels significantly below 10^{15} cm^{-3}, particularly at the time when most of these physics experiments were performed, the resulting $N \cdot L$ products were inevitably much larger than critical. In most of these experiments, therefore, it was the mature dipole mode that was observed, with various amounts of cathode fall added at the cathode.

Although this predominance of the mature dipole mode in physics experiments must be considered an accidental consequence of experimental and technological restrictions, historically the very simplicity of this mode made this accident a rather fortunate one. A propagating

---

[*]See, e.g., Refs. 31 and 34.

charge dipole all but required a static negative differential mobility for its explanation, and the uniform motion itself made a quantitative analysis of the probing data readily possible. For example, the determination that the GaAs $v(F)$ characteristic must have a flat valley is based on such data.[34,35] It is not surprising, then, that the overwhelming majority of the published papers that concern themselves with the space charge dynamices of the Gunn effect deal with this mode, rather than with transient phenomena such as more complicated modes.

A rather unfortunate by-product of this situation is the rather widely held misconception that most actual Gunn effect *microwave oscillators* (other than the LSA-mode devices described in Section 2. 5) exhibit mature dipoles. Few misconceptions could be farther from the truth. Most Gunn effect devices designed as microwave oscillators employ $N \cdot L$ products rather closer to the critical $N \cdot L$ product, typically not above 2×10^{12} cm^{-2}. For such $N \cdot L$ products the dipole should not be able to mature fully. An experimental determination of the operating mode, by potential probing of a 10-μ-thick device, is all but beyond the present state of the art, but the current wave shape of the Gunn oscillations of these devices is in any case rather different from that of the mature dipole mode. The latter exhibits a time-independent current, while the domain is in transit. As the domain reaches the anode, the field throughout the device rises again, as described earlier, leading to an appropriate current rise. As soon as the new dipole has been formed, and is maturing, the current collapses again to the mature domain value. The net result is a spiked current mode (Fig. 20) which is rich in harmonics but which contains only a relatively small fraction of the d-c imput power in any single harmonic, leading to a low-efficiency device. Since a high efficiency is typically one of the most important design criteria for a solid state microwave oscilla-

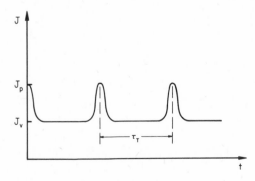

Fig. 20. "Spiked" current wave shape of the mature dipole mode.

tor, actual design procedures tend to *avoid* the mature dipole mode. In practice this takes the form of an $N \cdot L$ product rather close to the critical value, a consideration that in many cases is reinforced by the desire to obtain low thermal dissipation densities.

The fact that the mature dipole mode is, in its pure form, of little importance in *practical* Gunn effect microwave oscillators, in no way reduces its *conceptual* importance as a simple limiting case. Apart from the fact that there have been proposed certain nonmicrowave devices which fully utilize the mature dipole mode, most actual microwave devices operate in some more complicated fashion which is somewhere in between the pure accumulation mode* and a mature dipole mode. Any understanding of real devices will require an understanding of these two limits. For this reason, and for an understanding of the fundamental physics experiments, we shall present, in the following subsection, a mathematical theory of the mature dipole mode.

Mathematical Theory of Uniformly Propagating Dipole Domains — the Equal-Areas Rule. Let

$$j_x = qnv(F) - qD\frac{\partial n}{\partial x} \tag{62}$$

be the local electron conduction current density. The second term is a diffusion term; by writing the diffusion coefficient D outside the space derivative, we are implying that D is field-independent. This is, of course, an oversimplification. But by making it we are able to obtain an important approximate relationship, the so-called Butcher equal-areas rule,[36,37] between the domain propagation speed c, the velocity v_n, and the field F_n outside the domain and the maximum domain field, F_D.

If Eq. 62 is transformed into the u-coordinate system, Eq. 60, moving along with the domain with the velocity c, the amount qnc has to be subtracted from the current and we obtain

$$j_u = qn[v(F) - c] - qD\frac{dn}{du} \tag{63}$$

where we have replaced the partial derivative with respect to x, by a total derivative with respect to u, to indicate the fact that in this moving coordinate system the electron density will not be time-dependent.

---

*The LSA mode is considered here as a circuit–quenched form of a pure accumulation mode; see Section 2.5.

But in this case the divergence of the conduction current must vanish (there is no displacement current in this frame), i.e., the conduction current must be constant with position. Since far away from the domain the electron density will be constant and equal to N, and the electron velocity will be $v_n = v(F_n)$, this constancy requirement can be expressed as

$$n \cdot [v(F) - c] - D\frac{dn}{du} = N \cdot (v_n - c) \tag{64}$$

By making use of the similarly transformed Poisson equation,

$$\frac{dF}{du} = \frac{q}{\epsilon} \cdot (n - N) \tag{65}$$

the space derivative in Eq. 64 can be replaced by a field derivative:

$$n \cdot [v(F) - c] - \frac{qD}{\epsilon} \cdot (n - N) \cdot \frac{dn}{dF} = N \cdot (v_n - c)$$

Division by n, and integration from F_n to the maximum domain field F_D yields

$$\int_{F_n}^{F_D} [v(F) - c]\,dF - \frac{qD}{\epsilon}\left\{[n(F_D) - N] - N \cdot \ln\frac{n(F_D)}{N}\right\} = N \cdot (v_n - c)\int_{F_n}^{F_d} \frac{dF}{n} \tag{66}$$

Now the point of maximum domain field is, of course, a point of electrical neutrality,

$$n(F_D) = N \tag{67}$$

Thus the second and third terms on the left-hand side of Eq. 66 vanish. Furthermore, there are two ways to integrate from F_n to F_D: one through the AL, the other through the DL. The left-hand side of Eq. 66, not containing the electron density, is the same for both integrations. The integral on the right-hand side, however, differs drastically.

But Eq. 66 must, of course, hold in both cases, and this is possible only if the factor in front of the integral vanishes

$$c = v_n \tag{68}$$

i.e., the domain speed is equal to the electron velocity outside the domain, or the conduction current in the frame of reference of the moving domain itself vanished.

Equation 68 does not state what the common value of c and v_n is. A relationship for this is obtained directly from what is left of Eq.

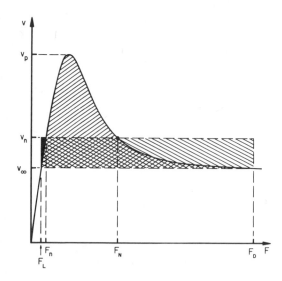

Fig. 21. Butcher's equal-areas rule. The oppositely cross-hatched areas must be equal.

66, after Eqs. 67 and 68 are inserted:

$$\int_{F_n}^{F_D} [v(F) - c]\, dF = 0 \tag{69}$$

This is Butcher's *equal-areas rule,* as illustrated in Fig. 21. According to Eq. 69 the two oppositely cross-hatched areas must be equal.

In most practical cases F_D is located far out along the flat valley portion of the $v(F)$ characteristic, and v_n and c are correspondingly close to the valley velocity. In this case it is useful to add, to *both* of the two areas which are required to be equal, the doubly cross-hatched area which is bordered by $v = v_n$, $v = v_\infty$, and the $v(F)$ characteristic. The trapezoidal area between $v = v_n$ and $v = v_\infty$ can, with negligible error, be replaced by a rectangle, giving Eq. 69 the simplified form,

$$\int_{F_L}^{\infty} [v(F) - v_\infty]\, dF = (v_n - v_\infty) \cdot (F_D - F_L)$$
$$\approx \mu_0 \cdot (F_n - F_L) \cdot (F_D - F_L) \tag{70}$$

where F_L is the low-field value opposite the valley (about 1 kV/cm for a v_∞ of 8×10^6 cm/sec), and μ_0 is the low-field mobility. (Note that the ascending portion of the $v(F)$ characteristic has a very nearly field-independent mobility right up to the peak velocity.) Equation 70 can be rewritten as

$$(F_n - F_L) \cdot (F_D - F_L) = F_A^2 \tag{71}$$

where

$$F_A^2 = \frac{1}{\mu_0} \int_{F_L}^{\infty} [v(F) - v_\infty]\, dF \tag{72}$$

is a quantity that is a property of the $v(F)$ characteristic only and does not depend on the details of the domain. An accurate value for the integral depends on exactly how $v(F)$ approaches the asymptotic value v_∞ and on the exact value of v_∞. If one assumes the analytical approximation of Eqs. 15 and 17, to Ruch and Kino's data, one obtains F_A = 2935 V/cm or roughly 3 kV/cm. There is little point in obtaining a more accurate value, in view of the only approximate validity of the entire Butcher approximation.

It is perhaps surprising to find that the two main results of our treatment, Eqs. 68 and 69, are independent of the magnitude of the (constant) diffusion coefficient. But, given this independence, it is clear at least that Eq. 68 must hold. For, in the case of no diffusion at all, the DL would deplete fully, and this means that no current could cross the DL.

This independence of our results from the magnitude of the diffusion coefficient is a consequence of the vanishing of the second term in Eq. 66, because of Eq. 67; our results could in fact have been obtained by neglecting the diffusion term in Eq. 62 in the first place. But it is readily seen that the diffusion term in Eq. 66 vanishes only because D was assumed to be field-independent, and D could therefore be pulled out of the integral with respect to F. For the much more difficult case of a field-dependent diffusion coefficient, the reader may refer to the work of Lampert,[37] who gives earlier references.

The Domain Shape. Whatever the quantitative limitations to the validity of Butcher's equal-areas rule, there is no doubt that it represents an excellent qualitative approximation, and as such it has numerous important consequences.

The first such consequence we shall consider here concerns the domain shape.[38] In the case of a flat-valley characteristic, the rule requires that the drift velocity at the point of highest field is always slightly less than the drift velocity outside the domain. Since the electron density at that point is equal to the outside electron density, and since the current must be constant with position, this can only mean that the discrepancy is made up by a diffusion current. But since a diffusion current at this point requires a density gradient, this implies that the AL and the DL are *not* separated by a neutral high field domain of finite width, but that they are directly adjacent to each other. In terms of a field versus distance distribution, this means that the domains are "triangular" rather than "trapezoidal" (Fig. 22), i.e., an AL region of increasing field is directly adjacent to a DL region of decreasing field, without a flat-topped region between the two space charge layers.

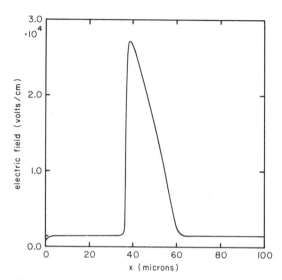

Fig. 22. Electric field versus distance for a mature dipole domain in a crystal with a donor density, $N = 10^{14}$ cm^{-3}.[33]

 The field rise on the AL side is usually much steeper than the field decrease on the DL side. This is a direct consequence of the fact that the space charge density in the AL can—and does—rise to values very much larger than the donor charge density, while in the DL it must of necessity stay below that value. Figure 23 shows an electron density profile obtained by computer simulation which clearly reveals this fact.[33] Clearly, this behavior means that most of the domain voltage drop occurs across the DL. The situation is very much analogous to the voltage distribution in a reverse-biased, very unsymmetrically doped pn junction, where most of the voltage drop occurs across the less heavily doped side. In the first order we can neglect the potential drop across the AL altogether; if we further assume full depletion of the DL, we again obtain Eq. 61 as a relationship between the maximum domain field and the domain voltage.

 Experimentally, it is not U_D that is given, of course, but the overall device voltage U, which is related to U_D by

$$U_D = U - \overline{F}_n L = (\overline{F} - F_n) \cdot L \tag{73}$$

where $\overline{F}$ is the bias field. Insertion into Eq. 61 and solution for $\overline{F}$ produces

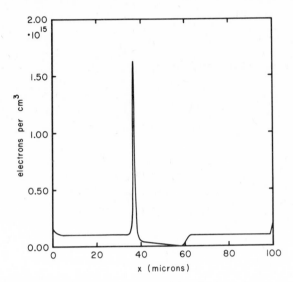

Fig. 23.   Electron density distribution corresponding to Fig. 22.

$$\overline{F} = F_n + \frac{\epsilon}{2qNL} (F_D - F_n)^2$$

$$= F_n + \frac{1}{2F_M} (F_D - F_n)^2 \qquad (74)$$

where

$$F_M = \frac{qNL}{\epsilon} = 1.45 \times 10^5 \, \text{V/cm} \times \left[\frac{NL}{10^{12} \, \text{cm}^{-2}} \right] \qquad (75)$$

is the maximum field that can be supported by the ionized donors in the entire device. For above-critical devices, $NL > 10^{12}$ cm^{-2}, F_M is very much larger than such fields as F_L, F_n, F_A, and F_p.

The two fields F_D and F_n in Eq. 74 are related to each other by the equal-areas rule. If we use the rule in its simple form Eq. 71 we can eliminate F_D from Eq. 74 and obtain

$$\overline{F} = F_L + \Delta F + \frac{1}{2F_M} \left[\frac{F_A^2}{\Delta F} - \Delta F \right]^2 > F_L \qquad (76)$$

We see that $\overline{F}$ depends on F_n only in the combination,

$$\Delta F = F_n - F_L \qquad (77)$$

This difference is always small compared to F_M, so that the ΔF term in the square brackets is small compared to all other terms and can be neglected:

$$\overline{F} = F_L + \Delta F + \frac{F_A{}^4}{2 F_M \Delta F^2} \tag{78}$$

According to Eq. 78 the bias field exhibits a minimum for

$$\Delta F = \Delta F_m = F_A \cdot \sqrt[3]{\frac{F_A}{F_M}} \tag{79}$$

$$\overline{F} = \overline{F}_m = F_L + \frac{3}{2} F_A \sqrt[3]{\frac{F_A}{F_M}} > F_L \tag{80}$$

For lower fields a mature diode domain cannot exist, at least not under our simplifying assumptions.* Actually, only that portion of the $\overline{F}$ versus F_n curve for which $\Delta F > \Delta F_m$, corresponds to physically realizable situations. As we cannot show here, solutions of Eq. 78 with $\Delta F < \Delta F_m$ are unstable against small space charge fluctuations, i.e., any small space charge fluctuation would grow until a solution with with $\Delta F > \Delta F_m$ develops.

Negative Domain Resistance. The result (Eq. 80) has an important consequence: mature dipole domains exhibit a negative differential resistance:[38-41]

$$\frac{d\overline{F}}{dj} = \frac{d\overline{F}}{d\Delta F} \cdot \frac{d\Delta F}{dj}$$

$$= \frac{1}{qN\mu_0} \left[1 - \frac{F_A{}^4}{F_M \cdot (\Delta F)^3} \right]$$

$$\approx - \frac{1}{qN\mu_0} \cdot \frac{F_A{}^4}{F_M \cdot (\Delta F)^3} \tag{81}$$

where the last line assumes $\Delta F \ll \Delta F_m$. The existence of such a negative resistance is experimentally well verified,[42] and in turn, it has several important consequences.

The first of these is the instability of any situation involving the simultaneous presence of more than one mature dipole in a device, such as in Fig. 24. In situations like this the larger domain will grow at the expense of the smaller one. Physically, what happens is the following: The electric field in the neutral region cannot be the equilibrium value for both the larger domain (which requires a smaller F_n) and the smaller one (which requires a larger F_n). But an in-between value of F_n will weaken the space charges in the smaller domain and

---

*Near this point, Eq. 70 can be a poor approximation, but this does not affect the qualitative validity of our argument.

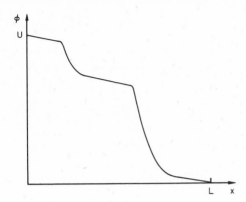

Fig. 24. A situation with more than one dipole domain is unstable.

further strengthen them in the large one, until only the larger domain is left.

A second consequence is the fact that the negative domain resistance can lead to spurious oscillations in the *bias* circuitry, unrelated to the domain transit-time oscillations themselves. Usually, these bias circuit oscillations are of a lower frequency. While in principle easily overcome, they have a considerably nuisance value in that they require considerable attention to the bias circuit design and they limit the design freedom of the circuit design.

Thim[42] has utilized this negative resistance to build an experimental microwave amplifier, using the negative resistance of the traveling dipole domain in a device that was simultaneous by oscillating in the mature dipole mode at a different frequency.

Avalanche Breakdown. Gunn effect devices, like *pn* junction, can break down by avalanche breakdown if the domain field exceeds the avalanche breakdown field F_B. While this phenomenon is not restricted to the mature dipole mode, it is particularly severe there, because the domain field reaches high values early in the oscillation cycle. The bias field at which breakdown sets in can easily be estimated from Eq. 74 by replacing F_D with F_B and neglecting the F_n terms which are much smaller than the F_B term:

$$\overline{F}_B = \frac{F_B{}^2}{2F_M} = \frac{\epsilon F_B{}^2}{2qNL} \tag{82a}$$

In GaAs, F_B is over 2×10^5 V/cm.[43] With this value we obtain numerically:

$$\overline{F}_B = 1.38 \times 10^5 \text{ V/cm} \times \left[\frac{10^{12} \text{ cm}^{-2}}{NL} \right] \tag{82b}$$

Clearly, for $N \cdot L$ products that are not large compared to the critical $N \cdot L$ product, the breakdown bias fields are high compared to the threshold field, a requirement for high-efficiency pulsed operation. However, it can be shown[6, 33] that these high breakdown fields can be realized only in GaAs crystals which are quite free of traps, and that traps can reduce the breakdown voltage to the point that breakdown sets in immediately upon exceeding the oscillation threshold.

Domain Collapse. Earlier in this section we found that there exists a minimum domain-sustaining voltage. If one inserts numbers into Eq. 80, making use of Eq. 75, one finds that for above-critically doped devices the domain-sustaining field always stays below F_p, the domain nucleation field in the limit of no-field enhancement at the cathode. For example, for $NL = 10^{12}$ cm^{-2} the second term on the right side of Eq. 80 amounts only to about 1.2 kV/cm, placing the sustaining field at about 2.2 kV/cm, below the threshold field. For $N \cdot L$ products that are more typical of *mature* dipole operation, say, 8×10^{12} cm^{-2}, $\overline{F}_m$ drops to about 1.6 kV/cm. In other words, once a domain has been nucleated, the bias field can be reduced almost to one-half the field necessary to nucleate the domain, without domain collapse. Under such circumstances, the domain will continue to travel toward the anode. However, once it has disappeared into the anode, no new domain will be nucleated as long as $\overline{F}$ stays below F_p. This behavior is indeed what is observed in devices with large $N \cdot L$ products, but it is important to recognize that it is predicated on the absence of a substantial cathode fall, due to imperfect cathode boundary conditions.

The existence of imperfect cathode boundary conditions, with a crossover point substantially beyond the threshold field, has two consequences:

1. A part of the total applied voltage now is consumed by the cathode fall voltage U_d in Fig. 19, and Eqs. 73 through 80 have to be revised to include this contribution.

2. The current density for uniformly propagating domains is now limited not by j_p but by the crossover current $j_x = qNv_x$.

According to the equal-areas rule this places a *lower* limit on the domain voltage U_D, which in general is higher than the domain-sustaining voltage estimated in Eq. 80.

Actually, the current density must stay significantly *below* the crossover current density because, for current densities approaching the crossover value, the cathode fall itself would occupy the entire device. But this necessity of keeping the cathode fall small further raises the minimum domain voltage.

A general theory of the magnitude of the domain-sustaining voltage, as a function of the crossover field, has not been carried out.
But we can easily illustrate our point by a simple example. Assume
a crossover field $F_x = 10^4$ V/cm. Judging from the fields that Gunn
has observed in some of his stationary cathode fails, [31] this is by no
means an extraordinarily large value. Assuming the analytical approximation Eq. 17 of Ruch and Kino's experimental $v(F)$ characteristic,*
we obtain a drift velocity of 1.16×10^7 cm/sec at this point, corresponding to a $\Delta F = F_n - F_L$ of 200 V/cm (see Eq. 77). Assuming further an $N \cdot L$ product of 2×10^{12} cm^{-2}, we obtain, from Eq. 75, an F_M of
2.9×10^5 V/cm. With these numbers, the last term on the right-hand
side of Eq. 78 becomes 3.5 kV/cm. Adding F_L and ΔF leads to about
4.7 kV/cm as the minimum bias field to sustain the domain alone, not
counting the cathode fall voltage U_d. But if the current were really
equal to the crossover current, the cathode fall by itself would require
a bias field equal to the crossover field, 10 kV/cm, not counting the
domain voltage U_D. The minimum bias field necessary in order to support *both* the cathode fall and the domain is somewhat hard to estimate—and its exact value really does not interest us here—but it is
clearly of the order of 15 kV/cm, far above the bias field needed to
nucleate the domains in the first place.

The fact that the domain-sustaining field is higher than the domain-
nucleating field has several important consequences:

1. Rather than lead to a uniformly propagating domain, as in Fig.
18, it leads to a shrinking domain, as in Fig. 19, or to domain collapse, as in Fig. 25.[18] In fact, both figures were obtained by computer
simulation of devices with crossover fields significantly beyond F_p.
In Fig. 19, F_x was about 7 kV/cm; in Fig. 25, 14 kV/cm. In both
cases $NL = 2 \times 10^{12}$ cm^{-2}, and $\overline{F} \approx 3.5$ kV/cm. It is obvious that, in the
case of Fig. 19, the bias field is at least close to being able to sustain
the domain. In Fig. 25 it is hopelessly inadequate. The above F_x
$= 10$ kV/cm example would lie somewhere in between the two cases.

2. The oscillation frequency is increased, as a consequence of
the fact that each domain collapses before reaching the anode, and a
new domain gets nucleated. In the case of Fig. 25, the calculated oscillation frequency was 20 GHz, which is four times the about 5 GHz
one would expect for a 20-μ-thick device. As the voltage is raised,
the domain collapse slows down and the oscillation frequency decreas-

---

*The experimental data by themselves scatter too much to be used directly
here.

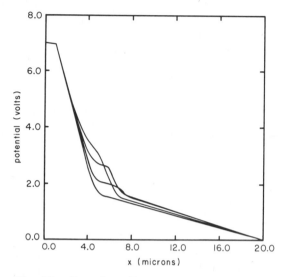

Fig. 25. Domain collapse, in a device similar to the one shown in Fig. 19, but with a crossover field of 15 kV/cm, at a bias field of 3.5 kV/cm. (After Kroemer.[18])

es. Such types of behavior are frequently observed experimentally.

3. The amplitude (i.e., the peak-to-valley ratio) of the oscillations, and thereby their efficiency, are drastically reduced because a new accumulation layer gets nucleated as soon as the current overshoots the crossover current, thereby dropping the current again.

From a practical point of view, this last consequence is perhaps the most important and the most undesirable one. It shows that, for high-efficiency devices, a crossover point close to the threshold field (or below) is absolutely essential. It also explains why most practical devices at present have efficiencies that are very much lower than should be theoretically possible.

2.4.3 The Role of Crystal Inhomogeneities

Dipole Formation for Positive Doping Gradients. Throughout our discussion of dipole domains, and to the extent that the question after the origin of the dipole formation arose, we have discussed this dipole formation as being a consequence of imperfect cathode boundary conditions. We must now turn to the second major mechanism of dipole domain formation, which can be operative even in devices with perfect cathode boundary conditions: crystal inhomogeneities. By this we mean such structural details as doping inhomogeneities, deviations

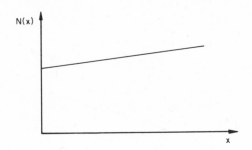

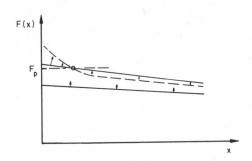

Fig. 26. The doping distribution in a crystal with a small positive doping gradient, and the resulting evolution of the field distribution, for bias field $\overline{F}$ approaching F_p.

from one-dimensional geometry, or anything else that would cause field nonuniformities in the device's active region, upon the application of a voltage. We shall concentrate here on doping nonuniformities, and we shall illustrate our point by considering the case of a crystal with perfect cathode boundary conditions, but with a (shallow) doping gradient, corresponding to a monotonic rise of the donor density from the cathode to the anode (Fig. 26). The electron density, of course, behaves in essentially the same way, i.e.,

$$\frac{\partial}{\partial x} n(x) > 0 \tag{83}$$

Assume now that a below-threshold bias field is gradually applied to the device. Because of the conductivity gradient throughout the structure, the resulting internal field will be nonuniform, with the highest field occurring near the cathode. The same is true for the drift velocity,

$$\frac{\partial}{\partial x} v[F(x)] < 0 \tag{84}$$

If the time required for the field to rise up to the threshold field is long compared to the low-field dielectric relaxation time, displacement currents can be neglected compared to conduction currents and

the conduction current will remain essentially divergence-free during
this rise, or

$$\frac{\partial n}{\partial t} = -\left[n \frac{\partial v}{\partial x} + v \frac{\partial n}{\partial x} \right] \approx 0 \qquad (85)$$

i.e., the velocity gradient and the density gradient terms cancel each
other out.

As the voltage is raised further, the field will exceed the threshold value, beginning at the cathode. At this point the drift velocity
near the cathode begins to decrease, to values less than those immediately downstream of the cathode, i.e., the velocity gradient term in
Eq. 85 reverses its sign. Since Eq. 83 is still valid, cancellation of the
two terms in Eq. 85 becomes impossible and we now obtain

$$\frac{\partial n}{\partial t} < 0 \qquad (86)$$

i.e., the electron density near the cathode, which is already less there
than it is further inside the sample, drops even lower. This, of
course, further increases the density gradient term Eq. 83, but more
important, it leads to a positive space charge ahead of the cathode.
This further increases the *field* at the cathode and thereby further decreases the velocity and the velocity gradient term, thus accelerating
the electron density decrease further. Clearly, we have here a space
charge runaway phenomenon typically for a negative mobility medium,
leading to the formation of a depletion layer ahead of the cathode by a
mechanism very similar to the mechanism downstream of a cathode
with imperfect cathode boundary conditions: the electron influx from
the cathode gets restrained below the rate at which electrons are removed from the cathode vicinity. The only major difference lies in
the effect of the boundary conditions on the formation of a subsequent
accumulation layer. In our present case of perfect cathode boundary
conditions there exists no crossover point and an AL can therefore
form as soon as the cathode field reaches F_p, i.e., as soon as the DL
itself forms. There will, therefore, be no voltage-dependent preos-
cillation cathode fall.

A weak positive doping gradient, combined with perfect cathode
boundary conditions, should therefore lead to a pure dipole mode.
Whether the dipoles will actually mature on their way to the anode is
again a question of the $N \cdot L$ product.

Negative Doping Gradients. If, in a device with a positive doping
gradient, the voltage polarity is reversed, one obtains a negative doping gradient. The highest prethreshold electric fields now occur on the

anode side; upon reaching threshold, an accumulation layer forms adjacent to the anode, the consequences of which are drastically different.

A DL, formed inside the device, enhances the electric field at the cathode and thereby leads to the injection of an AL from the cathode, as well as the formation of a dipole. This dipole can then grow as it propagates deeper into the device, leading to oscillations. By contrast, an AL formed inside the device *reduces* the field at the cathode. We must then distinguish between two possibilities:

1. This reduction in field is weak enough, and the rate of rise of the bias field, as well as the final value of the bias field, are large enough so that the formation of the "secondary" AL inside the device merely delays slightly the time during which F_c reaches F_p and a "primary" AL gets injected from the cathode.

2. The opposite conditions prevail and a primary AL does not form.

In case 1 we obtain essentially a pure accumulation mode, but with the difference that a weak secondary charge accumulation exists throughout the entire space between the primary AL and the anode. Since this entire space is biased in the negative mobility range, this is a highly unstable situation and these charges effectively attract each other, tending to coalesce into a single localized AL. As a result, the primary AL moves considerably faster than in the gradient-free pure accumulation mode, and the oscillation frequency is increased.

As pointed out in Section 2.4.1, the pure accumulation mode is inherently a high-efficiency mode for microwave oscillators but it is sensitive to the presence of doping gradients. As long as these gradients are weak, they will not appreciably affect the efficiency; but the *sign* of the gradient will affect the direction into which the breakup of the high-field domain of the pure accumulation mode goes. For positive gradients the breakup goes into the direction of a dipole mode with lower space charge layer velocities and frequencies. For negative gradients the AL widens, leading to higher space charge layer velocities and frequencies. In this context it is interesting to note that it has been observed that some of the highest-efficiency ($\cong 30\%$) Gunn effect microwave oscillators built to date[44] clearly fell into two groups that were made identically but which differed in oscillation frequency by a ratio of about 3:1. We suspect that the existence of weak residual gradients in the GaAs wafers are responsible for this behavior and that the sign of these gradients is different for the two groups.

In case 2 above, any oscillations can arise only from the internal nucleation, growth, disappearance, and renucleation of the secondary AL which is caused by the doping gradient. But since this secondary AL is nucleated close to the anode, it has very little space available in which to grow. Qualitatively speaking, the device behaves as though only that region near the anode, where the field is larger than F_p, acts as a Gunn effect diode, with the remainder of the crystal acting as a separate, parasitic series resistance. Since the "effective diode" is thinner than the overall device, and since it has a lower doping level, it has a lower $N \cdot L$ product, and if this $N \cdot L$ product is below-critical, no oscillations will take place. With increasing bias voltage, the negative mobility layer will, of course, grow thicker, and ultimately its $N \cdot L$ product should become above-critical. But even then, the residual positive mobility layer, rather than the heavily doped cathode region, will act as the effective cathode. Since the doping level of the positive mobility region merges continuously into that of the negative mobility region, this effective cathode can inject only very weak ALs. This weakens the reinjection feedback loop gain, however, and as a result the critical $N \cdot L$ product for the negative mobility portion of a negative gradient structure will be significantly larger than for structures with a heavily doped true cathode.

In the case of a relatively weak, negative gradient this usually means that oscillations will not set in until the bias voltage has been raised to the point at which the cathode field reaches F_p. As the gradient is increased, the bias field necessary for the onset of oscillations will increase. At the same time the field at the anode will increase for a given bias field; since a higher bias field is required for the onset of oscillations, it will increase faster than the oscillation threshold bias field itself. For sufficiently strong gradients the field at the anode end of the device may reach the avalanche breakdown field before the onset of oscillations, leading to the same destructive consequences already discussed on page 57. In fact, high-resistivity layers, adjacent to the anode contact, are a well-known cause of the destructive breakdown of Gunn effect devices.

In many practical cases, crystal inhomogeneities do not take the form of simple monotonic gradients, but of an alternation between positive and negative gradients. This can lead to very complicated space charge dynamics;[13] the interested reader is referred to the literature.

2.5 SPACE CHARGE DYNAMICS UNDER RESONANT LOADING

2.5.1 Space Charge Layer Delay and Quenching

Throughout Section 2.4 we dealt with space charge instabilities under fixed bias conditions. Actual Gunn effect microwave oscillators always operate in resonant circuits, i.e., under bias conditions containing a large a-c component. As long as the amplitude of the a-c component remains small enough so that the device stays above its threshold voltage during the entire oscillation cycle, nothing fundamentally new happens. To be sure, an exact calculation of the device behavior might be a matter that is as tedious as it is important. But it is basically a question of circuit engineering, not one involving additional physical principles.

The situation changes if the a-c amplitude drops the device voltage periodically below the threshold value, either for a time long enough for the space charge layers to reach the anode during that time, or to a value low enough for the space charge layers to collapse, or both. In either case, new space charge layers will not get nucleated until the device voltage reaches the threshold voltage again. This means, however, that the oscillation frequency is not determined by the transit time of the traveling space charge layers through the device but by the resonant frequency of the loading circuit. This remains true even if the circuit resonance frequency differs drastically from the transit-time frequency; in this way operation at frequencies drastically different from the transit-time frequency is indeed obtainable, particularly at frequencies that are large compared to the latter.

For what follows it will be important to distinguish between the two different reasons for the disappearance of the space charge layers.

1. If the space charge layers disappear into the anode during the below-threshold phase of each cycle, the resonant operation actually delays the nucleation of new space charge layers, as opposed to non-resonant operation. It is common to refer to this mode as *delayed space charge* or *delayed domain* mode.[45] The latter term implies the participation of high-field dipole domains, but the analysis of this mode of operation is the same for pure accumulation layers. This analysis is presented in Section 2.5.2.

2. If the space charge layers collapses inside the device because the device voltage drops below the value necessary to sustain the space charge layers, the oscillation frequency may be either below or above the transit-time frequency. The low-frequency case is of no particular interest: compared to a nonquenched delayed mode, it offers no advantages and it has, in fact, several disadvantages. The interesting

case, on which we shall concentrate here, is the quenched operation at frequencies that are larger than the transit-time frequency, potentially very much larger. In this case it is necessary to include the details of the space charge layer distribution of the particular mode involved. In its full generality this requires a computer treatment. We shall consider here only the two *limiting* cases, the quenched mature dipole mode with negligible cathode fall, [46] and the quenched pure accumulation mode, more commonly referred to as the LSA (limited space charge accumulation) mode. [47,48] Since the quenched mature dipole mode is the simpler of the two, we shall discuss it first, in Section 2.5.3 and then the LSA mode in Section 2.5.4.

2.5.2 The Delayed Space Charge Mode

In Fig. 27 we show what might be the device voltage U, as a function of time, across a device operating in the delayed space charge mode. For simplicity we have assumed a purely sinusoidal voltage and a d-c bias U_{dc} which is slightly larger than the threshold voltage U_{th}. The minimum voltage, U_m, is assumed to stay above the value U_s necessary to *sustain* any space charge layers, whatever that may be.

We obtain delayed space charge operation if

$$\tau_n < \tau_T < \tau_f = \frac{1}{f} \tag{87}$$

where τ_n is the length of the interval during which the device is biased above threshold, i.e., in a field range where the field *average* corresponds to a negative mobility.

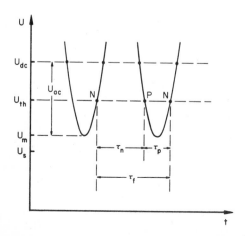

Fig. 27. The device voltage, as a function of time, for delayed space charge operation.

For a d-c bias field just equal to the threshold field, $\tau_n = \frac{1}{2}\tau_f$, indicating that for such a bias field the device can be circuit-tuned anywhere between $0.5f_T$ and $1.0f_T$.[45] While there is relatively little interest in the fact that these frequencies are lower than the transit-time frequency, there is considerable interest in being able to circuit-tune Gunn effect microwave oscillators over such a wide band in a large number of applications.

At a d-c bias higher than the threshold voltage the circuit tuning range gets narrower than $2:1$. It is possible to sustain oscillations even at d-c bias voltages below the threshold voltage if their amplitude is such that the voltage exceeds the threshold voltage during each cycle, and if the oscillations themselves generate enough power to be self-sustaining at this amplitude. By damping out the oscillations below this minimum amplitude, they can be extinguished; by triggering them again with a suitable transient they can be restarted. This bistable behavior can be utilized for memory purposes.[49]

2.5.3 The Quenched Mature Dipole Mode

In Section 2.4.2 it was stated a mature dipole with negligible cathode fall can be sustained at a voltage less than $U_{th} = F_p \cdot L$, the threshold voltage for its own nucleation, provided that the voltage stays above some sustaining voltage U_s, which was somewhat larger than $F_L \cdot L$, the excess over this lower limit depending on the $N \cdot L$ product (see Eq. 80). We now consider the events that take place if the voltage across the device dips below U_s during every oscillation cycle (Fig. 28). We assume that the oscillation period is shorter than the domain transit time,

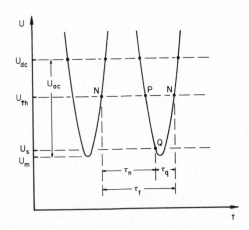

Fig. 28. The device voltage, as a function of time, for quenched domain operation.

$$\frac{1}{f} = \tau_f = \tau_n + \tau_q < \tau_T \qquad\qquad (88)$$

Under this condition the domains nucleated, during every oscillation cycle, at point N in Fig. 28, will still be present when the voltage reaches point Q, at which time they get quenched. By presenting the crystal with a resonant circuit of sufficiently high resonant frequency, it is possible to obtain Gunn effect oscillations at a frequency much higher than the transit-time frequency.[46]

This increase in oscillation frequency cannot, of course, be carried arbitrarily far, for the domain nucleation is neither for an instantaneous process nor does the domain adjust itself instantaneously to the voltage changes, including domain collapse. The speed of all of these processes depends on the doping level, and, in order to obtain a quenched operation of the mature dipole mode at frequencies significantly above the transit-time frequency, it is necessary that the $N \cdot L$ product be higher than for nonquenched operation. We shall not study the high-frequency limitations of the quenched dipole mode here, because domain quenching is a rather inefficient way to obtain higher frequencies than the transit time frequency.

As pointed out in Section 2.4.2 the mature domain mode is a comparatively inefficient mode to begin with; quenched operation reduces the efficiency further. There are two reasons for this.

1. Under quenched operating conditions the domain never propagates farther than a certain distance from the cathode, and the portion of the device that lies beyond this quenching distance acts purely as a positive series resistance. Not only does it stay in the positive mobility region during the time between quenching and renucleation, it returns to this region as soon as the domain has been renucleated. It is clear that such a series resistance, which is essentially always positive, does not contribute to the generation of oscillation energies but it dissipates such energy. Obviously, oscillations at the same frequency could be generated more efficiently by omitting this parasitic resistance, i.e., by making the semiconductor body shorter, so that its transit-time frequency would coincide with the desired oscillation frequency.

2. The second reason for the inefficiency lies in the fact that the quenching process itself reduces the amount of oscillation energy available from the device. Perhaps the easiest way to visualize this is to consider not a sinusoidal voltage wave shape, but a square wave, switching the device between a voltage below U_s and one above U_{th}. If we ignore transient effects, including the finite domain readjustment

time, the current then also exhibits a square wave.

In order for such an oscillation to correspond to power generation by the device (and delivery to some load), rather than power absorption, the current and the voltage wave must be in opposite phase. That is, the current at the quenching voltage must be higher than the current at the peak voltage. In other words, the line connecting the two operating points must correspond to a negative resistance. This is illustrated in Fig. 29, where the current through the device is shown, as a function of voltage, both in the absence and in the presence of a mature dipole domain.

Since the quenching voltage must be below the sustaining voltage, and since the sustaining current is typically significantly below the threshold current, it is clear that the oscillation power under quenched conditions is much less than the oscillation power available under non-quenched square wave drive, between the peak current, and the domain branch of the device current-voltage characteristic.

Under sinusoidal voltage conditions the situation becomes quantitatively much more complicated, but qualitatively our conclusions remain valid.

It is apparent from our square wave analysis, and from the de-

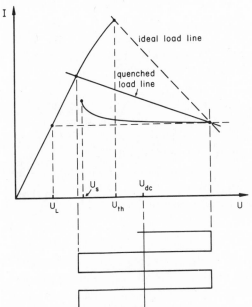

Fig. 29. The current-voltage characteristic of a mature domain, and the switching between the two branches of this characteristic, under quenching square wave drive.

pendence of the sustaining field on the $N \cdot L$ product, as given by Eqs. 75 and 80, that the current amplitude, relative to the d-c current, actually decreases with increasing $N \cdot L$ product, i.e., the oscillation power efficiency decreases. Conversely, the lower the $N \cdot L$ product, the closer will the sustaining field move to the threshold field, thus increasing the quenched mode efficiency. This, of course, rapidly leads to $N \cdot L$ products, for which the domains will no longer mature. The ultimate would be a mode without a DL at all, a quenched pure accumulation mode. We shall see in Section 2.5.4 that such a mode does, in fact, exhibit a much larger current amplitude, close to the situation indicated by the dashed load line in Fig. 29, and that it also eliminates the problem of a positive parasitic series resistance, mentioned above.

For the moment we wish to pursue another point. The narrow voltage range between U_L and U_s in Fig. 28 means that, under quenched operating conditions, the a-c voltage amplitude must all in a very narrow range; the question might arise as to how this amplitude is established. Fortunately, this amplitude is largely self-regulating. If the device would oscillate in a lossless and unloaded circuit, the amplitude would build up until the voltage reaches U_L during every cycle, at which point the power generated by the device becomes zero. As a-c power is extracted from the circuit, the oscillation amplitude drops, and the device now delivers power to the circuit. However, if the device is overloaded, the voltage will no longer drop below U_s and the entire quenched mode operation collapses, just at the point of maximum delivered power. As a result, quenched operation requires rather careful circuit design, a matter that goes beyond our scope here.

From a practical point of view, the main interest in the quenched domain mode lies in its use for those applications, like signal generators, for which tuning of the oscillation frequency over a wide frequency band is the primary consideration, rather than high power. The frequencies in such applications are frequently low enough so that they could readily be obtained by transit-time operation; but only by quenched operation is the desired wide band tunability achievable.

2.5.4 The LSA Mode

Quenched Accumulation Layers. We are now getting to what is undoubtedly the most important mode of quenched device operation, the so-called LSA mode.[47, 48] In our systematic terminology it might be called a quenched pure accumulation mode, but the term LSA mode is well established, and we shall follow this convention here.

The reaons for the importance of the quenched accumulation mode are twofold; they can best be illustrated by comparing this mode with the quenched dipole mode, as discussed in Section 2.5.3.

1. In the quenched dipole mode the field in that portion of the device that lies between the point of dipole quenching and the anode permanently stayed essentially in the region of positive mobility. By contrast, in the quenched accumulation mode, the field in this portion of the device stays biased in the range of negative differential mobility, except during the brief quenching and reinjection of the ALs. As we shall see presently, this field needs to drop only insignificantly below the threshold field during the quenching period, and this quenching period can be much shorter than in the quenched dipole mode. As a consequence the portion of the device that is never reached by the AL does not act as a parasitic *positive* series resistance but as a *negative* resistance. Such a negative resistance, of course, *generates* oscillation power rather than absorbs it; as a result frequencies that are much higher than the transit-time frequency can much more readily be achieved than in the quenched dipole mode.

2. A pure AL is much more readily quenched than a mature dipole, *particularly* during the earlier parts of its travel toward the anode, i.e., if the oscillation frequency is large compared to the transit-time frequency. This is readily seen by considering the magnitude of the field on the high-field (= anode) side of the AL, as a function of the position of the AL. If the AL is still close to the anode, the field on the anode side of the AL exceeds the instantaneous bias field by very little. Consequently, the bias field has to be dropped only slightly below the threshold field, in order also to drop F_a to a value below the threshold field. But under these conditions the entire crystal drops into the range of positive mobility and the AL disperses.* That the voltage has to be dropped only slightly below the threshold voltage to achieve quenching also implies that the current during quenching stays close to the threshold current. As a result the quenched accumulation mode has a very much higher efficiency than the quenched dipole mode *if* the oscillation frequency of the former is high compared to the transit-time frequency.

---

*Actually, the situation is more favorable. It is not necessary to drop $\bar{F}$ to the point that $F_a < F_p$, in order to quench the AL. A quantitative discussion of AL quenching goes beyond the scope of our discussion; our simple presentation shows that, even under our more stringent assumputions, the AL is readily quenched.

It is in this last case that we shall use the term LSA mode, rather than the quenched accumulation mode. Under LSA conditions, i.e., when the oscillation frequency is large compared to the transit-time frequency, the AL proceeds only such a short distance into the crystal that for some purposes its existence may, in a first-order treatment, be neglected altogether, and the behavior of the entire device may be treated as though the device were simply a bulk negative conductance. This fact greatly simplifies the theoretical and circuit analysis of the device.

Structual Requirements for SLA Devices. We must now turn to the detailed conditions that must be satisfied actually to obtain LSA oscillations. One of these conditions we have already mentioned: the need for an oscillation frequency that is large compared to the transit-time frequency,

$$f \gg f_T \tag{89}$$

or, to put it in structual terms: the need for a device length such that the transit-time frequency is small compared to the—presumably prescribed—oscillation frequency. But this is not a sharp criterion; in fact, the quenched accumulation mode as such is not characterized by this restriction. The condition Eq. 89 is required merely in the interest of high efficiency and, from a theoretical point of view, because under this condition many of the aspects of the device can be treated simply as though the device represented a bulk negative resistance.

Using this limit of device performance, Copeland, in a classical paper,[48] has given a quantitative analysis of the conditions for LSA operation. We outline his results here.*

The first set of requirements arise from the fact that, during the finite quenching time, no space charge will get *completely* quenched. In a "perfect" device the only space charge would be that of the primary AL. During quenching, this space charge will become attenuated to a very small but nevertheless finite residue. As the voltage goes above threshold again, this residue will find itself on the downstream side of the *new* primary AL, in the region of negative differential mobility, and it will grow again. Obviously, in order to obtain true quenched operation, the "regrown" residue of the old AL must stay weak compared to its original strength which it had before quenching. On the next quenching cycle this residue will then get quenched further,

---

*Our line of argumentation differs somewhat from that of Copeland, but the final results are the same.

rapidly decreasing in strength with every cycle. The entire situation is quite similar to the initial growth of a weak AL, as we discussed earlier, in Section 2.3.3. There we saw that the strength of the AL initially grows exponentially (Eq. 37), with a time constant equal to that dielectric relaxation time which corresponds to the absolute value of the negative differential mobility. In the present case the electric field does not stay inside the region of negative mobility, and the mobility becomes strongly time-dependent. Instead of Eqs. 37 and 38 we thus obtain, for the factor G by which the residue of the primary AL grows (or decays) during one cycle:

$$G = \exp\left[-\tau_f \cdot \left(\frac{1}{\tau_D}\right)\right] \tag{90a}$$

where

$$\left(\frac{1}{\tau_D}\right) = \frac{qN}{\epsilon \tau_f} \int_0^{\tau_f} \frac{dv}{dF}\, dt = \frac{qN\bar{\mu}}{\epsilon} \tag{91a}$$

Here,

$$\bar{\mu} = \frac{1}{\tau_f} \int_0^{\tau_f} \frac{dv}{dF}\, dt \tag{91b}$$

is the time-average electron mobility, on the downstream side of the primary AL, averaged over one full cycle. With Eq. 91, Eq. 90a can be rewritten

$$G = \exp\left[-\frac{N}{f} \cdot \frac{q\bar{\mu}}{\epsilon}\right] \tag{90b}$$

In order to obtain true quenched operation, G must of course be small compared to unity. Copeland has investigated this question quantitatively and has concluded that the plausible choice,

$$G < e^{-1} \tag{92}$$

is a reasonable requirement for LSA operation. Together with Eq. 90b this leads to

$$\frac{N}{f} > \frac{\epsilon}{q\bar{\mu}} \tag{93}$$

The right-hand side of Eq. 93 depends strongly on the d-c bias value, the a-c amplitude, and the wave shape of the electric field oscillations. For a simple symmetric square wave, operating between some point on the linear portion of the $v(F)$ characteristic and some point in the flat valley (Fig. 30), we would have $\bar{\mu} \approx 0.5\,\mu_0$. This would lead to the approximate requirement

$$N/f \gtrsim 10^3 \tag{94a}$$

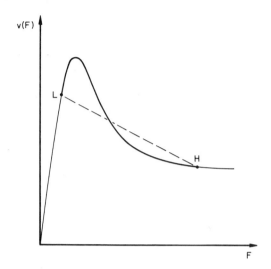

Fig. 30. Bulk negative resistor-style operation in the LSA mode.

sec/cm³. For sinusoidal operation the time spent on the positive mobility portion is very much shorter than for a symmetric square wave of the same amplitude, and a significant fraction of the total oscillation period is spent traversing those portions of the $v(F)$ characteristic that have a relatively large negative mobility. The net result is a drastic reduction in the average mobility. Copeland has considered this situation in considerable quantitative detail. He concludes that, under conditions in which the device *generates* net oscillation power, the time-averaged mobility will be less than about 0.02 μ_0, a fact that he expresses as the requirement

$$\frac{N}{f} > 2.10^4 \tag{94b}$$

sec/cm³. For smaller N/f ratios than given by Eqs. 94a and 94b the necessary degree of quenching would be obtainable only by increasing the a-c field amplitude on the positive mobility side of the $v(F)$ characteristic, so that no net oscillation power would be generated. For wave shapes intermediate between square wave and purely sinusoidal operation, the lower N/f limit will, of course, be intermediate between the two values given above.

In addition to the lower limit for the N/f ratio, there is also an upper limit. This can be seen as follows. The first Copeland condition, Eq. 92, makes no statement about the factor $G_n > 1$ by which any space charge residue grows during the *negative* mobility portion of each cycle, before it gets attenuated even more strongly, by G_p

$< 1/G_n$, during the subsequent *positive* mobility portion. Clearly, the larger the N/f ratio, the larger will be the amplitude with which the strength of the space charge residue oscillates. If one were dealing with a mathematically perfect, pure accumulation mode, the magnitude of G_n would not matter. But the pure accumulation mode is an idealization. Actual devices may have imperfect boundary conditions or macroscopic doping gradients; even in the most perfect structures, there will always be inherently unavoidable statistical fluctuations, due both to the randomness of the location of the individual impurity atoms and to noise.

All of these deviations from ideal pure accumulation mode behavior lead to the formation of additional, secondary, space charge layers downstream of the primary AL, in a region of negative differential mobility. Just like the primary AL, they get regenerated in full strength during every oscillation cycle. If now the growth factor G_n is large enough, these secondary space charge layers can grow to a strength comparable to the mature primary AL; they may, in fact, mature themselves. In either case, the existence of strong secondary space charge layers makes the quenching of *all* space charge layers more difficult, and thereby essentially destroys the advantages of LSA operation. This is quite obvious in the case of a strong secondary DL, such as it would be formed by imperfect cathode boundary conditions. Together with the primary AL it would clearly tend to form a dipole domain, and for a sufficiently large G_n we should obtain not the LSA mode, but the quenched dipole mode, with its quite inferior properties.

If the secondary space charge layer is a pure AL itself, as in the case of a strong negative doping gradient, the situation would be somewhat different, but also in this case the increased field nonuniformity throughout the device makes the space charge layer quenching more difficult.

If one assumes that imperfect cathode boundary conditions can, in principle, be avoided, the maximum permissible G_n is a function of the degree of nonuniformity of the semiconductor employed. Clearly, in sufficiently nonuniform crystals even no growth at all ($G_n = 1$) would cause sufficiently strong field nonuniformities to prohibit LSA-type operation. If one assume a 1% doping nonuniformity as achievable, a G_n value around 100 appears to be a reasonable value. Copeland's calculations were performed with the very similar limit,

$$G_n < e^5 \tag{95}$$

apparently arrived at by similar reasoning.

A G_n limit such as Eq. 95 can be translated into an upper limit for the N/f ratio through the relationship

$$G_n = \exp\left[-\frac{N}{f} \cdot \frac{q\bar{\mu}_n}{\epsilon}\right] \qquad (96)$$

which is similar to Eq. 91b, but where $\bar{\mu}_n$ represents the average mobility during the *negative mobility portion* of the oscillation cycle only. From Eqs. 95 and 96 we obtain the corollary to Eq. 93,

$$\frac{N}{f} < \frac{5\epsilon}{q|\bar{\mu}_n|} \qquad (97)$$

A reasonable low limit for $(\bar{\mu}_n)$ might be $0.01\ \mu_0$. Such a value is readily achievable under square wave drive; for sinusoidal drive it would require a high d-c bias field, in order to sweep through the region of high negative mobility as rapidly as possible. With this value the right-hand side of Eq. 97 becomes 2×10^5 sec/cm^3 and, together with Eq. 94b, this leads to Copeland's N/f range

$$2\times10^4 \frac{\text{sec}}{\text{cm}^3} < \frac{N}{f} < 2\times10^5 \frac{\text{sec}}{\text{cm}^3} \qquad (98)$$

a range that is in good agreement with experimental data by Copeland himself,[47,50] and by Eastman and Kennedy.[51]

It is perhaps interesting to note that an N/f ratio of 10^5 sec/cm^3 implies the same doping level, for a given frequency, as the doping level for a transit-time device for the same frequency, with an $N \cdot L$ product that is just critical $(\approx 10^{12}$ cm$^{-2})$. At least in the lower portion of the N/f range Eq. 98, LSA devices are therefore less heavily doped than transit-time devices.

Operating Conditions for LSA Devices; Their Limitations. In our discussion of the upper limit of the N/f ratio we noted that devices near this upper limit would operate in the LSA mode only for very high d-c bias fields. Copeland has estimated the minimum bias fields for a few sample values of the N/f ratio. For $N/f = 1\times10^5$ sec/cm^3 he finds 12 kV/cm, for 5×10^4 still about 6 kV/cm; only near the lower edge of the N/f range Eq. 98 does the minimum bias field approach the oscillation threshold itself. The LSA mode is thus a mode requiring an inherently harder drive than the various transit-time modes.

A second requirement for LSA operation is a proper load impedance, of a value suitable to connect the two desired terminal operating points in Fig. 30. Typical load resistances are between 10 and 100 times the low-field resistance of the device. For details we wish to refer to Copeland's paper.[48]

If both the d-c bias field and the load resistance are optimized, the LSA mode should be capable of remarkable efficiencies. For example, for $N/f = 4 \times 10^4$ sec/cm^3 Copeland has estimated a broad efficiency maximum of about 17% for a d-c bias field between 10 and 15 kV/cm. For lower N/f ratios the peak efficiencies drop rapidly; for higher ratios they increase only slightly.

For symmetric square wave drive the efficiency would be higher. For example, Copeland gives an example with an efficiency of 21.5%. But true square wave drive is all but impossible to achieve. It is true, however, that "squaring up" the wave shape by suitable tuning measures at the third harmonic is beneficial to the entire device operation, a fact already recognized by Copeland.

Regardless of the drive wave shape, there is some question as to whether those high efficiencies are compatible with *stable* device operation. Basically, they are obtained by pushing the lower oscillation limit (point L in Fig. 30) as close to the threshold field as is compatible with the quenching requirement Eq. 92. Obviously, a small—temporary or permanent—disturbance in the operating conditions such as a change in the load resistance or the bias voltage could destroy the quenching condition. In such a case LSA oscillations could not start and, if they are already underway, they would collapse, switching to lower-frequency, transit-time oscillations. If the disturbance is removed, the oscillations should switch back into the LSA mode. But as Copeland has already shown, the starting of LSA oscillations takes many transit-time cycles.* In fact, the more marginally the quenching condition is satisfied, the slower is the switchover, and it might not take place voluntarily at all. In order to assure stable and reliable LSA operation it might well be necessary to operate under conditions other than those leading to the predicted maximum efficiencies.

All of these efficiency limitations are related to the need for quenching. Although the LSA mode has a much higher efficiency than the quenched or unquenched dipole mode, or than the quenched pure accumulation mode operating near the transit-time frequency, it is nevertheless not as potentially efficient as an unquenched pure accumulation mode or a "hybrid mode" which is close to the latter. We saw in Section 2.4.1 that, for the pure accumulation mode, the current peak-to-valley ratio does approach the velocity PVR. Also, the d-c bias field is restricted only by breakdown considerations, which makes possible high voltage PVR's, further increasing the efficiency. Indeed,

---

*This, in itself, is a limitation of the LSA mode.

the highest efficiencies ever reported for Gunn effect oscillators, around 30%, were reported for devices that almost certainly operated not in the LSA mode,[44] but probably in a mode very close to the pure accumulation mode.

It is therefore probably clear that the LSA mode, despite its many merits, is not a universal mode, ideal for all application; its usefulness lies primarily in those applications for which its principal feature, its high f/f_T ratio, represents an important advantage. There are two such applications: (a) the generation of very high frequencies and (b) the generation of very high powers. They will be discussed below.

LSA *Applications*. It probably is obvious that the LSA mode is particularly suited to the generation of frequencies that are large compared to, say, 10 GHz. Transit-time devices for such frequencies would require device thicknesses small compared to 10 μ. Such thin layers are difficult to prepare with good uniformity. By employing the LSA mode, much thicker devices can be used, which are easier to construct, despite the high homogeneity requirements of the LSA mode.

Perhaps a more profound reason for the LSA mode in such applications is the following: in a transit-time device the thickness decreases inversely proportional with increasing frequency. Furthermore, since the $N \cdot L$ product must be kept in some reasonable range, the proper doping level increases proportional to the frequency. The combination of reduced thickness and increased doping implies that the power per unit area remains the same, but that the device impedance, for a given amount of power, decreases inversely proportional to the square of the frequency. In order to keep the device impedance at some reasonable level, the device area, and thereby the device power, must be reduced, inversely proportional to the square of the frequency, or by 6 dB per octave. This is, of course, a result that is familiar from most other electronic devices as well. LSA devices are one of the few exceptions: even though their proper doping, too, increases proportional to the frequency, the device thickness does not have to be reduced. In fact, it might be increased to compensate for the (linear) impedance decrease due to the increased doping. By increasing the length proportionately to $\sqrt{f}$ and by decreasing the area inversely proportionately to $\sqrt{f}$, both the impedance and the power can be kept constant, and the reduction in area at the same time compensates for the increasing skin effect. As one might expect, this type of scaling cannot be carried on arbitrarily far; ultimately several new limitations set in.[48] To the extent that these are fully understood, their discussion goes beyond the scope of our presentation. For our purposes, it is sufficient to have shown why the LSA mode is very much superior to the transit-time modes for the generation of very high frequencies. In fact, almost all Gunn effect work at frequencies above, say, 40 GHz appears to have employed this mode.

Besides pushing to the highest frequencies, the LSA mode can also be utilized to push to the highest powers, at comparatively low frequencies, say, below 10 GHz. In a transit-time device an increase in power requires an increase in *area* which quickly drops the device impedance below useful levels. In the LSA mode the power can be increased by increasing the device *length* instead, which *increases* the device impedance. By increasing both the device length and the device area in proportion, very high power can be achieved for a fixed impedance. Because of skin effect limitations this approach is particularly effective at lower frequencies; the highest rf pulse power achieved to date, about 6 kW, was achieved at around 1.8 GHz, using this principle.[17]

The ultimate limitation in this case is, of course, a thermal one. The larger the volume of the device, the longer it will take to remove any given amount of heat from the device. Since the total amount of heat generated is actually proportional to the volume, such high rf powers can be generated only in the form of very short pulses of low-duty cycle, with both the pulse length, and even more, the duty cycle decreasing rapidly with increasing power.

This situation is aggravated by two of the inherent limitations of the LSA mode: (a) its quenching limitations on efficiency, and (b) its sensitivity to resistivity gradients,[52] even if they are not caused by doping gradients but by temperature gradients. The efficiency limitations imply that the amount of heat dissipated per unit rf power generated is actually larger than in the more efficient nonquenched accumulation or hybrid modes. The sensitivity of the LSA mode to resistivity gradients implies that the LSA mode can actually tolerate only a smaller temperature rise inside the device than those other, less sensitive modes.

It is clear from these considerations that the LSA route to high powers is a practical one only if the requirements in pulse duration and duty cycle are quite low, and the requirements on power so high that a non-LSA device would face impedance limitations. For power levels that can be achieved with transit-time devices of a practicable impedance level, the use of such devices is more practical.

REFERENCES

1. J. B. Gunn, "Microwave Oscillations of Current in III-V Semiconductors," *Solid State Commun.*, 1, 81–91 (September 1963).
2. J. B. Gunn, "Instabilities of Current in III-V Semiconductors," *IBM J. Res. Develop.* 8, 141–159 (April 1964).
3. B. K. Ridley and T. B. Watkins, "The Possibility of Negative Resistance in Semiconductors," *Proc. Phys. Soc. (London)*, 78, 293–304 (August 1961).
4. C. Hilsum, "Transferred Electron Amplifiers and Oscillators," *Proc. IRE*, 50, 185–189 (February 1962).
5. B. K. Ridley, "Specific Negative Resistance in Solids," *Proc. Phys. Soc. (London)*, 82, 954–966 (December 1963).
6. J. A. Copeland, personal communication.
7. L. A. Eastman, "Gunn and LSA Oscillators—Capabilities and State of the Art," *Electro-Technology*, 85, 25–28 (February 1970).
8. H. Kroemer, "Theory of the Gunn Effect," *Proc. IEEE (Correspondence)*, 52, 1736 (December 1964).
9. D. E. McCumber and A. G. Chynoweth, "Theory of Negative-Conductance Amplification and of Gunn Instabilities in 'Two-Valley' Semiconductors," *IEEE Trans. Electron Devices*, ED-13, 4–21 (January 1966).
10. J. G. Ruch and G. S. Kino, "Transport Properties of GaAs," *Phys. Rev.*, 174, No. 3, 921–931 (Oct. 15, 1968).
11. H. Ehrenreich, "Band Structure and Electron Transport of GaAs," *Phys. Rev.*, 120, 1954–1963 (Dec. 15, 1960).
12. G. Reddi, personal communication.
13. H. Kroemer, "Non-Linear Space-Charge Domain Dynamics in a Semiconductor with Negative Differential Mobility," *IEEE Trans. Electron. Devices*, ED-13, 27–40 (January 1966).
14. B. W. Hakki and S. Knight, "Microwave Phenomena in Bulk GaAs," *IEEE Trans. Electron Devices*, ED-13, 94–105 (January 1966).
15. H. W. Thim, M. R. Barber, B. W. Hakki, S. Knight, and M. Uenohara, "Microwave Amplification in a DC-biased Bulk Semiconductor," *Appl. Phys. Lett.* 7, 167–168 (September 1965).
16. W. Shockley, "Negative Resistance Arising from Transit Time in Semiconductor Diodes," *Bell System Tech. J.*, 33, 799–826 (July 1954).
17. H. Kroemer, "Generalized Proof of Shockley's Positive Conductance Theorem," *Proc. IEEE*, 58, 1844–1845 (November 1970).
18. H. Kroemer, "The Gunn Effect under Imperfect Cathode Boundary Conditions," *IEEE Trans. Electron Devices*, ED-15, 819–837 (November 1968).
19. H. Kroemer, "Detailed Theory of the Negative Conductance of Bulk Negative Mobility Amplifiers, in the Limit of Zero Ion Density," *IEEE Trans. Electron Devices*, ED-14, 476–492 (September 1967).
20. H. Kroemer, "Negative Conductance in Semiconductors," *IEEE Spectrum*, 5, 47–56 (January 1968).
21. B. K. Ridley, "The Inhibition of Negative Resistance Dipole Waves and Domains in n-GaAs," *IEEE Trans. Electron Devices*, ED-13, 41–43 (January 1966).
22. B. W. Hakki, "Amplification in Two-Valley Semiconductors," *J. Appl. Phys.*, 38, 808 (February 1967).

23. S. Mahrous and P. N. Robson, "Small-Signal Impedance of Stable Trans-ferred-Electron Diodes," *Electronics Lett.*, 2, 107–108 (March 1966).

24. A. L. McWhorter and A. G. Foyt, "Bulk GaAs Negative Conductance Am-plifiers," *Appl. Phys. Lett.*, 9, 300–302 (Oct. 15, 1966).

25. P. N. Robson, G. S. Kino, and B. Fay, "Two-Port Microwave Amplification in Long Samples of GaAs," *IEEE Trans. Electron Devices (Correspondence)*, ED-14, 612–615 (September 1967).

26. M. T. Vlaardingerbroek, G. A. Acket, and P. M. Boers, "Reduced Build-up of Domains in Sheet-Type Gallium Arsenide Gunn Oscillators," *Phys. Lett.* 28A, 97 (November 1968).

27. G. S. Kino and P. N. Robson, "The Effect of Small Transverse Dimensions on the Operation of Gunn Devices," *Proc. IEEE*, 56, 2056–2057 (November 1968).

28. K. Kumabe, "Suppression of Gunn Oscillations by a Two-Dimensional Effect," *Proc. IEEE*, 56, 2172–2173 (December 1968).

29. S. Kataoka, H. Tateno, and M. Kawashima, "Suppression of Travelling High-Field-Domain Mode Oscillations in GaAs by Dielectric Surface Load-ing," *Electronics Lett.*, 4, 48–50 (February 1969).

30. R. H. Dean, "Optimum Design of Thin-Layer GaAs Amplifier," *Proc. IEEE*, 57, 1327–1328 (July 1969).

31. J. B. Gunn, "Properties of a Free, Steadily Travelling Electrical Domain in GaAs," *IBM J. Res. Develop.* 10, 300–309 (July 1966).

32. W. K. Kennedy, "Negative Conductance in Bulk Gallium Arsenide at High Frequencies," MS thesis, Cornell University, Ithaca, N.Y., September 1966.

33. H. Kroemer, unpublished work.

34. J. S. Heeks, "Some Properties of the Moving High-Field Domain in Gunn Effect Devices," *IEEE Trans. Electron Devices*, ED-13, 68–79 (January 1966).

35. J. B. Gunn, "On the Shape of Traveling Domains in Gallium Arsenide," *IEEE Trans. Electron Devices*, ED-14, 720–721 (October 1967).

36. P. N. Butcher, "Theory of Stable Domain Propagation in the Gunn Effect," *Phys. Lett.*, 19, 546–547 (Dec. 15, 1965).

37. M. A. Lampert, "Stable Space-Charge Layers Associated with Bulk Nega-tive Differential Conductivity: Further Analytic Results," *J. Appl. Phys.*, 40, 335–340 (January 1969).

38. P. N. Butcher, W. Fawcett, and C. Hilsum, "A Simple Analysis of Stable Domain Propagation in the Gunn Effect," *Brit. J. Appl. Phys.*, 17, 841–850 (July 1966).

39. J. A. Copeland, "Electrostatic Domains in Two-Valley Semiconductors," *IEEE Trans. Electron Devices (Correspondence)*, ED-13, 189–191 (January 1966).

40. G. S. Hobson, "Small-Signal Admittance of a Gunn-Effect Device," *Electronics Lett.*, 2, 207–208 (June 1966).

41. B. W. Knight and G. A. Peterson, "Domain Velocity, Stability, and Im-pedance in the Gunn Effect," *Phys. Rev. Lett.*, 17, 257 (August 1966).

42. H. W. Thim, "Linear Microwave Amplification with Gunn Oscillators," *IEEE Trans. Electron Devices*, ED-14, 517–522 (September 1967).

43. H. Kressel and G. Kupsky, "The Effective Ionization Rate for Hot Carriers in GaAs," *Int. J. Electronics*, 20, 535–543 (October 1966).

44. J. F. Reynolds, B. E. Berson, and R. E. Enstrom, "High–Efficiency Transferred Electron Oscillators," *Proc. IEEE*, 57, 1692–1693 (September 1969).

45. P. N. Robson and S. N. Mahrous, "Some Aspects of Gunn Effect Oscillators," *Radio Electronic Engr.*, 30, 345–352 (December 1965).

46. J. E. Carroll, "Oscillations Covering 4 Gc/s to 31 Gc/s from a Single Gunn Diode," *Electronics Lett.*, 2, 141 (April 1966).

47. J. A. Copeland, "A New Mode of Operation for Bulk Negative Resistance Oscillators," *Proc. IEEE (Correspondence)*, 54, 1479–1480 (October 1966).

48. J. A. Copeland, "LSA Oscillator Diode Theory," *J. Appl. Phys.*, 38, 3096–3101 (July 1967).

49. T. Hayashi, "Three-Terminal GaAs Switches," *IEEE Trans. Electron Devices*, ED-15, 105–110 (February 1968).

50. J. A. Copeland, "CW Operation of LSA Oscillator Diodes," *Bell System Tech. J.*, 46, 284–287 (January 1967).

51. W. K. Kennedy and L. F. Eastman, "High Power Pulsed Microwave Generation in Gallium Arsenide," *Proc. IEEE*, 55, 434–435 (March 1967).

52. J. A. Copeland, "Doping Uniformity and Geometry of LSA Oscillator Diodes," *IEEE Trans. Electron Devices*, ED-14, 497–500 (September 1967).

Chapter 3

AVALANCHE DIODE OSCILLATORS

W. J. Evans

Bell Telephone Laboratories, Incorporated
Murray Hill, New Jersey

3.1 INTRODUCTION

In the short time since the discovery of microwave oscillations in an avalanching p-n junction diode by Johnston, De Loach, and Cohen,[1] the avalanche diode oscillator has developed very rapidly into a practical microwave generator. The basic theoretical explanation for the origin of this phenomenon had been given earlier by Read,[2] although in terms of a somewhat more complex structure. Since 1965 a large body of literature has developed on this subject and a great deal of progress has been made in the theoretical analysis, although Read's original analysis has been proven remarkably accurate. At the same time the technology has advanced rapidly, so that 1000 W of pulsed rf power at 1 GHz and 1 W of continuous wave rf power at 50 GHz are now possible.

In this chapter some of the general features of avalanche diode devices will be reviewed. We begin with a discussion of the physics of Impatt (an acronym from impact-ionization avalanche transit time) oscillators with the emphasis on the limitations on power and efficiency. The study of large-signal operation of Impatt diodes leads naturally into the study of the high-efficiency or Trapatt (trapped plasma avalanche triggered transit) mode of operation. Again we shall emphasize the limitations on diodes operating in this mode. Because the circuit plays a very important role in Trapatt operation, some time will be spent on the interaction between the diode and circuit in Trapatt oscillators. In Section 3.4 oscillator design will be considered. Here we shall discuss the problems of diode design, diode fabrication,

thermal design, and circuits. In Section 3.5 the applications of avalanche diodes will be discussed. We shall consider methods of constructing power oscillators and amplifiers, and methods of oscillator
modulation. Finally, some information on diode failure will be covered and the latest state-of-the-art results for avalanche diodes will
be given. Topics that are not covered include the areas of device
fabrication and the detailed mathematical analysis of avalanche diodes.
The latter is given in excellent papers by Read,[2] Misawa,[2a] and Gummel and Scharfetter.[3]

3.2 IMPATT PHYSICS

3.2.1 Review

Before discussing the large-signal behavior of Impatt diodes in detail, the basic principles of operation first described by Read will be
reviewed.[2] Figure 1a shows the n^+pip^+ Read diode structure. If this
diode is reverse biased to breakdown, the field profile is that shown
in Fig. 1b. If the a-c voltage shown in Fig. 1c is added to the d-c
voltage, the peak field at the n^+p junction initially increases and background carrier density at the junction is greatly multiplied by impact
ionization. Later in the cycle the decreasing diode terminal voltage
and the space charge of the generated carriers cause the field at the
junction to drop below its average value and multiplication stops. The
carriers generated by this process quickly separate under the influence
of the high field, with the electrons moving into the n^+ region and the
holes drifting at saturated velocity in the intrinsic region. The drifting hole current (i.e., the current injected by the avalanche in Fig.
1d) induces in the external circuit the current shown in Fig. 1e. If
we compare Figs. 1c and 1e, it is clear that the diode exhibits a negative resistance at its terminals and that a transit time for the holes
of approximately one-half of the period of the voltage wave form
yields the largest negative resistance.

3.2.2 Small-Signal Characteristics

The first-order analysis of the Read diode is relatively simple
because the avalanche multiplication and the time delay due to carrier
drift can be separated. Gilden and Hines[4] have made one additional
assumption to simplify the small-signal analysis, and that is that the
transit angle of the drift region (i.e., the time required for carriers
to cross the intrinsic region compared to the oscillator period) is

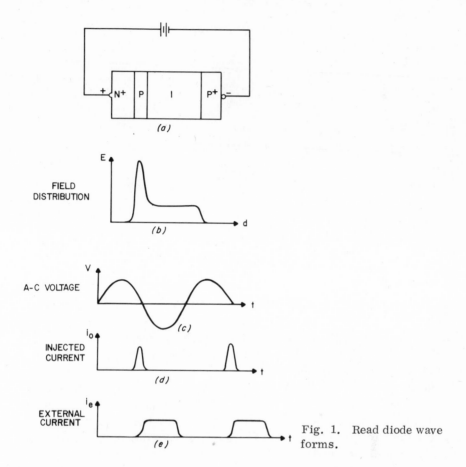

FIELD
DISTRIBUTION

A-C VOLTAGE

INJECTED
CURRENT

EXTERNAL
CURRENT

Fig. 1.   Read diode wave
forms.

small. This assumption simplifies the expression obtained for the
small-signal diode impedance and permits one to characterize the
diode in terms of a simple equivalent circuit. Figure 2 shows the
lumped circuit representation for this case and the diode impedance
versus frequency. Note that the diode impedance is negative above a
characteristic frequency ω_a, called the avalanche frequency. Gilden
and Hines have shown that the avalanche frequency is proportional to
the square root of the d-c bias current. This current dependence of
the impedance has been found to be true both theoretically and experimentally for a wide range of impurity profiles.

A numerical analysis of the Read structure by Gummel and Scharfetter[3] has removed many of the assumptions made in Read's analysis.
Figure 3 shows their results for the small-signal diode admittance at
several values of d-c current. The admittance is seen to pass very

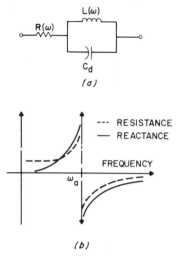

(a)

(b)

Fig. 2. (a) Equivalent circuit and (b) small-signal impedance for Read diode. (After Gilden and Hines.[4])

close to the origin, corresponding to the impedance plane pole shown in Fig. 2. Because of the similarity of these results with Fig. 2, we shall continue to use the notion of an avalanche frequency and define it as the frequency at which the diode susceptance changes sign (and the diode conductance is negative).

The Read diode concept is very useful because it is easy to analyze (to the first order at least) and yields a good qualitative understanding of the origin of the negative resistance. Unfortunately, it is not easy to fabricate this structure. However, it was discovered experimentally[1] that diode profiles of considerably less complexity will generate microwave power. Figure 4 shows the impurity concentration and field for a reverse-biased 6 GHz germanium Impatt diode. The diode is constructed by forming a lightly doped epitaxial p region on a heavily doped p^+ substrate. The junction is then formed by an n^+ diffusion to a depth of about 3 μm. The calculated admittance for this structure is shown in Fig. 5. The admittance for this structure is seen to be very similar to the admittance of the Read diode. The n^+pp^+ diode has a larger negative Q (i.e., ratio of diode susceptance to conductance) for a given current than the Read diode. Large-signal studies have shown that generally a diode with smaller negative Q is preferred for maximum power and efficiency. This disadvantage to the n^+pp^+ structure is offset to some extent by the behavior of the negative conductance at higher-current densities. Figure 6 shows the diode conductance for several field profiles versus the d-c current. The frequency is held constant. The PIN diode is seen to have less negative conductance than the Read diode at low current, but its conduc-

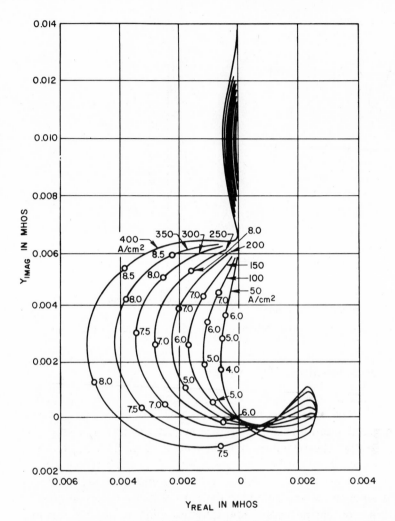

Fig. 3. Admittance of Read diode. Selected frequencies, in GHz, are marked off on each curve. (After Gummel and Scharfetter,[3] copyright 1966, and the American Telephone and Telegraph Co., reprinted by permission.)

tance continues to increase at higher currents. Thermal considerations dictate which profile is preferred for a given application; these factors will be discussed in greater detail in Section 3.3. Generally, increasing frequency requires smaller diode areas, however the thermal spreading resistance increases only inversely as the square root of the diode area. This permits very high current densities to be used at higher frequencies so that the n^+pp^+ structure is preferred for continuous wave operation of millimeter wave diodes.

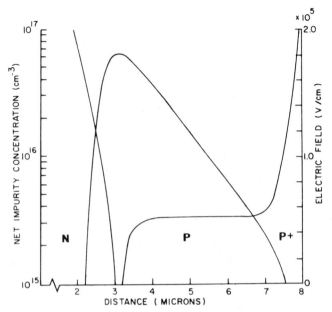

Fig. 4.   Doping and field profiles for experimental germanium diodes.

3.2.3 Large-Signal Characteristics

A number of authors have given large-signal analyses of Impatt oscillators.[2,5-7] The most accurate and complete of these was given by Scharfetter and Gummel and is again a numerical solution (see Appendix). Basically, the program that they used performs a simultaneous time-dependent solution of Poisson's equation, the current equations, and the continuity equations for a one-dimensional structure. Included in the program are accurate expressions for electron and hole mobilities, ionization rates, and thermal generation and recombination. The solution is obtained by subdividing the diode into approximately 100 mesh points and applying specified terminal conditions. The program then obtains the solution for the field and carrier densities at successive time steps.

Because of the complexity of this program, it is in general not practical to examine many different diode structures and circuits. Nevertheless, the program is sufficiently accurate so that in many cases computer experiments can be done more economically than laboratory experiments. The program is also an extremely valuable tool for improving our conceptual understanding of Impatts.

Figure 7 shows a typical solution for a Read diode carrying a d-c current and no a-c component. If an rf voltage is applied to this diode,

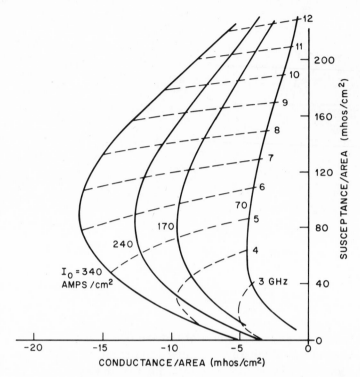

Fig. 5. Small-signal, admittance plane plot for a germanium N^+PP^+ diode from Gummel-Scharfetter program. (Courtesy of C. A. Brackett.)

the change in diode admittance with rf amplitude can be found. Typical results are shown in Fig. 8. Note that generally the diode susceptance increases with increasing rf voltage amplitude, whereas the diode negative conductance decreases. These results show efficiencies as high as 18%, although neither oscillator power nor efficiency is saturated at that point. It is difficult with this program to determine the maximum efficiency or power output because these quantities depend on the proper termination of the harmonics of the fundamental, and these constraints are not known a priori. The rf voltage and particle current wave forms are shown in Fig. 9. These should be compared to the simplified wave forms for a Read diode shown in Figs. 1c and 1e.

3.2.4 Limitations on Efficiency

Space Charge Effects. In the last section, the general large-signal behavior of the Read diode was shown to be approximately that expected from the qualitative discussion in Section 3.2.1. In this sec-

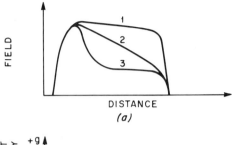

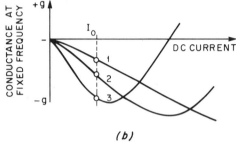

Fig. 6. Dependence of diode conductance on field profile. (a) Field
profiles; (b) conductance versus d-c
current.

tion we shall discuss in more detail some of the factors that reduce the
efficiency of Impatt diodes. Probably the most important omission
from the simple model of Section 3.2.1 is the effect of the space
charge of the mobile charge carriers.

Figure 10a is a plot of the field profile and carrier densities in an
oscillating Impatt diode at a particular point in time. The profile
shown here is for a 6 GHz n^+pp^+ epitaxial germanium diode. The plots
are made for the particular point in the rf voltage and current wave
forms indicated by the squares in Fig. 10b. The carriers shown have
been generated by avalanche breakdown in the high-field region on the
left. The holes then drift to the right and the electrons to the left.

One of the saturation mechanisms of this mode of operation is
seen in the depression of the field by the space charge of the avalanche-
generated carriers. The reduction in field turns off the avalanche
prematurely and thus reduces the phase delay provided by the avalanche. This in turn reduces the negative conductance of the oscillator.

Another saturation mechanism is illustrated in Fig. 11. Here the
effect of the space charge on the field can be seen after generation
has stopped and the holes have drifted into the lower-field region of
the diode. Again the space charge is depressing the field so that, to
the left of the carrier pulse, the carrier drift velocity may drop below
saturation and some carriers may be trapped for a short time in the

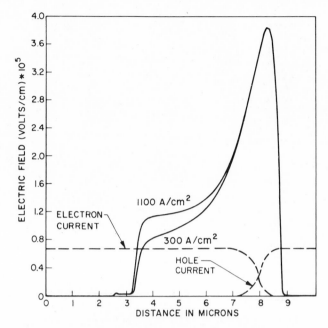

Fig. 7.   Electric field versus distance in d-c steady-state Read diode. (After Scharfetter and Gummel. [7])

low-field region.   This change in carrier velocity is not in itself a loss process, as will be seen in the discussion of the Trapatt mode. However, it does sufficiently change the terminal current wave forms so that the power generated at the transit-time frequency is reduced.

Epitaxial Layer Resistance.   In Fig. 8 it was shown that the negative conductance of the Impatt decreases with increases in rf voltage amplitude.   At the point of maximum efficiency the diode conductance thus becomes quite small and any loss will cause the efficiency to drop.   A loss that can easily occur in the fabrication of diodes is caused by the resistance of the epitaxial layer.   Figure 12a from Kovel and Gibbons[8] shows a diode structure that breaks down before the field reaches the substrate.   This leaves a short length of unswept epitaxial material having a relatively high resistivity ($\sim 1\,\Omega$-cm).   Figure 12b shows the effect of this unswept layer on the diode efficiency. The diode diameters used in these experiments are such that 2 μm of unswept material corresponds to about 1 ohm of series resistance.   It is apparent that small resistive losses near the diode are very detrimental to efficient oscillator operation.   From this experiment we also conclude that the circuit loss should be kept small.

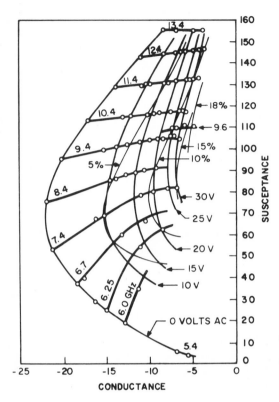

Fig. 8. Diode admittance (susceptance versus conductance) as a function of frequency and a-c voltage amplitude, with the resultant efficiency indicated. Current density 200/A-cm². (After Scharfetter and Gummel.[7])

Reverse Saturation Current. Read[2] has shown that one would expect oscillator efficiency to decrease with increasing reverse saturation current. Misawa[9] has made some quantitative calculations of this effect. Figure 13 shows the effect of saturation current on the diode admittance for a Read diode. The saturation current reduces the negative conductance and thus the efficiency of the oscillator. The efficiency of an Impatt diode will therefore be reduced with increasing junction temperature; this effect is usually observed in continuous wave oscillators just before burnout. The effect of reverse saturation current (which is caused by thermal generation of carriers), although important at high temperatures, should not degrade oscillator performance under normal operating conditions. However, there are two very interesting conditions that result in an "effective" saturation current and have a pronounced effect on the oscillator efficiency.

Misawa has shown that high-field diffusion can lead to an effective saturation current in Impatt diodes.[10] Figure 14 shows the structure analyzed by Misawa. According to Read's theory, carriers are generated by impact ionization in the high-field region and, in this

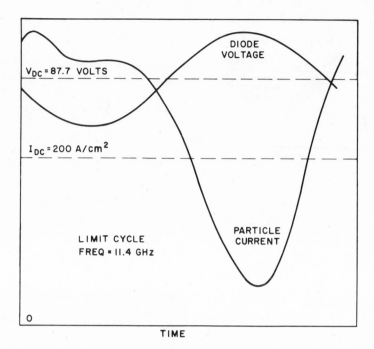

Fig. 9. Diode voltage and particle current versus time for 1 cycle of steady-state oscillation. (After Scharfetter and Gummel.[7])

case, electrons drift to the right and holes to the left. However, the avalanche region is very narrow and the carrier concentration may be very high. Therefore, the carrier density gradient is large, giving rise to a significant diffusion current. By writing down the current equation for one carrier, we can estimate the magnitude of the electron current j_n

$$j_n = qn\mu E + qD_n \frac{\partial n}{\partial x} \tag{1}$$

where q = electronic charge
   n = electron density
   μ = electron mobility
   E = electric field
   D_n = diffusion constant.

   Now we would like to know under what conditions the diffusion component can exceed the drift component. Therefore set $j_n = 0$ and note that for high fields $\mu E = v_s$, the saturated drift velocity. Then Eq. 1 gives

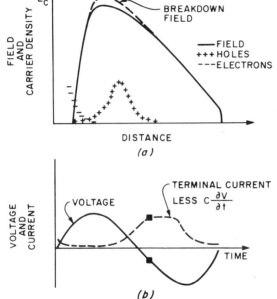

Fig. 10. Large-signal Impatt field and carrier densities during the avalanche portion of the rf cycle.

$$n = n_0 e^{-x v_s / D_n} \qquad (2)$$

If we assume that D_n is independent of field[11] and has a value of 26 cm^2/sec, and v_s a value of 10^7 cm/sec, $D_n/v_s = 0.026$ μm. Thus, for a sharply defined junction region, a relatively large number of electrons may diffuse into the p^+ region, where the field is very small. These carriers are stored in the p^+ region until the avalanching ceases and the carrier density at the junction drops. The stored carriers then diffuse and drift back into the high-field region. This current is very similar in effect to a reverse saturation current, and it reduces the phase delay of the avalanche buildup on the next Impatt cycle. This in turn reduces the oscillator efficiency. Figure 15 from Misawa[10] shows this minority carrier storage current versus time for the structure shown in Fig. 14. The electron current density is calculated at the boundary of the high-field region, and the positive electron current indicates electrons that have diffused out of the high-field region and are flowing into the p^+ end region. The negative current flowing after the electron density has reached a maximum represents current flowing back into the high-field region.

Another source of minority carriers in the heavily doped end region was discovered by Decker, Dunn, and Frost.[12] They observed that germanium diodes made from identical slices of $n^+ p p^+$ germanium

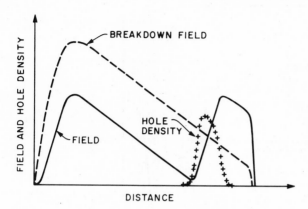

Fig. 11.   Large-signal Impatt field and carrier densities during the drift portion of the rf cycle.

but contacted under varying conditions showed a substantial variation in microwave performance.   In order to separate the influence of the diode contacts from the junction region of the diode, they constructed the device shown in Fig. 16.   Ideally, the two top contacts are ohmic, however; in fact, they form a pair of back-to-back Schottky diodes with most current in the n^+ region being carried by electrons.   However, some holes are injected into the n region by the forward-biased Schottky barrier.   Because the thickness of the n^+ region is small compared to the diffusion length for the holes, they diffuse into the high-field region created by the reverse bias V_2.   The current that flows in the high-field region I_2 is thus due to the injection of minority carriers into the n^+ region (V_2 is well below that required for avalanche generation).   The ratio of I_2 to I_1 is defined as the injection ratio γ for the contact.   Figure 17 shows the oscillator efficiency versus injection ratio for three different samples used in the experiments.   As before, the loss of efficiency can be explained by simply postulating an increase in the saturation current which corresponds to the injected carriers. (This is only an approximate model, since the saturation current amplitude does not depend on the terminal current amplitude, as does the injected current.)

Skin Effect.   As the operating frequency of Impatt oscillators is increased above 50 GHz, the nonuniform distribution of current in the diode caused by skin effect in the substrate may become a problem. De Loach[13] has considered this problem and found that it may indeed be an important effect in millimeter wave diodes.   Figure 18 shows a simplified drawing of a diode with a substrate of length l.   In the ex-

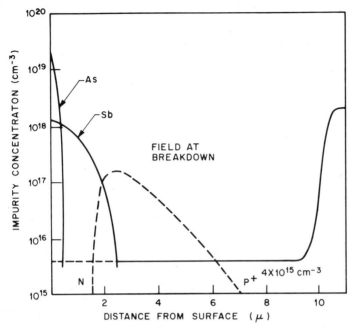

Fig. 12*a*. 6-GHz germanium Impatt diode impurity distribution and field profile. (After Kovel and Gibbons.[8])

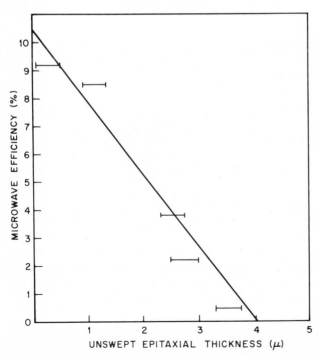

Fig. 12*b*. Microwave efficiency versus unswept epitaxial thickness. (After Kovel and Gibbons.[8])

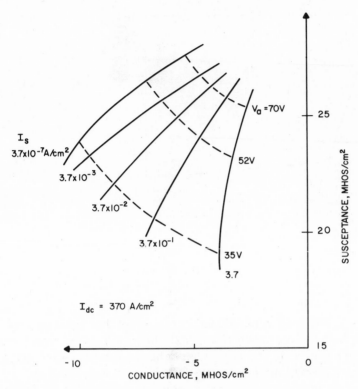

Fig. 13. Change of admittance with amplitude of signal voltage, V_a, for five values of saturation current. (After Misawa.[9])

treme case the current is primarily confined to flow within a distance δ of the surface of the substrate. Thus the effective resistance of the substrate is increased (because only a fraction of it is carrying current). Furthermore, there will be a voltage drop across the radius of the diode near the junction. Figure 19 shows this voltage drop versus the diode diameter for several current densities at 10 and 100 GHz in a 0.001 Ω-cm substrate. Since d-c current densities of 10^5 A/cm^2 in 100 GHz diodes are not uncommon and reactive currents may be 10 times this value, there can be a significant voltage drop across the diode diameter. This in turn will lead to a nonuniform current distribution in the diode and an effective series resistance that reduces efficiency.

Ionization Rates. In very narrow diodes, the field required for avalanche breakdown may exceed 10^6 V/cm. At this field intensity, the slopes of the ionization rates are a function of field decrease. Thus less current modulation is obtained with a given RF voltage and the effi-

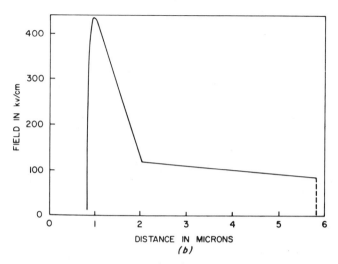

Fig. 14.   10 GHz silicon $p^+n\nu n^+$ diode field profile in breakdown.   (After Misawa.[10])

ciency is reduced.   For abrupt junction silicon diodes this limit is reached for diodes that break down at approximately 6 V and operate as 200 GHz oscillators.   In a Read diode, where the avalanche region is intentionally made narrow, the frequency limit is somewhat lower. Conversely, in a $P\nu N$ diode the avalanche region is larger and the frequency limitation somewhat above 200 GHz.

Although it is difficult to put all of the limitations discussed above into a large-signal analysis of Impatt diodes, it does seem probable that efficiency will decrease rapidly above 100 GHz.

3.2.5   Limitations on Power

Power Impedance Product.   In the last section we discussed some of the limitations on diode efficiency.   If it is assumed that there is some maximum efficiency for an Impatt diode, the question of the maximum available power remains to be answered.   This problem has been discussed by De Loach[15] for Impatt oscillators; we give a simplified version of his analysis.

The power impedance product for an Impatt diode is obtained by calculating the maximum power that can be imparted to the mobile charge carriers by the field.   For a semiconductor of length l this value is given by

$$P = I_{max} V_{max} = (qn v_s A) \cdot (E_c l) \tag{3}$$

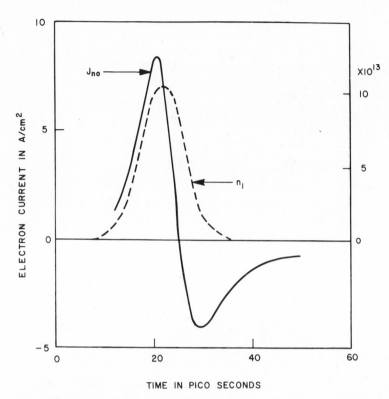

Fig. 15. Computed electron density, n_1, and electron current, J_{n0}, at the boundary between high-field and neutral regions. (After Misawa.[10])

where q = electronic charge

n = number of carriers

v_s = saturated drift velocity

A = device area

E_c = critical field for avalanche breakdown.

The maximum current is determined by the number of carriers moving at the maximum drift velocity, v_s. The maximum voltage is determined by the field that causes avalanche breakdown.

The number of charge carriers that can participate is limited by the carrier space charge through Poisson's equation,

$$\frac{\partial E}{\partial x} \cong \frac{E_c}{l} = \frac{qn}{\epsilon} \qquad (4)$$

Therefore Eqs. 3 and 4 give

$$P = \epsilon E_c^{2} v_s A \qquad (5)$$

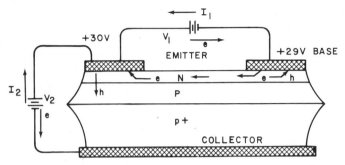

Fig. 16. Test structure and carrier flow diagram. (After Decker, Dunn, and Frost.[12])

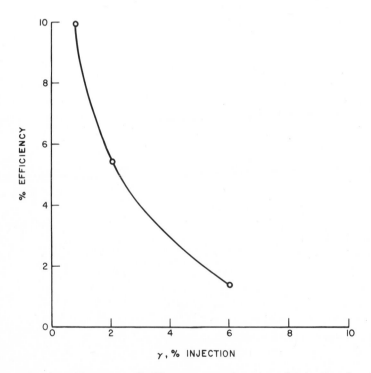

Fig. 17. Maximum rf efficiency as a function of injection level. (After Decker, Dunn, and Frost.[12])

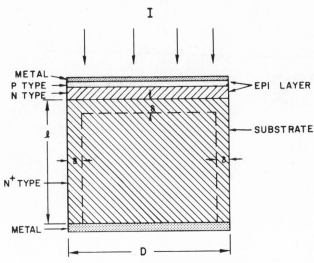

Fig. 18. Cross-sectional view of cylindrical diode model. (After De Loach, Jr.[13])

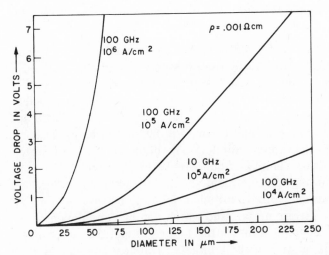

Fig. 19. The voltage drop from diode center to diode edge (for 0.001 Ω-cm substrate material) versus diode diameter, where J is assumed uniform. (After De Loach, Jr.[13])

117

Noting that the device capacitance C is

$$C = \frac{\epsilon A}{l} \tag{6}$$

and the transit-time frequency f is

$$f = \frac{v_s}{2l} \tag{7}$$

Eq. 5 becomes

$$P \cdot x_c = \frac{E_c^2 v_s^2}{4\pi f^2} \tag{8}$$

where $x_c = 1/2\pi f C$.

If we assume that we are limited to some minimum circuit impedance (e. g. , by circuit loss), Eq. 8 predicts that the maximum power which can be given to the mobile carriers decreases as $1/f^2$. This relationship is found to be approximately true experimentally (see Fig. 72, a plot of power output versus frequency). The agreement between these results and Eq. 8 does not necessarily confirm the theory, however, since a similar conclusion might be drawn from thermal considerations alone, as will be shown in the next section. Low-frequency, continuous wave devices do not exhibit this kind of behavior because of thermal spreading resistance.

Thermal Limitations. The continuous wave performance of Impatt oscillators is, of course, limited by the power that can be dissipated in a semiconductor chip. In Section 3.4, the thermal design of oscillators is considered in detail. In this section, the equations which describe the thermal resistance of a semiconductor are given. Some appreciation of this problem is gained from Fig. 20, which compares a human hair and a 50-GHz Impatt diode. The diode shown is capable of dissipating a d-c power density of 500, 000 W/cm^2. Figure 21 shows the typical mounting arrangement for an Impatt diode. The cylindrical semiconductor wafer is attached (usually by thermocompression bonding) to a larger metal stud which can be considered a semifinite heat sink. Several intervening layers of metal are usually present to improve the quality of the contact. The heat is generated in the high-field region in the semiconductor. The thermal resistance thus consists of two components. The first is the sum of the series-connected resistances of the silicon and metallic layers. This resistance is inversely proportional to the area of the wafer. The second component is the thermal spreading resistance of the heat sink. This resistance is inversely proportional to the diameter of the wafer. The

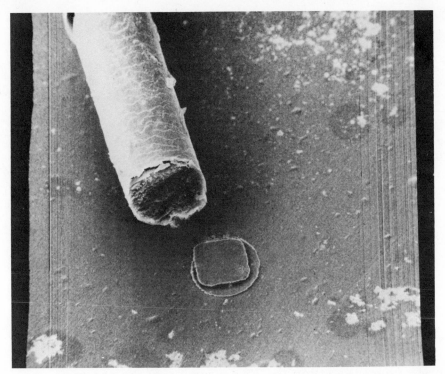

Fig. 20. 50-GHz Impatt diode and a human hair. (With permission of Bell Laboratories Record, copyright 1966, and the American Telephone and Telegraph Co.)

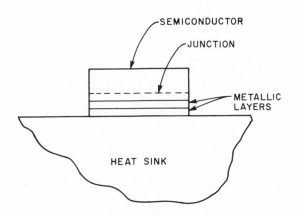

Fig. 21. Diode mounting.

energy that can be dissipated in the diode must equal the heat energy
that can be transmitted to the heat sink. Therefore,

$$P = \frac{\Delta T}{R_T} \qquad (9)$$

where P is the power dissipated in the diode, ΔT is the temperature
difference between the junction and the heat sink, and R_T is the total
thermal resistance which is given by[16]

$$R_T = \frac{4}{\pi d^2} \sum \frac{l_i}{\sigma_i} + \frac{1}{2 d \sigma_c} \qquad (10)$$

where d = the wafer diameter

   l_i = the thickness of each of the materials between the junction
        and the heat sink

   σ_i = the thermal conductivity of each of these materials

   σ_c = the heat sink thermal conductivity.

At high frequencies the diode diameter becomes small and the
first term of Eq. 10 may dominate. This is particularly true if the
diode is mounted so that the heat must pass through a substantial
amount of substrate material. Thus, from Eq. 9, it is apparent that
for this case the dissipated power is proportional to the diode area.
Comparing this with Eq. 5, we conclude that thermal resistance can
also result in a $1/f^2$ limitation. Which of these effects will dominate
depends on the details of the device fabrication.

3.2.6 Noise

Avalanche Noise. There are several sources of noise in Impatt
oscillators; the first that we shall consider is the noise generated in
the avalanche process. The large-signal noise problem, i.e., the
noise in the sidebands of an oscillator, has not been analyzed for
Impatt oscillators. For most applications this is probably the most
interesting problem. However, work on the small-signal noise in
Impatt diodes has been done by Hines[17] and Gummel and Blue.[18] Their
results are directly applicable to small-signal amplifier noise and give
us a good deal of insight into the factors that influence oscillator noise.

In the Gummel and Blue model, charge is assumed to be generated
randomly in time and position in the diode. The amount of charge generated is $q\gamma$, where q is the electronic charge and γ the generation rate
$\alpha_n v_n n + \alpha_p v_p p$. The next step is to calculate the voltage induced by this
charge in the external circuit. The effect of space charge that is
neglected for the individual events is included in the process of calculating the transfer impedance relating the generated charge to the

terminal voltage. Thus the individual events are influenced by the average value of the space charge.

Using this approach and the small-signal assumption, the noise generation of diodes with arbitrary field profiles can be calculated numerically. If additional assumptions used by Read are made, an analytical solution for the noise can be obtained. Figure 22 shows the ratio of mean square diode noise current to conventional shot noise, $2qI_{dc}\Delta f$ for a Read diode. At low frequencies and high d-c current, the noise amplitude is constant and decreases with increasing d-c bias. Above the avalanche frequency the noise decreases rapidly. Haus and Adler[19] have defined the noise measure for an amplifier to be

$$M \equiv \frac{F - 1}{1 - (1/G)} \qquad (11)$$

where F is the noise figure and G the amplifier gain. The optimum value for M in negative conductance amplifiers is given by[20]

$$M = \frac{\overline{i}^2}{4KTg} \qquad (12)$$

where $\overline{i}^2$ = the noise current

K = Boltzmann's constant

T = the temperature of source

g = magnitude of the negative conductance.

Figure 23 is a plot of this optimum noise measure for a Read diode versus frequency. There are several points of interest in this result. The first is that the noise measure decreases with increasing d-c current. The second is that the minimum noise is achieved only over a small band of frequencies. Finally, we note that the rapid decrease in noise above the avalanche frequency results in a minimum noise measure at about twice the transit-time frequency. However, the negative conductance in this region is small, so that it would be difficult to obtain high gain in a practical amplifier.

Using a numerical solution, Gummel and Blue obtained the noise measure for more typical diode structures and material parameters. Figure 24 gives the noise measure for abrupt junction silicon and germanium diodes having an assumed parasitic series resistance of 0 and 1 Ω. These results are quite similar to those of Fig. 23; however, they indicate that silicon is somewhat noisier than germanium. The difference in the germanium and silicon results is apparently a result of the unequal ionization rates for holes and electrons in silicon. For high-gain amplification Eq. 11 shows that the noise measure equals the noise figure. Thus the noise measure of about 30 dB for

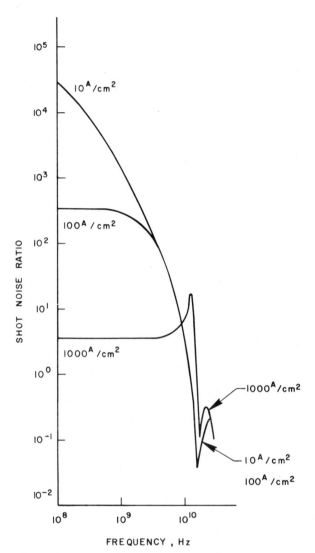

FIG. 22. Short-circuit mean-squared noise current, normalized to shot noise mean-squared current, associated with total d-c current. (After Gummel and Blue.[18])

germanium diodes is consistent with measured noise figures of the same value.

Although the small-signal calculations provide some insight into noise generation in Impatt oscillators, they are not complete enough to describe oscillator noise adequately. Among other problems, the

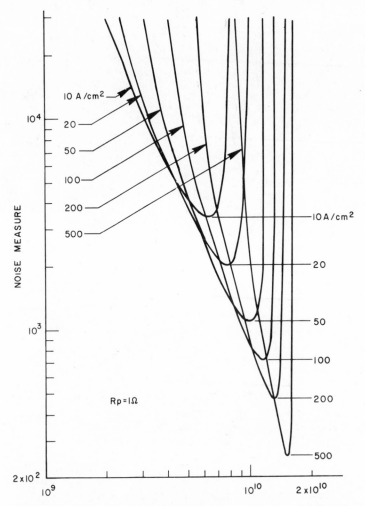

Fig. 23. Noise measure for parasitic resistance of 1 Ω and various current densities. (After Gummel and Blue.[18])

large-signal nonlinearities of an oscillating diode result in strong up-conversion of low-frequency noise to the oscillator sidebands. Calculations on the up-conversion gain available in an oscillating diode show this to be a potentially large source of sideband noise.[21] (As shown in Fig. 22, the noise power available at the lower frequencies is considerable.) This contribution to oscillator noise can be reduced by proper bias circuit design.

Circuit Effects. Another source of noise in Impatt oscillators, or

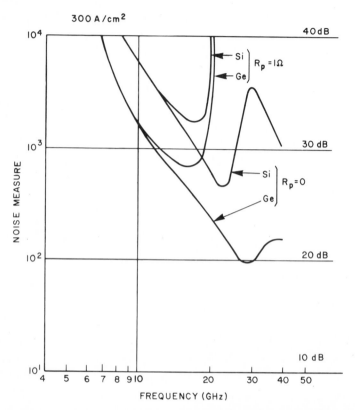

Fig. 24. Noise measure for germanium and silicon diodes at 300 A/cm² for parasitic resistance of 0 and 1 Ω. (After Gummel and Blue.[18])

any negative conductance oscillator, is caused by the rf circuit which is used to resonate the device. In Fig. 10 we showed a typical admittance plane plot for an Impatt diode. We can also plot on this figure the admittance as a function of frequency for the circuit. For a singly tuned circuit the admittance plane plot would appear as shown in Fig. 25. Stable oscillation occurs (i.e., the amplitude is not growing or decaying) if the device admittance curve for some rf voltage amplitude and the circuit curve intersect at the same frequency. (Other conditions must be met for stable operation in a singly tuned circuit,[22] and the case in which multiple harmonics are present.[23])

However, circuits, and particularly microwave circuits, may exhibit multiple resonances near the desired operating frequency. Using a low-frequency analog circuit Kenyon[24] has investigated the effect of a double-tuned circuit on the Impatt spectrum. Figure 26 shows the

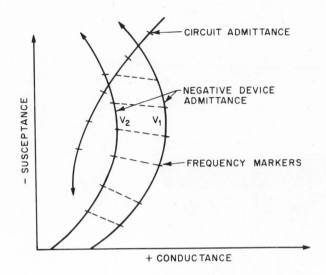

Fig. 25. Circuit admittance and negative device admittance. Arrow is in the direction of increasing frequency; broken lines connect points of constant frequency.

spectrum of an oscillator when the device and circuit curves intersect at nearly zero angle. In this case the system is indeterminant and the frequency not defined.

3.2.7 Improvements to Impatt Oscillators

Second Harmonic Tuning. Experiments with germanium Impatt diodes have shown that in some cases there is a strong dependence of oscillator efficiency on the circuit impedance at the second harmonic of the oscillator. Figure 27a shows experimental results obtained by Swan,[25] indicating the improved power output with harmonic tuning. Figure 27b, however, shows that the improvement is not so great if compared to the normal mode (i.e., without second harmonic tuning) output at higher frequencies. In fact, on a Pf^2 basis the normal mode is somewhat better. Thus it appears that second harmonic tuning improves the range of efficient operation but does not enhance the maximum efficiency greatly.

Theoretical analysis of this problem is quite involved; however, the general features of the experimental results have been obtained by several authors.[23,26] A simple qualitative explanation due to J. L. Blue follows.[6] Saturation of the rf power output with increasing bias current occurs in the normal mode (at lower frequencies) because the space charge of the carriers injected from the avalanche zone reduces the phase delay in the avalanche. However, this phase delay can be regained if the peak in the voltage wave form can be delayed (without changing the phase of its fundamental component). This is accomplished by the addition of a second harmonic voltage, as shown in Fig. 28.

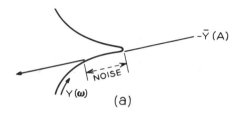

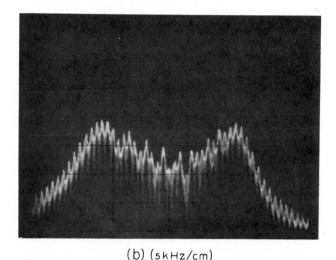

(b) (5kHz/cm)

Fig. 26. (a) Admittance plane plot for device and circuit at an unstable inter-section; (b) broadband noise output for unstable tuning condition. (After Kenyon,[24] copyright 1970, and the American Telephone and Telegraph Co., reprinted by permission.)

Double-Drift Impatts.[27] Conventional avalanche diode oscillators (p^+nn^+ or n^+pp^+) are essentially single-drift-region devices in that transit-time delay (for the Impatt mode) occurs in a single region of one impurity type. The double-drift-region structure (p^+pnn^+) is essentially two complementary avalanche diode oscillators in series. The obvious result is that power output per unit area and impedance on a per unit area basis are both essentially doubled. Hence the structure has a power-impedance product approximately a factor of 4 greater than a comparable single-drift-region unit. It has, in addition, some more subtle advantages: The "two diodes" share a single high field, centrally located avalanche region and have less parasitic series loss than that associated with a comparable single-drift-region unit. This should lead to considerable improvement in oscillator efficiency and device Q.

The first-order improvement expected with a double-drift-region structure is illustrated by Fig. 29. The heavy lines in the electric field plot and the structure schematic represent a single-drift-region unit for reference. The lighter lines correspond to the additional field profile and structure associated with the change to a double-drift-region unit. The double-drift-region unit will have a centrally located high-field avalanche region and two drift spaces, one for holes and one for electrons. The microwave negative resistance would be essentially doubled on a per unit area basis. (Detailed computer calculations verify this conclusion.)

Since ultimate power output of Impatt oscillators is limited by microwave circuit impedance limitations, the double-drift-region unit can have essentially double the area of a corresponding (in frequency) single-drift-region unit and maintain the same microwave impedance level. The ultimate power output is thus expected to be increased by a factor of four. This is readily seen from the schematic of the double-drift-region unit of Fig. 29. For the same impedance level, the double-drift-region unit has four times the working volume of a single-drift-region unit. Series resistance associated with the contacting n^+ and p^+ regions will correspondingly be reduced by the increase in area. Since only one avalanche region is required for both drift regions, increased efficiency is also expected. Increased efficiency is expected for two reasons: first, the additional d-c voltage in the double-drift-region structure will result essentially from the second drift space. In particular, it has been shown[7] that the avalanche region and drift region voltages are essentially equal for a silicon p on n structure, and therefore the total d-c voltage for such a double-drift-region device may be only 50% greater than that of the single-drift-region unit. Second, the central location of the avalanche region, away from a heavily doped contact region, greatly reduces the minority carrier storage effects discussed in Section 3.2.4.

Small-signal calculations of the admittance of single- and double-drift-region silicon units are presented in Fig. 30. These calculations were performed with material parameters appropriate for silicon at a temperature of 300 °C and a current density of 10,000 A/cm^2. As can be seen by the represented frequencies, the structures correspond to millimeter wave units. Such structures have been chosen for preliminary studies because of the expected advantages of double-drift-region units at millimeter wave frequencies. Lower frequency units will be limited in power output by thermal rather than circuit impedance considerations.

The detailed numerical analysis for the admittance of the double-drift-region unit, as shown in Fig. 30, reveals an important advantage over a single-drift-region unit not obvious from the simplified description given previously. The additional depletion region width reduces the susceptance by approximately a factor of 2, as expected. However, the reduction in negative conductance is somewhat less than a factor of 2. The result is a considerable improvement in small-signal negative Q.

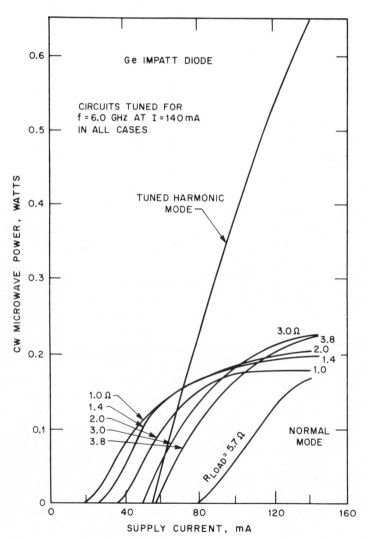

Fig. 27a. Normal-mode and tuned-harmonic mode operation of a germanium Impatt. (After Swan.[25])

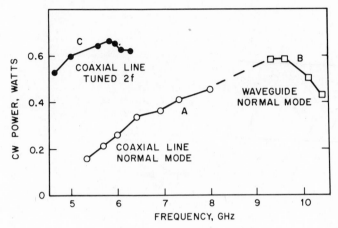

Fig. 27*b*. Experimental continuous wave output power for a germanium Impatt at 140-mA supply current. (After Swan.[25])

3.3 TRAPATT PHYSICS

3.3.1 Review of Trapatt Theory

A high-efficiency mode of operation using avalanche diode oscillators was first obtained in silicon diodes by Prager, Chang, and Weisbrod[28] and in germanium by Johnston, Scharfetter, and Bartelink.[29] This mode is characterized by efficient conversion of d-c to rf energy (up to 60%), operation at frequencies well below the transit-time frequency of operation, and a significant change in the d-c operating point when the diode switches into the mode. In 1968 Scharfetter, Bartelink, Gummel, and Johnston[30] proposed an explanation for this mode of operation. The explanation that they proposed, which was given the acronym Trapatt, was suggested by computer experiments performed with a complex program capable of modeling semiconductor devices rather accurately. Subsequent modeling of a complete diode-circuit system has shown that the original proposal presents an accurate description of the physical phenomena that give rise to the high-efficiency mode of operation. The computer experiments have also suggested where a number of simplifying assumptions can be made and, as a result, simplified analyses of the Trapatt oscillator that provide a more quantitative understanding of its operation have been devised.[32-34]

In order to understand the operation of the high-efficiency mode, it is necessary to recall the comments made in Section 3.2 on the ef-

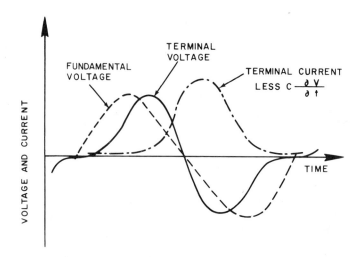

Fig. 28. Impatt wave forms with second harmonic voltage present. (After Blue,[6] copyright 1969, and the American Telephone and Telegraph Co., reprinted by permission.)

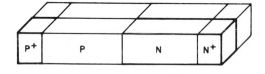

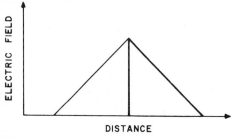

Fig. 29. Electric field profile and structure (schematic) for single-drift-region (heavy lined only) and double-drift-region Impatt diodes.

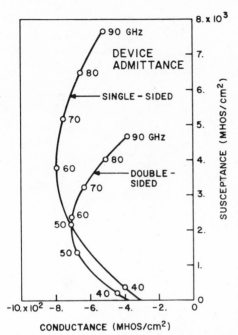

Fig. 30. Small-signal admittance of single- and double-drift-region silicon millimeter wave Impatt diodes. (Calculations for abrupt junction P^+-N-N^+ and P^+-P-N-N^+ units. Impurity concentration of all lightly doped regions is 5×10^{16} cm^{-3}. The N and P-N layers are 0.7 μm and 0.6 and 0.6 μm, respectively.)

fects of space charge on Impatt oscillator efficiency. Figures 10 and 11 showed the effects of space charge suppression of the avalanche and carrier trapping in the drift region. In the Trapatt mode these two effects play an important role. Figure 31 shows the situation that can occur when an Impatt diode is operated in a high Q circuit, where large rf voltages can develop. The field and carrier densities are shown at a point in time just after the diode voltage has reached a maximum. In this case, the large overdrive in voltage causes the generation of enough charge to reduce the field to nearly zero on the left. However, the field ahead of these carriers is still above the critical field, and avalanche generation begins to occur on the right. Thus an avalanche zone has been created which travels through the diode, leaving the diode filled with a plasma of electrons and holes and nearly zero electric field. Because of the low-field condition the plasma is trapped and is extracted at the boundaries by space charge limited flow.

The terminal characteristics of this mode for a full cycle are shown in Fig. 31b. Initially, there is a current that charges the depletion layer capacitance to a voltage of about twice the breakdown voltage. This initiates the traveling avalanche zone which drops the voltage to almost zero. The particle current is very high during this low-voltage

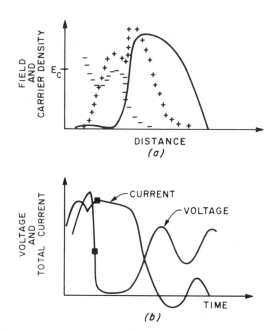

Fig. 31.   Trapatt wave forms.   (a) Field and carrier density;   (b) voltage and current.

state.   After the plasma has been removed from the diode, the current drops to a low average value and the voltage returns to near the break-down value.   Since the voltage is low while the current is high and vice versa, this is a very efficient mode of oscillation that has been calcula-ted to be capable of efficiencies of over 60% in silicon and 50% in germanium. [34]

The Trapatt mode occurs in avalanche diodes primarily because it is possible in some circuits rapidly to drive the diode to a voltage well in excess of that required for avalanche breakdown.   It is the in-herent delay in the avalanche process that permits this to occur.   How-ever, at some point in time after the application of the excess voltage, the multiplication of carriers begins to occur very rapidly.   Because the field is now well above the breakdown value (by 50 to 100%) the multiplication greatly exceeds that required to sustain the terminal current of the diode.   The excess carriers that are generated are there-fore sufficient to cause the field in the diode to collapse, leaving the diode filled with a plasma and in a near zero-voltage state.   If the terminal current is maintained, the plasma is removed from the diode and the diode voltage recovers to its breakdown value.   The time re-quired to remove the plasma depends on the density of the plasma and the magnitude of the recovery current. However, a recovery time of several times the transit time of carriers moving at the saturated drift velocity is typical.

A simple circuit that can be used for the Trapatt mode consists of a transmission line with the diode mounted at one end and a low-pass filter at the other end. The low-pass filter is placed approximately one-half wavelength away from the diode and presents a short circuit to all the harmonics of the Trapatt oscillation frequency. On the load side of the filter some tuning is required to match the diode impedance to the lead impedance. Figure 32 shows a simple circuit model of the circuit just described. Since we are primarily interested in the high-frequency components of the Trapatt wave form that are responsible for triggering the diode, the low-pass filter has been replaced by a short circuit. The inductance and shunt capacitance are the series lead inductance and the package capacitance of the diode. The capacitance may also include any additional capacitance due to Impatt tuners or lumped capacitance near the diode.

If the diode current is divided into two parts, the displacement

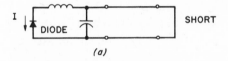

(a)

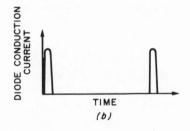

(b)

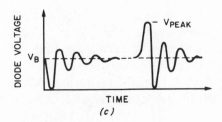

(c)

Fig. 32. Simplified circuit analysis. (a) Simplified circuit model; (b) diode conduction current versus time; (c) diode voltage versus time.

current and the conduction current, it is then possible to calculate the circuit response to an assumed conduction current wave form. Figure 32b shows a simplified conduction current wave form for a low-frequency Trapatt oscillator. The high-conduction current state occurs during the formation of the Trapatt plasma state. Figure 32c shows the calculated voltage response to such a current. Note that the diode voltage rapidly drops to a value near zero volts. This negative-going voltage pulse propagates down the line to the short circuit provided by the low-pass filter. The −1 reflection coefficient causes a positive voltage pulse to be reflected back toward the diode. The voltage pulse that returns to the diode therefore has an amplitude of about twice the breakdown voltage of the diode. This pulse then produces an avalanche shock front in the diode and the field again collapses.

3.3.2 Computer Simulation of a Trapatt Oscillator

In the original computer experiments by Scharfetter et al.[30] the response of an avalanche diode to a step in diode terminal current was calculated. However, it remained to be demonstrated that the results obtained by those experiments explained the experimental results. Since the low-frequency (2 to 3 GHz) coaxial circuits used in the experimental germanium oscillator could be modeled with reasonable accuracy, it was decided that a complete Trapatt oscillator could be simulated on the computer. Therefore, an experimental oscillator was set up and careful measurements made of oscillator performance and mechanical adjustments (i.e., tuning slug positions). A particular operating point was chosen which, although not the point of highest efficiency, was representative of Trapatt operation and not critically dependent on tuner positions or diode area.

Shown in Fig. 33 is the beginning of the turn-on transient, shortly after bias pulse has been applied. The upper curve is the computer simulation, and the lower curve the experimentally recorded response. Both wave forms are taken at the load. The experimental curve has been shifted downward by 20 V for clarity. The computer simulation continued to follow the experimental wave form quite closely, with structure similar to that shown here for approximately 16 nsec, at which time the simulated and experimental responses grew rapidly to a high-efficiency oscillation. The calculated diode voltage and the diode current wave forms are shown Figs. 34 and 35.

In Fig. 34 diode terminal voltage is plotted with a scale running from 0 to 100 V versus time over 0.5 nsec. The fundamental is at 2.38 GHz, with a three-to-one ratio between the Impatt frequency and the fundamental. The voltage swing is from approximately 93 to 2 V.

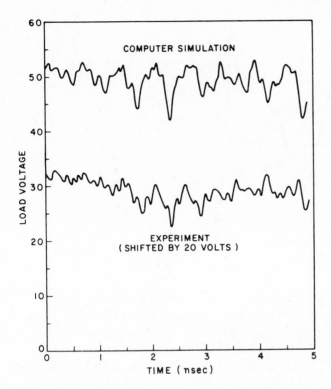

Fig. 33. Load voltage versus time during the turn-on transient of the 2.5-GHz oscillator.

The terminal current shown in Fig. 35 indicates that, while the voltage is low (the trapped plasma state), the current is maintained high by the circuit. Over the first roughly 0.2 nsec the voltage was very low. Also note that during the Impatt period, which follows the trapped plasma state, the current component at the fundamental is quite low, while the voltage is swinging about the breakdown value. The efficiency calculated for this oscillation period is 24.7% at 2.4 GHz and 25.7% over all frequencies; the average current is slightly less than the measured value, indicating that with a very slight tuning adjustment the computer experiment should agree almost exactly with the laboratory experiment. Table I summarizes the result of the simulation.

3.3.3 Analysis of Trapatt Oscillators

The computer simulation confirms the original hypothesis that a traveling avalanche zone and the subsequent trapped plasma state are responsible for the high-efficiency oscillations observed experimentally. Thus the simplified wave form shown in Fig. 36, which contains

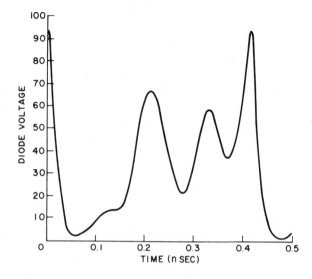

Fig. 34. Diode voltage versus time for the steady-state Trapatt oscillation.

these basic effects, can be used to obtain analytical expressions for oscillator performance. The analysis proceeds by calculating the diode response to a step in the terminal current. The response is divided into time intervals in which different simplifying assumptions can be made. During the first interval, the diode voltage rises to a value well in excess of breakdown, and little conduction current flows. During the next interval an avalanche shock front forms and the diode field

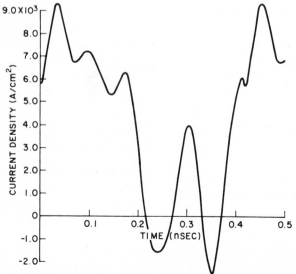

Fig. 35. Diode current versus time for the steady-state Trapatt oscillation.

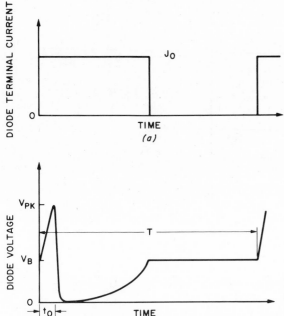

Fig. 36. Diode wave forms for computer experiments. (*a*) Applied current; (*b*) voltage response.

collapses. During this time the conduction current is large and a plasma is formed in the diode. Finally, during the recovery part of cycle, the plasma is removed and the diode voltage returns to the breakdown value. After the voltage reaches breakdown, the "Trapatt" part of the cycle is complete and, as the simulation results showed, a number of Impatt oscillations may occur. For simplicity we assume the square wave of current shown and calculate the efficiency at the fundamental frequency.

Charging. During the charging part of the Trapatt cycle,[35] the current is very low and the diode voltage increases linearly with time. The time delay in the buildup of the avalanche current can be calculated

TABLE 1

	Computer Simulation		Experimental Oscillator	
Frequency	2.38	GHz	2.38	GHz
d-c voltage	34.75	V	34	V
d-c current	453	mA	475	mA
Power output	3.9	W	5.3	W
Efficiency	25	%	33	%

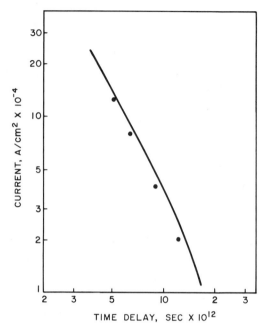

Fig. 37. Time required for diode voltage to reach the maximum value versus amplitude of the applied current step.

analytically if a number of simplifying assumptions are made. Here we shall simply state those assumptions and give the results of the calculation. The electron and hole drift velocities are assumed equal; and the ionization rates for holes and electrons are also assumed equal. Diffusion is neglected. Figure 37 shows the calculated time delay versus the amplitude of the current step for a 50-GHz Impatt diode. Shown also is the time delay calculated numerically for a 50-GHz silicon Impatt diode. Comparison of the two results indicate that the simple model accurately describes diode behavior during this part of the cycle. These results are important because the magnitude of the time delay determines the peak voltage amplitude. This, in turn, determines whether a given diode structure can be operated in the Trapatt mode.

Avalanche Shock Front. The next part of the Trapatt cycle is the creation of an avalanche shock front in the diode.[36] Referring to Fig. 31, the field profile and carrier densities during the zone transit part of the cycle are shown. In the region on the right, the carrier densities are small, so that the terminal current J_T is carried by the displacement current, i.e.,

$$J_T = \epsilon \frac{\partial E}{\partial t} \tag{13}$$

However, for an abrupt junction diode with a uniform doping N_a, Poisson's equation gives

$$\frac{\partial E}{\partial x} = \frac{q N_a}{\epsilon} \tag{14}$$

Thus we conclude that $E(x, t)$ must have the form $E(x - v_z t)$, where v_z, the zone velocity, is given by

$$v_z = \frac{\overline{J_T}}{q N_a} \tag{15}$$

This equation shows how we can overcome the normal space charge limitation in Impatt diodes. Recall that the space charge generated by the avalanche depresses the field and limits the generation. However, if a terminal current of more than $J = v_s q N_a$ is applied to the diode, the avalanche zone moves faster than the carriers and the avalanche is not restricted by the carrier space charge.

The form of the field given above represents a traveling wave in the diode; we can replace the time derivative in the continuity equations by $-v_z d/dz$. (See Appendix, Eqs. A12 and A13.) This substitution simplifies the continuity equations and, using the assumptions listed in the preceding section, an analytical expression for the magnitude of the current multiplication through the avalanche zone can be obtained.

The result of this calculation is

$$J_{cm} = J_{cs} e^{J_{cm} - J_{cs}/J_T} \tag{16}$$

where J_{cs} is the particle current ahead of the zone and J_{cm} the particle current behind the zone, in the region where $\partial E/\partial t$ is negative. This equation is interesting, since it indicates that the current multiplication through the avalanche is independent of the exact form of the ionization rate $\alpha(E)$. Furthermore, it shows that the excess multiplication increases with decreasing J_{cs}. Thus the number of carriers that remain after the last Impatt cycle determines the amplitude of the overmultiplication. Values for J_{cm} are typically 10 times J_T; the charge density remaining in the diode after a zone transit is about 10 times N_a.

Recovery. After the avalanche shock front passes through the diode, the diode is filled with a plasma of electrons and holes and the field is quite low.[33] In order to obtain the highest frequency of opera-

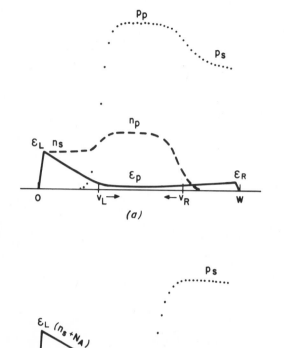

Fig. 38.   Recovery transient.   Hole concentration, electron concentration, and electric field versus distance. (a) Plasma velocity period; (b) saturated velocity period. (After Scharfetter,[34] copyright 1970, and the American Telephone and Telegraph Co., reprinted by permission.)

tion from the diode, it is necessary to remove the plasma as quickly as possible. Figure 38 shows the carrier densities and field profiles for two points in time during the recovery period. Initially, the plasma fills most of the diode; however, since the plasma density is quite uniform, diffusion is negligible and the drift component, determined by the low-field mobility is still dominant. In the regions between the low-resistivity ends and the plasma, the current is the result of carrier drift at saturated velocity and displacement current due to the increasing field. Thus, if the terminal current is too high, the fields in these depleted regions increase rapidly and may exceed the field required for avalanche breakdown. If this happens, the diode does not return to its high voltage state and Trapatt operation is inhibited.

The boundaries of the plasma move toward the center of the diode with velocities which depend on the low-field mobility of the carriers, i.e.,

$$V_L = \mu_P E_P$$

and

$$V_R = \mu_n E_P$$

where E_P = the field in the plasma
μ_P and μ_n = the hole and electron mobilities
V_L and V_R = the plasma boundary velocities
as defined in Fig. 38. When the two boundaries meet, the field rapidly exceeds that required to saturate the drift velocities, and the remaining carriers are removed at their scattering limited velocity.

Trapatt Operation. The simplified pictures of the various parts of the Trapatt cycle[34] can be combined to yield some quantitative analytical results for a Trapatt oscillator. As shown in Fig. 36, the procedure is to assume a square wave of diode terminal current and to calculate the diode voltage response for each of the intervals discussed above. A numerical Fourier analysis of the resulting wave form then gives the oscillator power, efficiency, and impedance. For example, Fig. 39 (from Scharfetter[34]) shows the theoretical power versus frequency behavior that one might expect from this mode of operation. Using this approach, efficiencies as high as 60% for silicon diodes are predicted. However, the simple square wave assumption for the terminal current is surely not the optimum; thus even higher efficiencies may be realized experimentally.

3.3.4 Trapatt Circuit Studies

In Section 3.2 on Impatt physics, the circuits used with the diode to make an oscillator were not discussed. For single-frequency Impatt oscillators it is reasonable to separate the study of the device and circuit because the diode can be represented by relatively simple circuit models. These models usually represent the diode in terms of a lumped circuit model (e.g., Fig. 2) which is valid over a limited frequency range. However, when an avalanche diode is operated in the Trapatt mode, a number of harmonically related frequencies are present and important to the operation of the oscillator.[37] Thus a simple, single-frequency representation of the oscillator is not adequate. In this section, the interaction of the diode and the circuit will be examined.

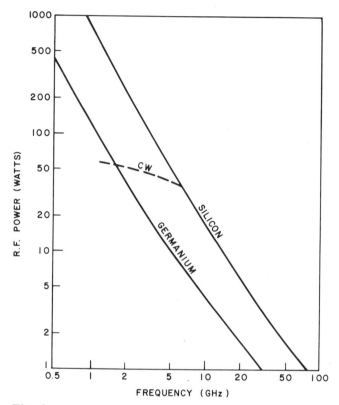

Fig. 39. Power output versus frequency. (After Scharfetter,[34] copyright 1970, and the American Telephone and Telegraph Co., reprinted by permission.)

Although there is presently no frequency domain model for a diode operating in the Trapatt mode, the simple time domain model discussed above can be used to predict a considerable amount about oscillator performance. As shown in Fig. 32 the diode can simply be thought of as a switch that is closed by the triggering pulse confined to the transmission line between the diode and the low pass filter. The switch opens a short time after it is closed because the plasma in the diode is exhausted. Thus the frequency of operation should depend on the transit time of the triggering pulse between the diode and the filter. Figure 40 shows this to be the case for experimental oscillators.

In many experimental oscillators, the Trapatt frequency is a factor of 10 below the transit-time or Impatt frequency. In these cases, the conduction current through the diode is not a square wave but, as shown in Fig. 32, a pulse of short duration. Therefore, the charge

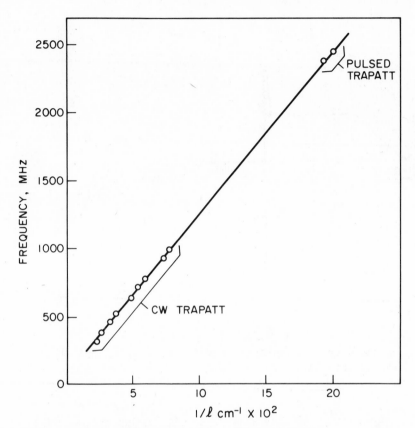

Fig. 40. Oscillator frequency versus filter position.

that flows through the diode for this short period of time must come from the capacitance in the immediate vicinity of the diode. The capacitance that can be discharged during that time consists of the lumped capacitance near the diode and the capacitance of a length of line which depends on the duration of the high-current state. Therefore the total capacitance can be written as

$$C_T = C_0 + \frac{\tau_P}{Z_0} \tag{17}$$

where C_0 = the lumped capacitance near the diode

τ_P = the length of time that the high current or trapped plasma state exists

Z_0 = the characteristic impedance of the coaxial air line.

After the plasma has been cleared from the diode, the diode voltage is nearly back to its breakdown value. However, the capacitance near the diode can recover only as quickly as the line can recharge it.

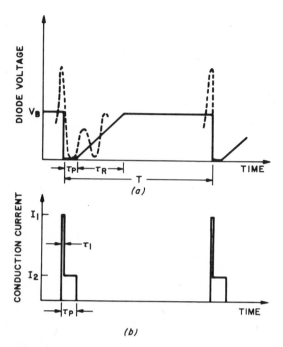

Fig. 41. Trapatt voltage and current wave forms. (a) Diode voltage versus time; (b) diode current versus time.

The diode voltage will thus have a shape similar to that shown by the broken line in Fig. 41a. The sinusoidal variations shown there are a result of ringing of the circuit inductance and capacitance near the diode. To a first approximation, it may be assumed that the average value near the diode increases linearly after the high-current state and stops when it reaches the breakdown voltage. This is a reasonable assumption, since the capacitance near the diode is recharged by a transmission line. For this model the conduction current will have the shape shown in Fig. 41b. Initially, a large current flows which discharges the lumped capacitance near the diode. This is followed by a lower current that discharges a short length of the transmission line. This type of current wave form has been found in computer simulations of a 500-MHz oscillator.

Neglecting the initial spike of voltage that initiates the traveling avalanche, we can calculate the amplitudes of the fundamental components of voltage and current. Although this is a simple model of the oscillator, it has been found to give results that compare favorably with the experimental data. Figure 42 shows the calculated power

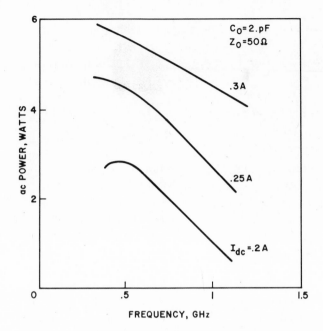

Fig. 42. Alternating current power versus frequency for values of d-c bias current.

versus the frequency for a 50-V germanium diode. Table 2 compares results calculated from Fig. 42 to experimental results for a 50-V germanium diode. The excellent agreement indicated in Table 2 is principally a result of the low frequency of operation. This oscillator was operating at the fourteenth subharmonic of the Impatt frequency, so that the assumed wave forms are not very sensitive to the details of the high-frequency components.

The operation of the Trapatt mode is thus seen to depend very strongly on the circuit around it, e.g., it is found that high-efficiency

TABLE 2

	Calculation	Experimental Results
a-c power	2.5 W	2.6 W
d-c voltage	41.5 V	38.9 V
Efficiency	30 %	34 %
Diode resistance	$-33.7\ \Omega$	$-35.5\ \Omega$
Diode reactance	16.7 Ω	19.0 Ω
Frequency	468 MHz	462 MHz

oscillations are possible even though the trapped plasma state lasts only a small fraction of the total period (such as 1/20). This occurs because the circuit recovery time is much longer than the diode recovery time.

3.3.5 Computer Experiments

The simple model of Trapatt oscillators given above is very useful for circuit design and performance calculations.[35] However, very little was said about the diode design. The analysis assumes that a trapped plasma state can be excited and that the diode will generate sufficient power in the harmonics to make up for circuit losses, regenerating the required triggering voltage. In this section the diode requirements are considered in more detail.

The computer program referenced above[7] has been used to investigate the response of a number of diode profiles to the square wave in the terminal current shown in Fig. 36. As pointed out in Section 3.3.1 above, the circuit is capable of periodically driving the diode voltage to approximately twice its breakdown value. Thus the voltage in excess of the normal breakdown voltage that is required to create an avalanche shock front in a given diode is an important parameter.

A simple argument that shows how the excess diode voltage depends on the resistivity of the high-field region in the diode is illustrated in Fig. 43. Figure 43 shows field profiles for a nearly *PIN* and an abrupt junction diode for d-c breakdown. If the avalanche region is

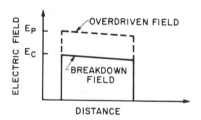

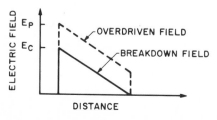

Fig. 43. Comparison of field profiles for punch-through factors of 5 and 1.

confined to a small fraction of the diode, then it is approximately the peak field, E_p, in the diode that determines when the avalanche multiplication will be large enough to start the field collapse. From Fig. 43 it is clear that the same increment terminal voltage is required for the field to reach E_p, (which is typically $1.5\,E_c$) in the PIN diode as in the abrupt junction case (provided the field in the abrupt junction diode just reaches the substrate at breakdown). Thus the peak voltage relative to the breakdown voltage is much less for the PIN diode. Alternatively, the PIN diode can be thought of as requiring much less triggering energy (relative to the stored energy at breakdown) to sustain Trapatt operation.

The results shown in Figs. 45 to 47 were calculated in the following way. A numerical solution for the field and carrier densities is obtained for a given diode structure at a reverse bias current of $1\ \text{A/cm}^2$. A step in the terminal current is then applied to the diode with the initial conditions determined by the $1\ \text{A/cm}^2$ solution. The solution is allowed to time evolve until the voltage returns to breakdown in the recovery part of the cycle. This point is defined as one-half of the oscillator period, and the Fourier coefficients for the d-c and fundamental components of the voltage wave form are calculated numerically. The instantaneous $i \cdot v$ product is integrated over one period to find power input to the diode. Thus, for each drive current, the efficiency, frequency, impedance, and peak diode voltage are obtained.

Before presenting the results of these calculations it will be helpful to define several parameters used in the analysis. The first is the punch-through factor F. Figure 44 (from Scharfetter and Gummel)[17] shows that this factor is obtained by extrapolating the field versus the distance for a punched-through diode of width W (biased to breakdown) to the point of zero field. The ratio of the extrapolated width to the diode width is the punch-through factor F. A second parameter is the overvoltage V_0. This is defined as the maximum diode voltage occurring during the Trapatt cycle normalized to the diode breakdown voltage. Finally, it is useful to define a current density $J_c = qNv_s$, where N is the depletion layer impurity concentration for electrons or holes (cf. Eq. 15).

Figure 45 shows a plot of the efficiency versus the d-c current density (normalized to J_c) for two germanium diodes. The field profile for these diodes is shown in the inset in the figure. Diode 1 is a typical N^+PP^+ 6-GHz germanium Impatt with a doping of about 4×10^{15} cm^{-3} in the p-region. The punch-through factor for this diode is 1. Diode 2 is also a germanium diode but it has been scaled up in frequency by a factor of 4. It also has a punch-through factor of 1, and a doping density of $2 \times 10^{16}\ \text{cm}^{-3}$. (Note that the doping density is more

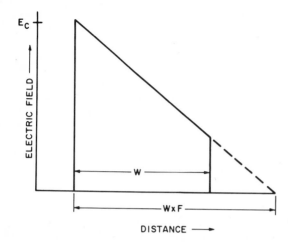

Fig. 44. Electric field versus distance for a typical abrupt junction diode defining the punch-through factor F.

than a factor of 4 larger than diode 1. This occurs because the peak field for avalanche breakdown increases as the doping density is increased.） Note that the d-c current given in the figure is one-half the drive current J_0 shown in Fig. 36 because the current wave form is assumed to be a square wave. It is found that the Trapatt frequency does not depend strongly on the d-c current, since a large value of J_0

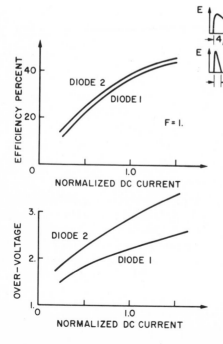

Fig. 45. Germanium Trapatt efficiency and overvoltage versus $d-c$ current (normalized to qNv_s). Diode 1 has a breakdown voltage of 50 V and a maximum Trapatt frequency of ~1.8 GHz. Diode 2 has a breakdown voltage of 19 V and a maximum Trapatt frequency of ~4.5 GHz.

creates a larger plasma but also extracts the plasma at a high rate. Therefore, in each figure, a nominal Trapatt frequency is given.

Also in Fig. 45, the overvoltage is plotted as a function of the d-c current. It can be seen that there is a significant difference between the two diodes with regard to this parameter. For example, at an efficiency of 35%, diode 1 requires an overvoltage of about 2.15 times its breakdown voltage which is consistent with the comments of Section 3.3.1. However, diode 2 requires an overvoltage of 2.75 times its breakdown voltage. The expected difference in performance for these two diodes is likely to be even greater in practice, since the Impatt frequency for diode 2 is about 25 GHz and the obtainable circuit Q will be lower. Thus the higher-frequency circuit will develop less overvoltage.

Figure 46 compares silicon diodes of the same width (1.0 μm) but having different values of F. It is immediately apparent that increasing the punch-through factor results in a significant reduction in the required overvoltage, as predicted from the simple argument above. At 35% efficiency, for example, the required overvoltages are 2.4, 2.0, and 1.7 for punch-through factors of 1, 2, and 5, respectively. A further advantage of diodes with high punch-through factors is that a given efficiency can be achieved with less current (recall that the plots are shown for current normalized to qNv_s, and N decreases approximately as $1/F$). For example, at 35% efficiency the d-c current

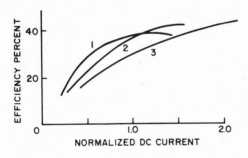

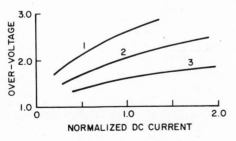

Fig. 46. Silicon Trapatt efficiency and overvoltage versus d-c current (normalized to qNv_s) for diodes with a depletion layer of 1 μm and different punch-through factors. For diode 1, $F = 1$, $V_B = 30$ V and $f_0 \cong 8$ GHz; for diode 2, $F = 2$, $V_B = 38$ V and $f_0 \cong 6$ GHz; and for diode 3, $F = 5$, $V_B = 45$ V and $f_0 \cong 5$ GHz.

densities required are 18, 000, 10, 000, and 6000 A/cm² for $F = 1$, 2, and 5, respectively. Finally, the ultimate efficiency with larger values of F is greater. It can be seen from Fig. 46 that, for $F = 1$, the efficiency saturates at $J_{dc} = 1$. This happens because of the premature avalanche breakdown that occurs in the recovery part of the cycle and is discussed elsewhere.[6] These results are summarized in Fig. 47.

The results of computer analysis show that the overvoltage problem can be avoided by choosing diodes with a larger punch-through factor and accepting some trade-off in frequency. However, this conclusion is valid only for pulsed operation. If continuous wave operation is desired, the loss in negative conductance (see Fig. 6) at the transit-time frequency cannot be made up by increased current density. The problem is simply that the requirements for optimum Trapatt operation diverge from those for efficient Impatt operation as the frequency is increased.

A solution to this overvoltage problem can be found in modification of the diode structure. Shown in Fig. 48 is a structure that has been fabricated by ion-implantation techniques and has been successfully operated as a continuous wave Trapatt oscillator in the 4 to 6-GHz range.[38] In this structure a double-sided design has been employed. The right-hand side has been designed as an efficient Impatt oscillator whose function is to provide the large Impatt oscillations that are required to start the Trapatt mode. The left-hand side is optimized for Trapatt operation in the 4 to 6-GHz range. This side of the diode has a punch-through factor of about 5. Calculation of the diode Q at the frequency of maximum negative conductance (which according to Ref. 7 is a good indication of large-signal performance) indicates that this

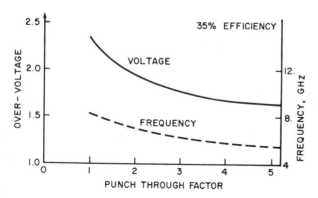

Fig. 47. Overvoltage and Trapatt frequency versus punch-through factor for diodes of Fig. 46 at 35% efficiency.

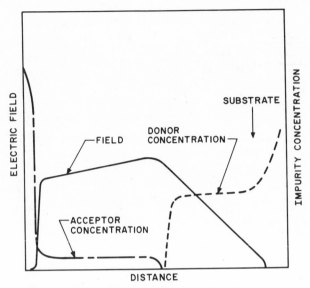

Fig. 48. Asymmetric field profile for high-frequency continuous wave Trapatt operation.

diode is as good at its transit-time frequency as a conventional single-sided Impatt.

3.4 OSCILLATOR DESIGN

3.4.1 Materials

In this section, some details of oscillator design will be given, starting with the choice of material. For reference, Table 3 gives some of the relevant properties of the three most common semiconductor used for avalanche diodes.[39] From the discussion in Section 3.2, Eq. 5 showed that the maximum power which can be absorbed by the carriers is proportional to $E_c^2 v_s$. Thus the data in Table 3 indicate that silicon and GaAs are comparable, while germanium is at least a factor of 4 lower in Pf^2 product. As pointed out in Section 3.2, at lower frequencies this difference is not seen in continuous wave oscillators because of thermal resistance. If the thermal drop in the semiconductor can be kept small, silicon and GaAs are comparable in power handling ability. However, in the double-drift Impatts, for example, the heat is produced some distance from the contact. Thus the superior thermal conductivity of silicon makes a

TABLE 3

	Ge	Si	GaAs
Dielectric constant	16	11.8	12.9
Energy gap (300 °C)(eV)	0.66	1.12	1.43
Intrinsic carrier concentration (cm^{-3})	2.4×10^{13}	1.4×10^{10}	1.1×10^{7}
Thermal conductivity (W/cm C°) (300 °K)	0.64	1.45	0.46
Thermal diffusivity (cm^2/sec) (300 °K)	0.36	0.9	0.24
Saturated drift velocity (300 °K) Electrons Holes (cm/sec)	$\sim 7 \times 10^{6}$ $\sim 6.5 \times 10^{6}$	1×10^{7} 1×10^{7}	$\sim 7 \times 10^{6}$ $\sim 7 \times 10^{6}$
Drift mobility[a] Electrons Holes (cm^2/V–sec)	3900 1900	1500 600	8500 400
Breakdown field (V/cm)[b]	$\sim 2 \times 10^{5}$	$\sim 3.5 \times 10^{5}$	$\sim 4 \times 10^{5}$

[a]For lightly doped material.
[b]For an abrupt junction diode with an impurity concentration of 4×10^{15} cm^{-3}.

substantial difference in the operating temperature for a given power input. In cases in which the diodes are operated at very high temperatures, GaAs has the advantage because the intrinsic carrier density remains negligible to higher temperatures. The high mobility of GaAs reduces to some extent the series loss in the substrate of a diode; however, thinning techniques are available that can be used to remove most of the substrate material.[45]

Table 4 summarizes the noise data that are available for Impatt devices.[40] The equal ionization rates for holes and electrons in germanium and gallium arsenide result in a substantially lower noise generation for devices made with these materials. The small-signal noise measure data are generally in agreement with the theoretical analysis by Gummel and Blue,[18] although the value for GaAs has not been as accurately calculated because of the uncertainty in some of the material parameters for GaAs.

The oscillator FM noise is obtained from the expression given by Josenhans,[40]

$$\Delta f_{rms} = \frac{f_0}{Q_{ext}} \left(\frac{k T_0 M \overline{B} \, \overline{W}}{P_0} \right)^{1/2} \tag{18}$$

TABLE 4. STATE OF THE ART: NOISE MEASURE
OF IMPATT DEVICES [a,b]

Value of $M = \dfrac{T}{T_0}$, $T_0 = 290$ °K		
	Small–Signal	Large–Signal
Silicon	40 dB	55 dB
Germanium	30 dB	40 dB
GaAs	25 dB	35 dB

[a]Courtesy of J. G. Josenhans.[40]
[b]The amplifier and oscillator noise in this table are
for a lossless circuit at the frequency corresponding
to the maximum oscillator efficiency, without harmonic tuning.
   Amplifier noise measure is given directly by the
M in the small–signal column.
   The bias circuit is assumed terminated to give minimum oscillator noise.

where $f_0 =$ the oscillator frequency
   $Q_{ext} =$ external circuit Q
      $k =$ Boltzmann's constant
      $T_0 =$ load temperature
      $M =$ the large–signal noise measure from Table 4
   $\overline{BW} =$ measurement bandwidth
      $P_0 =$ oscillator power output.
   The AM noise is given by

$$\left.\frac{N}{C}\right|_{SSB} = \frac{2kT_0 M\overline{BW}/P_0}{S^2 + 4Q_{ext}^2 f_m^2 / f_0^2} \tag{19}$$

where S is a large–signal saturation parameter (nominally about 2 for
Impatt oscillators) and f_m the separation of the measurement from the
carrier. Near the carrier the noise in silicon Impatts is found to be
approximately independent of the frequency separation from the carrier. Thus Impatt oscillators exhibit somewhat lower $1/f$ noise than
is observed in many oscillators.

   The FM noise can be reduced by using a stabilizing cavity that
raises the external Q of the oscillator.[41] The AM noise can be reduced by connecting multiple diodes in series or parallel, since the rf
voltages are correlated and the noise voltages are not. In both cases
the noise is reduced by operating at the highest-power output, although
this conclusion does not remain valid if the signal amplitude is large
enough to cause the field in the diode to drop below that required to

maintain saturated velocities. In any event it is clear from Table 4 that the lowest noise can be achieved with GaAs.

The data given in Table 4 are obtained for oscillators operating near the point of optimum efficiency. However, as shown in Fig. 23 lower-noise measures can be obtained when the operating frequency is twice the transit-time frequency. Unfortunately, this corresponds to a point of very low negative conductance, and thus an undesirable operating point for most circuits.

3.4.2 Impurity Profile

Impatt. In Section 3.2, Fig. 6, it was shown how the negative conductance for diodes with different profiles depends on current density. From this plot we conclude that the Read diode would be potentially superior at low d-c current densities. However, the abrupt junction diode is easier to fabricate and exhibits a higher conductance at higher-current densities. Furthermore, for diodes operating in the millimeter-wave part of the spectrum, diode area (for constant impedance) decreases approximately as $1/f^2$ and the operating voltage as $1/f$, whereas the thermal dissipation is proportional to the square root of the diode area. Therefore, the current density can be increased by approximately f^2. The result is that the abrupt junction diode can be operated as a continuous wave oscillator at maximum efficiency in the millimeter wave spectrum. (These comments on scaling will be discussed in more detail in the next section.) The advantages of the abrupt junction diode were first pointed out by Misawa and the reader is referred to his paper[42] for a more complete description.

The abrupt junction diode has several other advantages at high frequencies. The avalanche zone is not as sharply defined as it is in a Read diode; hence there is less minority carrier storage due to diffusion (cf. Section 3.2.4). In addition the broader avalanche region (approximately 33% of the total diode width) requires a smaller peak field. This in turn reduces the peak field that would occur in a very narrow avalanche region (ref. Section 3.2.4). Of course, further improvement is obtained if the double-drift design discussed in Section 3.2 is used.

Design information for abrupt junction Impatts can be obtained from Figs. 49 and 50. Figure 49 (from Sze[39]) gives the diode breakdown voltage and depletion layer width as a junction of doping concentration. From this plot and the saturated drift velocity for a given material and carrier type, the transit-time frequency can be calculated. Since saturated velocities are not in general accurately known (and are a function of temperature), and since Impatts may be operated

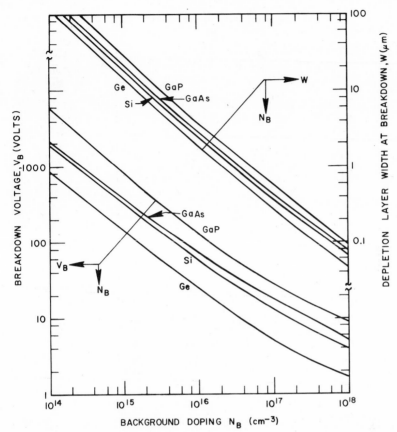

Fig. 49. Depletion layer width and breakdown voltage versus doping for abrupt junction diodes. (After Sze.[39])

some distance either side of the transit-time frequency, such a calcu-lation is only approximate. Figure 50 (from Dunn[43]) therefore should be used only as a rough guide line. The curves shown have been plotted for a transit angle of 0.75 π. The broken line in the 100 to 300-GHz range for silicon reflects the data of Bowman and Burrus[14] and indi-cates the change from abrupt junction diodes to more graded junctions with higher-breakdown voltages to avoid tunneling.

Trapatt. The correct impurity profile for Trapatt oscillators cannot be as accurately defined as the Impatt profile because, in this mode, the diode may operate at almost any subharmonic of the transit-time frequency. However, the simplified square wave assumption does provide a good estimate of the highest frequency at which good efficien-

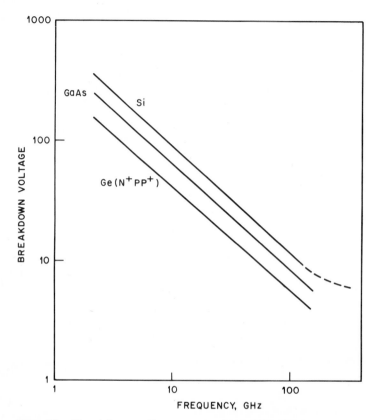

Fig. 50.   Breakdown voltage versus Impatt frequency.

cy can be obtained. Therefore, Fig. 51 has been calculated using the methods discussed in Section 3.3.5 above. The frequencies are only approximate because they depend to some extent on the magnitude of the bias current. Since higher-frequency diodes must be designed with larger punch-through factors to keep the triggering energy at an acceptable value, the results have been given for various values of punch-through factor.

3.4.3 Scaling Law

Although a detailed analysis to predict power output and efficiency is rather difficult, some very useful information can be obtained by simply scaling experimental results at one frequency to other frequencies. In doing this one assumes that the same physics are operative at both frequencies (e.g., if tunneling is not considered at low frequen-

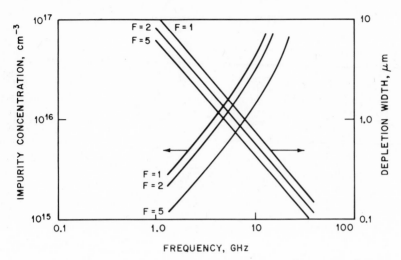

Fig. 51. Impurity concentration and depletion width versus frequency for Trapatt diodes.

cies, it is also neglected for higher frequencies). This type of analysis has been done by Scharfetter.[44] Here we will give only a few of the results of his analysis. Figure 52 shows the expected power versus frequency behavior for abrupt junction silicon Impatts on copper heat sinks. This curve is obtained by scaling well-characterized experimental results at 50 GHz.[45] The following assumptions are made in the scaling process. First, the thermal impedance for diodes at 50 GHz is scaled according to Eq. 10, which in scaling form can be written as

$$R_T = R_S\left(\frac{D_0}{D}\right) + R_A\left(\frac{D_0}{D}\right)^2 + R_f\left(\frac{D_0}{D}\right)^2\left(\frac{W}{W_0}\right) \tag{20}$$

where D_0/D = reference diameter over diameter
  W/W_0 = depletion thickness over reference (this factor, appearing in the third term, accounts for the relative variation in the length of the thermal path in the silicon wafer)
  R_S = spreading resistance of the reference
  R_A = thermal resistance of metallic layers
  R_F = thermal resistance of the silicon wafer.

Second, the diode negative conductance and susceptance are scaled by the relations,

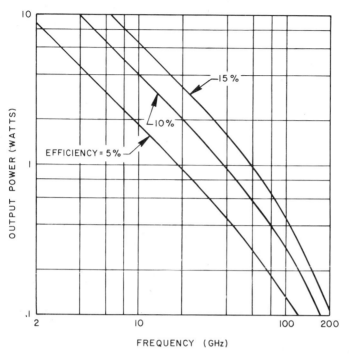

FREQUENCY (GHz)

Fig. 52. Power output versus frequency for silicon Impatt oscillators on copper heat sinks with efficiencies of 5, 10, and 15%. (After Scharfetter.[44])

$$\frac{G}{G_0} = \left(\frac{I}{I_0}\right)\left(\frac{V_0}{V}\right) \tag{21}$$

and

$$\frac{S}{S_0} = \left(\frac{f}{f_0}\right)\left(\frac{C}{C_0}\right) \tag{22}$$

where I = bias current

V = d-c voltage

C = depletion layer capacitance.

The inverse relationship between conductance and the d-c voltage is obtained by assuming that the conductance is approximately inversely proportional to the a-c voltage and that at maximum efficiency the ratio of a-c voltage to d-c voltage is constant.

The breakdown voltage for abrupt junction diodes scales approximately as

$$\frac{V}{V_0} = \left(\frac{W}{W_0}\right)^{0.82} \tag{23}$$

and

$$\frac{W}{W_0} = \frac{f_0}{f} \tag{24}$$

Finally, because of skin effect loss, the minimum microwave circuit impedance is scaled by

$$\frac{R}{R_0} = \left(\frac{f_0}{f}\right)^{0.5} \tag{25}$$

Using these equations, the experimentally observed power at 50 GHz is scaled to other frequencies. Figure 53 shows the predicted bias currents required to obtain the power outputs shown in Fig. 52. The broken line represents the space charge limitation in the drift region (see Section 3.2.5). Finally, Fig. 54 shows the diameters to be used for any frequency. Scharfetter has carried out these calculations for a number of different cases; the reader is referred to his paper for more detail.

In using these results, one must keep in mind that one is extrapolating from given experimental results by rather simple laws. For example, at frequencies above 100 GHz one must be very careful to keep circuit losses within Eq. 25, since surface roughness is a major problem. Diode fabrication is also more difficult, so that the results of Fig. 52 should be considered an optimistic scaling of the 50-GHz results. On the other hand they should not be considered as an upper

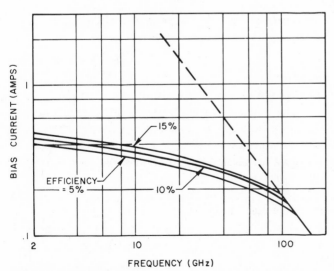

Fig. 53. Operating current versus frequency for silicon Impatt oscillators on copper heat sinks with efficiencies of 5, 10, and 15%. (After Scharfetter.[44])

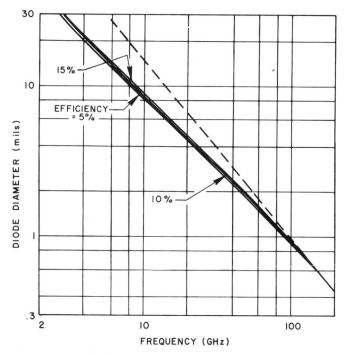

Fig. 54. Diode diameter versus frequency for silicon Impatt oscillators on copper heat sinks with efficiencies of 5, 10, and 15%. (After Scharfetter.[44])

limit, since improvements at the reference frequency (e.g., the use of double-drift diodes) can also be achieved at other frequencies.

3.4.4 Thermal Design

In Section 3.2.5 it was pointed out that Impatt power output is severely limited by heat dissipation. Equation 10 summarizes the various contributions to diode thermal resistances; a simple calculation shows that, if diodes are mounted junction-side down and have been fabricated with the junction close to the surface, the thermal resistance is rather fundamental and not a great deal can be done about it. Swan[46] has suggested that type IIA diamond be used for the heat sink because it has about five times the thermal conductivity of copper at 0 °C. In practice this much improvement is not realized, since devices are usually operated at 200 to 300 °C where the difference is less than a factor of 2.

There are several geometrical ways to reduce the thermal spreading resistance. The simplest is to mount a number of small diodes in parallel. Swan[16] has obtained good results on up to 5 diodes mounted in parallel on the same header. Another possibility is the ring diode geometry described by Marinaccio.[47] In this case the diode is formed in the shape of a ring. The improvement in thermal resistance of the ring structure over that of a disk of equal area has been determined by Garlinger and Stover[48] and is shown in Fig. 55.

In practice the experimental evaluation of various thermal designs is greatly facilitated if the junction temperature as a function of the d-c input power can be simply estimated. One method of doing this is to observe the change in d-c voltage with input power.[49] The breakdown voltage V_B of a diode is given approximately by

$$V_B = V_{BO}(1 + \beta \Delta T) \tag{26}$$

where β is approximately constant. The change in temperature is

$$\Delta T = R_T I_{dc} \cdot V_d \tag{27}$$

and the diode voltage at a given current and temperature is

$$V_d = V_B + r_{sc} I_{dc} \tag{28}$$

where r_{sc} is the space charge resistance. The r_{sc} is given approximately by[39]

$$r_{sc} = \frac{x^2}{2\epsilon A v_s} \tag{29}$$

where x = the drift region width, about 0.66 times the depletion width of the diode for an abrupt junction diode

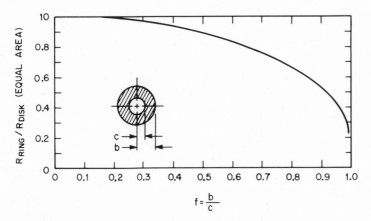

Fig. 55. Relative thermal impedance of a ring structure, compared to a disk of equal area. (After Garlinger and Stover.[48])

A = diode area

v_s = saturated drift velocity

The combining of Eqs. 26, 27, and 28 gives

$$V_d = V_{BO} + \beta V_{BO} R_T I_{dc} V_d + r_{sc} I_{dc} \tag{30}$$

If the thermal measurements are made at low currents (such that $\Delta T \leqq 50\,°C$) then in the second term V_d can be replaced by V_{BO}:

$$V_d \cong V_{BO} + \beta V_{BO}{}^2 R_T I_{dc} + r_{sc} I_{dc} \tag{31}$$

The value of β can be determined for a given type of diode by measurements of the breakdown voltage in an oven. The r_{sc} can be determined approximately from Eq. 29 or by pulsed measurements of V_d versus I_{dc}.

Thus the thermal resistance can be estimated by simple d-c measurements. The result can be used to calculate the approximate junction temperature. In addition to the assumptions made above, an additional error is present in estimating junction temperatures. This occurs because the variation in breakdown voltage given by Eq. 26 results in a nonuniform current and temperature in the plane of the junction.

3.4.5 Circuit Design

Impatt Circuits. In principle Impatt circuits should be simple to design. The small-signal diode impedance is relatively easy to calculate and the diode resistance is negative over a limited band of frequencies. Thus an oscillator can be made by resonating the diode reactance and transforming the load impedance to match the diode negative resistance. Since the negative resistance is band limited, no attention must be paid to broadband resistive loading to suppress parasitic oscillations. In fact, if low power outputs and efficiency are acceptable, these conclusions are valid in practice. However, at higher-power-output levels a number of problems develop. First, recall that the diode negative conductance decreases with increasing rf voltage amplitude (Fig. 8).* This requires a larger transformation of the load impedance and at the same time increases the importance of circuit loss. Second, the large amplitude rf voltage results in the creation of negative resistance in the diode over a much wider frequency spectrum. This requires careful design of the out-of-band cir-

---

*The negative resistance also decreases with rf voltage amplitude, since $R = G/(G^2 + B^2)$ and $B \gg G$, where $B \cong$ constant (see Fig. 8).

cuit characteristics, particularly in the bias connection. Third, the proper tuning of the harmonic components of the wave form have a significant effect on oscillator efficiency. Thus, while the designer may not wish to take advantage of the improvement in efficiency afforded by harmonic tuning, it is still important to be aware of the circuit impedances at harmonic frequencies, since some tuning conditions can reduce the power output.[23] Finally, at higher-input powers the designer is constrained to place the diode in a position where the heat can easily be removed.

One of the first circuits used by De Loach and Johnston[50] for microwave Impatt oscillators is shown in Fig. 56. The matching function is accomplished with a raised cosine taper to reduced height wave guide, and the diode is resonated by a sliding short. The impedance transformation from the full-height wave guide to the diode impedance is quite large. A coaxial circuit designed by Iglesias, requiring an impedance transformation from only 50 Ω, is shown in Fig. 57.[51] This circuit can be operated quite successfully with only a single-quarter-wavelength tuning slug. Improved matching is expected as additional tuning slugs are added; however, the improvement found experimentally was somewhat more than that expected on the basis of single-frequency operation. Swan[25] showed that this improvement resulted from second harmonic tuning of the diode. The coaxial circuit is also useful because it is relatively easy to model for analytical studies. Figure 58 is a circuit designed for single-frequency operation.[52] At the resonant frequency of the cavity, very little power is lost in the absorber, whereas at all other frequencies the diode is resistively loaded. Again, this circuit has the advantage of being well characterized.

At millimeter wave frequencies a circuit that has been used with good success is shown in Fig. 59.[45,53] The cap which is placed over the diode forms a localized radial line cavity, keeping circuit losses to a minimum. Although the cap forms the primary resonance, the sliding short permits tuning over about a 20% band.

Trapatt Circuits. In contrast with Impatt circuits, Trapatt circuits are not simple in principle, the reason being that a number of frequency components must be considered in any design. Some progress can be made, however, by using the time domain approach of Section 3.3. In Section 3.3 a simple model of the Trapatt was used which consisted of a diode, some capacitance and lead inductance near the diode, and a transmission line with a low-pass filter at a distance of $\lambda/2$ from the diode. Figure 60 shows a specific realization of this simple circuit model;

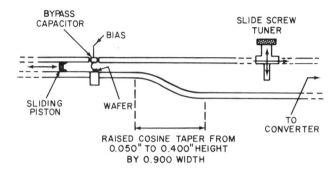

Fig. 56. Reduced height wave guide circuit. (After De Loach, Jr., and Johnston.[50])

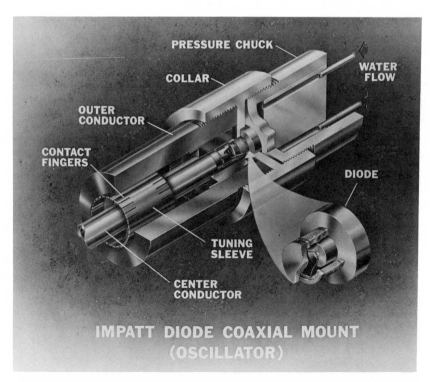

Fig. 57. Coaxial Impatt circuit. (Courtesy of D. E. Iglesias.)

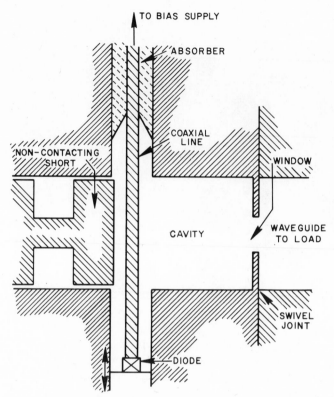

TO BIAS SUPPLY

ABSORBER

COAXIAL
LINE

NON-CONTACTING
SHORT

WINDOW

CAVITY

WAVEGUIDE
TO LOAD

SWIVEL
JOINT

DIODE

Fig. 58. Coax-wave guide cavity circuit. (After Magalhaes and Kurokawa.[52])

the data in Table 5 show that this is all that is required for efficient
Trapatt operation. The circuit consists of a coaxial air line with the
diode mounted in one end. Contacting the diode package is a 1-pF
lumped capacitor, formed by a disk on the center conductor. A distance down the line is a commerical low-pass filter. Following the
low-pass filter is a commercial, slotted coaxial line tuner. The experimental data in Table 5 show that, generally speaking, as fewer
harmonics of the Trapatt oscillation are confined to the high Q cavity,
the shape of the pulse reflected by the filter is degraded and the power
output decreases. The lumped capacitor, series lead inductance, and
diode capacitance also form a low-pass filter in this circuit. Care
must be taken that none of these elements is so large that the triggering pulse is degraded before reaching the diode. This requires that the
lead inductance and the lumped capacitance resonate at the Impatt frequency or higher.

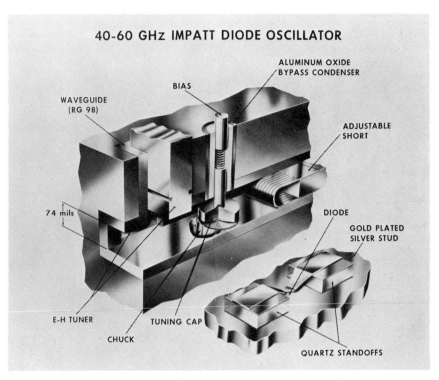

40-60 GHz IMPATT DIODE OSCILLATOR

Fig. 59. Radial line cavity. (Courtesy of D. E. Iglesias.)

TABLE 5. EXPERIMENTAL DATA FOR CIRCUIT SHOWN
IN FIG. 60.

$I_{dc} = 200$ ma

Low-Pass Filter Cutoff	Power, 500 MHz, in Watts	Power, 1000 MHz
700 MHz	2.7	—
1000 MHz	1.7	—
1500 MHz	2.5	1.4
3000 MHz	1.3	1.0
4000 MHz	0.9	0.9
4 Slug filter	2.6	1.3

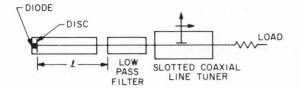

Fig. 60. Simplified Trapatt circuit.

The results shown in Table 5 were obtained for a Trapatt oscillator operating at about the fourteenth subharmonic of the transit-time frequency. With this much frequency separation, the time domain approach to circuit design works very well. As the subharmonic number is lowered, this approach is less accurate and should be replaced by a frequency domain approach. Nevertheless, the time domain analysis has given good quantitative results at subharmonic numbers as low as 5. Since our comments in Section 3.3.4 indicate that the efficient continuous wave operation at high frequencies is limited to operation at about this subharmonic number, the time domain approach will probably remain useful.

The simple transmission line delay for the triggering pulse is, of course, only one way of achieving the desired triggering. An equivalent approach is to measure the circuit impedance seen by the diode in a transmission line circuit and then to reconstruct this impedance with lumped elements. This has been done successfully at low frequencies by Reynolds *et al.*[54] This approach is particularly useful for low-frequency pulsed oscillators, since it reduces the circuit size and only a few harmonics have to be matched because the diodes are relatively easy to trigger.

3.5 APPLICATIONS

3.5.1 Power Oscillators

In Section 3.2 it was shown that transit-time oscillators are limited by a Pf^2 relationship. However, these conclusions are valid only for devices whose length is limited by the transit time. When devices are placed in series, as for example in the double-drift Impatt, this limitation no longer applies. Thus, in principle, one could expand the double-drift concept to make "diodes" that contained many avalanche and drift regions all in series. The thermal resistance limits this approach for continuous wave devices to the simple double-drift diode with one avalanche region. It is possible, however, to connect a num-

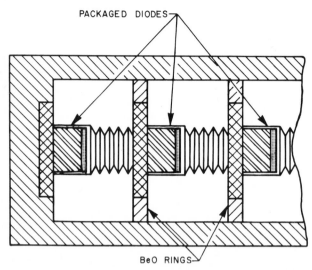

PACKAGED DIODES

BeO RINGS

Fig. 61. Heat sinking series diodes with beryllium oxide rings. (After Magalhaes and Schlosser.[56])

ber of individual diodes by providing a parallel heat path and series electrical connection. Josenhans[55] has made such an oscillator by forming metallized islands on type IIA diamond. Diodes are bonded to each island and the series connected with small wires.

Another method of series combination has been reported by Magalhaes and Schlosser[56] and is shown in Fig. 61. In this structure, packaged diodes are series connected along the center conductor of a coaxial line. Beryllium oxide rings (with a thermal conductivity of ~ 2 W/cm-C$^\circ$) provide a shunt heat path to the outer conductor. In both of these techniques the power output increased linearly with the number of diodes.

The techniques discussed above combine diodes within a small fraction of a wavelength and thus the resulting package can be treated as a single diode that oscillates in a single mode. Further improvement in power output can be obtained by various combining schemes for multiple oscillators. In combining circuits care must be taken to avoid spurious modes of oscillation. Such a circuit, devised by Rucker, is shown in Fig. 62.[57] The stabilizing resistors suppress spurious oscillations and maintain both diodes in phase. Rucker has used this circuit successfully to combine up to 5 diodes connected radially about the points X. Another possibility has been reported by Tatsuguchi[58] and is shown in Fig. 63a. Stabilization is achieved in this circuit by the ex-

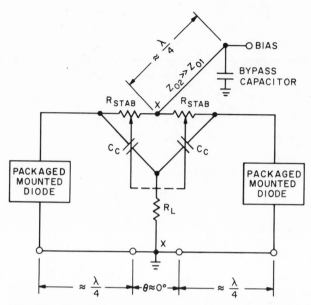

Fig. 62. Equivalent circuit of two-diode oscillator. (After Rucker.[57])

tensive use of circulators. The capacitors permit adjustment of the phase of the locking signal and also enable the circuit to combine outputs of unequal amplitude and phase. The circuit clearly depends on having low-loss circulators; Fig. 63b shows the loss in power as a function of circulator loss. Using a stripline fabrication, Tatsuguchi has reported on a power combiner with circulators having a forward loss of less than 0.2 dB.

3.5.2 Amplifiers

In many cases amplification is preferred to direct generation, particularly when low FM noise is desirable. Reflection-type amplification from the diode negative resistance has been demonstrated by Napoli and Ikola.[59] Small-signal amplification can be obtained across the band of device negative resistance. Instantaneous broadband amplification is possible if the circuit is designed to match the device impedance across this band.[60] However, broadbanding is more difficult under large-signal conditions because of the variation in diode impedance with signal amplitude. Hines[61] has considered the general cases of negative resistance power amplification, and some of his results are summarized in Fig. 64. This figure shows the normalized power

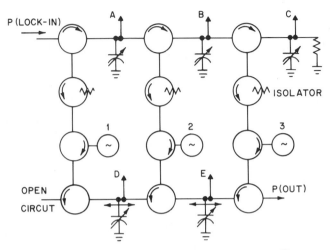

Fig. 63a. Power combiner. (After Tatsuguchi.[58])

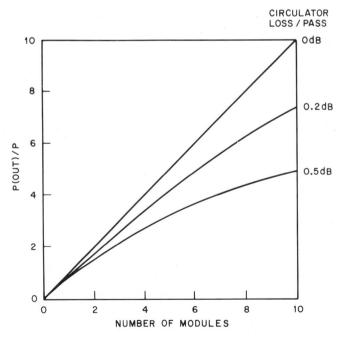

Fig. 63b. Power out over power of a single module versus number of modules for various values of circulator loss per pass. (After Tatsuguchi.[58])

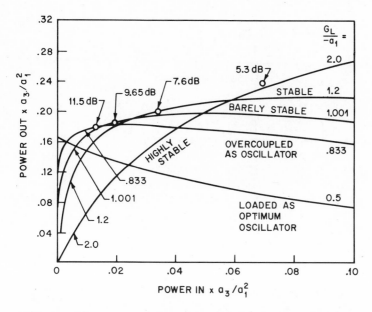

Fig. 64. Input–output characteristic for a negative resistance device with cubic nonlinearity. (After Hines.[61])

out versus power in for a device that has a simple cubic nonlinearity in the I-V characteristic (i. e., $i = a_1 v + a_3{}^3 v^3$). The results thus do not accurately describe avalanche diode oscillators, but many of the qualitative conclusions are applicable. First, for stable amplification, i.e., for large-load conductances, the power output saturates with increasing input power. At large-input powers the device might better be described as a power adder and, except that it does not oscillate with the input removed, it operates in much the same way as the power combiners discussed above. On the other hand, the lightly loaded device oscillates with no input and operates as a locked oscillator. Hines' analysis also predicts that, under large-signal conditions, the amplifier gain is not a maximum at the center frequency but has a response similar to a double-tuned circuit. This prediction has been confirmed experimentally for Impatt amplifiers by Noisten.[62]

High-efficiency mode amplifiers were first reported by Prager, Chang, and Weisbrod.[63] The characteristics of amplification in this mode are similar to the Trapatt mode oscillator. In general, the d-c to rf conversion efficiency is high, above 25%, at frequencies well below the transit-time frequency. Figure 65 shows the circuit for a typical 500-MHz continuous wave germanium Trapatt amplifier; Fig.

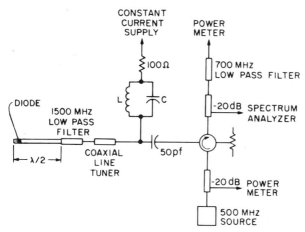

Fig. 65. 500-MHz amplifier circuit.

66 shows the power and efficiency versus input power for this circuit.[64]
Amplification is obtained by simply detuning the circuit so that it does
not oscillate.   This type of circuit is quite noisy because, in order to
create a low-frequency negative resistance in the abrupt junction
diodes used, the circuit must be tuned very close to the threshold for
oscillation.   Lower-noise amplification is possible if the circuit is al-

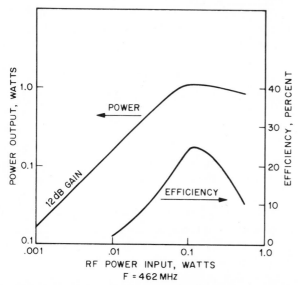

Fig. 66. Power output and efficiency, power out-power in/d-c power, for a
continuous wave Trapatt amplifier.

lowed to oscillate in the Trapatt mode at one frequency and amplify parametrically at a second frequency.[64, 65]

3. 5. 3 Modulation

There are a number of ways to modulate Impatt sources. The three that are reviewed here are direct deviation,[66] varactor,[66] and path length modulation.[67] In Section 3. 2 it was pointed out that the avalanche frequency is proportional to the square root of the d-c bias current. Thus, if the bias current is modulated, the diode reactance changes and the oscillator frequency can be changed. For practical systems there are a number of difficulties with this method of modulation. First, the power output changes with bias current, so that the frequency modulation is accompanied by amplitude modulation. Second, the diode reactance is a function of rf amplitude, as well as bias current. Therefore, the deviation is a complex function of the bias current for which no simple analytical model is available. For example, large-signal effects can cause the operating frequency to decrease with increasing bias current, whereas the small-signal models predict an increase.[5]

Impatt oscillators can also be modulated by including in the resonating circuit a varactor diode. Figure 67 shows such a circuit for a 50-GHz Impatt oscillator.[66] The Impatt diode is mounted in a resonant-iris Sharpless wafer which fits in the wave guide as shown. This mounting scheme exhibits a relatively low Q, so that a large-frequency deviation can be obtained. The varactor diode is similarly mounted and placed close to the Impatt diode. Lee and Standley obtained a frequency deviation of several GHz around 56 GHz with this circuit. One difficulty with frequency deviation using varactor diodes is that the loss in the varactor reduces the oscillator power output. By reducing the coupling between the varactor and the diode, more power is obtained but with a loss in tuning sensitivity.

A third method of modulating an Impatt oscillator is the path length modulator shown in Fig. 68.[67] In this circuit the phase of the rf wave form is digitally modulated by varying the path length of the rf wave by 180°. The receiver then compares adjacent pulses, giving an output of 0 or 1, depending on whether the two pulses are in phase or out of phase. This type of modulation has the advantage of separating the modulation function from the oscillator and allows each to be optimized for its own function. Clemetson *et al.* have reported on a 50-GHz modulator with an overall insertion loss of 1. 2 dB, operating at a bit rate of 300 Mb.

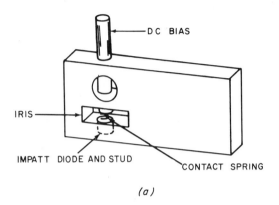

(a)

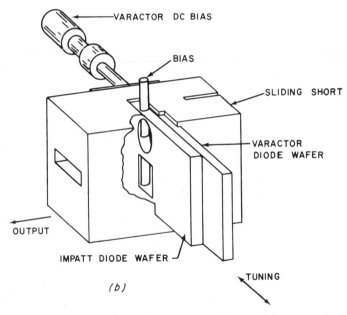

(b)

Fig. 67. Varactor-modulated Impatt oscillator. (*a*) Impatt diode wafer; (*b*) oscillator mount. (After Lee and Standley.[66])

3.5.4 Frequency Conversion

The highly nonlinear behavior of an oscillating Impatt diode suggests that an Impatt may operate as a self-pumped parametric device.
In fact, spurious oscillations in the first Impatt oscillators were identified by De Loach and Johnston as parametrically generated oscillations.[50] It is possible to suppress these parasitic oscillations by proper

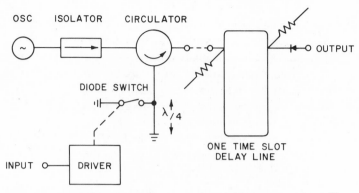

Fig. 68. Path length modulator. (After Clemetson *et al.*[67])

circuit design and to use the device negative resistance to amplify,[68] or in conjunction with the diode nonlinearity to amplify and convert, the frequency.[69] An analysis[21] of the parametric effects that occur in an oscillating diode shows that, when a large-signal rf voltage is present at one frequency, the diode exhibits strong parametric effects. Thus the diode conductance at one frequency depends on the termination of various idler frequencies. For example, the lower sideband converter in which we consider conversion from f_1 to f_2 (where f_1+f_2 $=f_{Impatt}$) exhibits negative input conductance at f_1 for a wide range of terminations at f_2. The conclusions that are drawn from this analysis must remain rather qualitative because of the assumptions of the analysis. A quantitative theory that accurately describes the parametric processes is difficult to devise because of the multiple frequencies and nonlinearity of the device. Thus, for example, it has not been demonstrated whether or not parametric frequency conversion can be done efficiently.

3.5.5 Diode Failure

The maximum power output and efficiency in an Impatt oscillator is typically obtained at a very high d-c power density that often results in diode failure.[70] Several types of thermal failure can be identified. The simplest case occurs when the d-c input power times the calculated thermal resistance gives a temperature change that causes the diode to exceed some known failure temperature (e.g., for a silicon diode with gold contacts, the gold-silicon eutectic point). This type of failure is observed in small-area, millimeter wave Impatt diodes.[45]

A similar failure is observed in high-efficiency Trapatt oscillators but occurs while tuning the oscillator. When a diode switches into the Trapatt mode, the d-c voltage drops significantly and the rf output power is a substantial part of the d-c input power. Therefore, in an oscillator that employs a constant current bias supply, considerably more power is dissipated in the nonoscillating state than in the oscillating state and failure may occur when the diode is tuned out of oscillation. Diodes that can switch into the Trapatt mode can also be destroyed if there is excessive shunt capacitance in the circuit. When a diode switches into the low-voltage state, the shunt capacitance discharges through the diode, which may result in thermal failure.

Some information on diode failure can be obtained from scanning electron micrographs of the diodes after burnout. However, before discussing these micrographs in detail, it will be helpful to have in mind some of the steps used in fabricating these diodes. Figure 69 shows the sequence of steps used to make the 10-GHz silicon Impatt diodes. Note that there are three silicon etching processes. The first is used to define the mesa; the second to etch apart the individual wafers; and the third, done after bonding, to tailor the junction capacitance. Figure 70 is an electron micrograph of a silicon diode that failed when operated as a Trapatt oscillator, and was etched, recov-

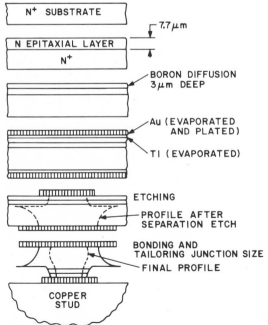

Fig. 69. Diode fabrication steps.

Fig. 70. Scanning electron micrograph of the junction region of a diode, showing two gold whiskers. Secondary electron image obtained with a stage tilt of 75° and a 30-kV beam.

ered, and burned out a second time while operating as a Trapatt os-cillator. The figure shows that two gold whiskers were formed by this experiment. The first whisker indicates the location of the sur-face of the diode at the time of the first failure. The second whisker shows the location of the second burnout relative to the surface. The shape of the diode surface near the junction was found to influence this type of failure. Figure 71 shows a diode that survived to nearly twice the current density of the diode shown in Fig. 70. The failure has the same form, a whisker of gold shorting the junction, but the shape of the surface is different.

Diodes have also been burned out with pulsed (using a pulse width of less than 0.5 μsec) bias. In these cases, the diodes do not short but rather exhibit a somewhat erratic reverse characteristic. Diodes burned out in this manner can be subsequently shorted by applying a relatively low level (< 100 mA) d-c bias. Evidently then, the gold whiskers are formed by a capillary growth of a filament of gold in the surface of the diode. An estimation of the heat generated as a result of the resistivity of a gold filament and the heat lost from the filament

Fig. 71. Closeup of the junction region of a diode without capacitance tailoring etch. Scanning electron micrograph, obtained with a 30-kV beam and stage tilt of 81°, using the secondary electron image.

because of heat conduction into the silicon indicates that such a gold filament would grow to approximately the observed size. Thus whatever the mechanism of failure is, it is not necessarily responsible for the final form of the gold whiskers.

3.5.6 State of the Art

To conclude this review, a summary of the present state of the art is given in Figs. 72 and 73. The highest Pf^2 result on these curves is the pulsed result of Liu, Risko, and Chang.[71] These results exhibit some characteristics of Trapatt; however, the efficiency is low (2 to 3%) and the output is at the transit-time frequency. The highest Pf^2 continuous wave results were obtained at 50 GHz with the double-drift structure. A d-c to rf conversion efficiency of 14%[81, 82] has been obtained with this structure. The continuous wave data points without notation are single-drift units on copper heat sinks at room temperature. All data shown are for silicon diodes; however, at lower frequencies GaAs has been demonstrated to be as good as or better than silicon.[74, 83]

State of the art data for Trapatt oscillators are summarized in

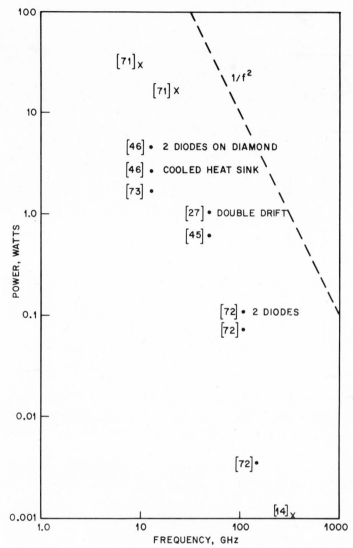

Fig. 72. Impatt state of the art (● cw, × pulsed).

Fig. 73. The numbers in parentheses indicate oscillator efficiency. The high-pulsed power output and low frequency are characteristic of Trapatt. The continuous wave result at about 4.3 GHz was obtained with the double-sided Trapatt diode described in Section 3.3.

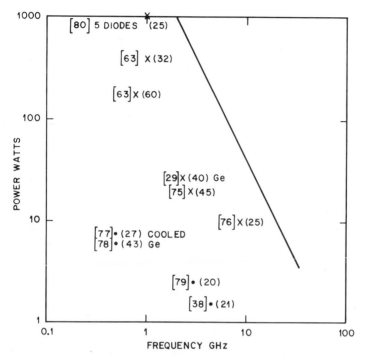

Fig. 73. Trapatt state of the art (● cw, × pulsed).

ACKNOWLEDGMENTS

The author would like to acknowledge the contributions of numerous colleagues at Bell Telephone Laboratories. In particular, he has drawn heavily from the work of B. C. De Loach, T. Misawa, and D. L. Scharfetter; and has benefited by many helpful discussions with R. Edwards.

APPENDIX

A relatively accurate mathematical description of avalanche diode oscillators can be obtained by a simultaneous solution of the set of one-dimensional equations given below.

Equations Covering Diode Operation

1. Continuity equations:

$$\frac{\partial p}{\partial t} = g - \frac{1}{q}\frac{\partial J_p}{\partial x} \tag{A1}$$

$$\frac{\partial n}{\partial t} = g + \frac{1}{q}\frac{\partial J_n}{\partial x} \tag{A2}$$

2. Poisson's equation:

$$\frac{\partial E}{\partial x} = \frac{q}{\epsilon}(p - n + N_d - N_a) \tag{A3}$$

3. Current equations:

$$J_p = q\mu_p p E - kT\mu_p \frac{\partial p}{\partial x} \tag{A4}$$

$$J_n = q\mu_n n E + kT\mu_n \frac{\partial n}{\partial x} \tag{A5}$$

4. Avalanche generation:

$$g_r = \frac{1}{q}(\alpha_n(E)|J_n| + \alpha_p(E)|J_p|) \tag{A6}$$

5. Shockley-Read-Hall generation and recombination:

$$g_d = \frac{pn - n_i^2}{\tau_{p0}(n + n_1) + \tau_{n0}(p + p_1)} \tag{A7}$$

6. Avalanche coefficients:

$$\alpha_{p,n} = A_{p,n}\, e^{(-b_{-p,n}/E)} \tag{A8, 9}$$

7. Mobility:

$$\left(\frac{\mu_0}{\mu}\right)_{n,p}^2 = 1 + \left(\frac{N_d}{N_0/S + F}\right) + \left(\frac{(E/A)^2}{E/A + F}\right) + \left(\frac{E}{B}\right)^2 \tag{A10, 11}$$

where A, B, and F are chosen to fit experimental mobility data. A complete solution of these equations must be obtained numerically, as done by Gummel and Scharfetter.[3] For analytical solutions a number of simplifying assumptions are usually made; these have been discussed above. A useful set of equations is obtained if we assume the following:

1. Neglect diffusion.
2. $v_n = v_p = v_s$ (not valid for the recovery portion of the Trapatt cycle).
3. $\alpha_n = \alpha_p = \alpha$.
4. $g_d = 0$.

Then, we have

$$\frac{1}{v}\frac{\partial J_p}{\partial t} = -\frac{\partial J_p}{\partial x} + \alpha(J_p + J_n) \tag{A12}$$

$$\frac{1}{v}\frac{\partial J_n}{\partial t} = \frac{\partial J_n}{\partial x} + \alpha(J_p + J_n) \tag{A13}$$

$$J_p = qv_s p \tag{A14}$$

$$J_n = -qv_s n \tag{A15}$$

Further assumptions on the form of $\alpha(E)$ are often made to simplify the analysis.

REFERENCES

1. R. L. Johnston, B. C. De Loach, Jr., and B. G. Cohen, "A Silicon Diode Microwave Oscillator," *Bell System Tech. J.*, 44, 369–372 (February 1965).
2. W. T. Read, "A Proposed High-Frequency Negative Resistance Diode," *Bell System Tech. J.*, 37, 401–446 (March 1958).
2a. T. Misawa, "Negative Resistance in *P-N* Junctions under Avalanche Breakdown Conditions," *IEEE Trans. Electron Devices*, ED-13, 137–151 (January 1966).
3. H. K. Gummel and D. L. Scharfetter, "Avalanche Region of Impatt Diodes," *Bell System Tech. J.*, 45, 1797–1828 (December 1966).
4. M. Gilden and M. E. Hines, "Electronic Tuning Effects in the Read Microwave Avalanche Diode," *IEEE Trans. Electron Devices*, ED-13, 169–175 (January 1966).
5. W. J. Evans and G. I. Haddad, "Large-Signal Analysis of Impatt Diodes," *IEEE Trans. Electron Devices*, ED-15, 708–717 (October 1968).
6. J. L. Blue, "Approximate Large-Signal Analysis of Impatt Oscillators," *Bell System Tech. J.*, 48, 383–396 (February 1969).
7. D. L. Scharfetter and H. K. Gummel, "Large-Signal Analysis of a Silicon Read Diode Oscillator," *IEEE Trans. Electron Devices*, ED-16, 64–77 (January 1969).
8. S. R. Kovel and G. Gibbons, "The Effect of Unswept Epitaxial Material on the Microwave Efficiency of Impatt Diodes," *Proc. IEEE*, 55, 2066–2067 (November 1967).
9. T. Misawa, "Saturation Current and Large Signal Operation of a Read Diode," *Solid-State Electronics*, 13, 1363–1368 (October 1970).
10. T. Misawa, "Minority Carrier Storage and Oscillation Efficiency in Read Diodes," *Solid-State Electronics*, 13, 1369–1374 (October 1970).
11. T. W. Sigmon and J. F. Gibbons, "Diffusivity of Electrons and Holes in Silicon," *Appl. Phys. Lett.*, 15, 320–322 (Nov. 15, 1969).
12. D. R. Decker, C. N. Dunn, and H. B. Frost, "The Effect of Injecting Contacts on Avalanche Diode Performance," *IEEE Trans. Electron Devices*, ED-18, 141–146 (March 1971).

13. B. C. De Loach, Jr., "Thin Skin Impatts," *IEEE Trans. on Microwave Theory and Technique*, MTT-18, 72–74 (January 1970).

14. L. S. Bowman and C. A. Burrus, "Pulse-Driven Silicon *p-n* Junction Avalanche Oscillators for the 0.9 to 20 mm Band," *IEEE Trans. Electron Devices*, ED-14, 411–418 (August 1967).

15. B. C. De Loach, Jr., "Recent Advances in Solid-State Microwave Generators," *Advances in Microwaves*, Vol. 2, Academic Press, New York, 1967.

16. C. B. Swan, T. Misawa, and L. P. Marinaccio, "Composite Avalanche Diode Structures for Increased Power Capability," *IEEE Trans. Electron Devices*, ED-14, 584–589 (September 1967).

17. M. E. Hines, "Noise Theory for the Read Type Avalanche Diode," *IEEE Trans. Electron Devices*, ED-13, 158 (January 1966).

18. H. K. Gummel and J. L. Blue, "A Small-Signal Theory of Avalanche Noise in Impatt Diodes," *IEEE Trans. Electron Devices*, ED-14, 569–580 (September 1967).

19. H. A. Haus and R. B. Adler, *Circuit Theory of Linear Noisy Networks*, John Wiley and Sons, New York, 1959.

20. B. C. De Loach, Jr., "The Noise Performance of Negative Conductance Amplifiers," *IRE Trans. Electron Devices*, ED-9, 366 (July 1962).

21. W. J. Evans and G. I. Haddad, "Frequency Conversion in Impatt Diodes," *IEEE Trans. Electron Devices*, ED-16, 78–87 (January 1969).

22. K. Kurokawa, "Some Basic Characteristics of Broadband Negative Resistance Oscillator Circuits," *Bell System Tech. J.*, 49, 1937–1956 (July-August 1969).

23. C. A. Brackett, "Characterization of Second-Harmonic Effects in Impatt Diodes," *Bell System Tech. J.*, 49, 1777–1810 (October 1970).

24. N. D. Kenyon, "A Lumped-Circuit Study of Basic Oscillator Behavior," *Bell System Tech. J.*, 49, 255–272 (February 1970).

25. C. B. Swan, "Impatt Oscillator Performance Improvement with Second-Harmonic Tuning," *Proc. IEEE (Lett.)*, 56, 1616–1617 (September 1968).

26. W. E. Schroeder and G. I. Haddad, "Effect of Harmonic and Subharmonic Signals on Avalanche-Diode Oscillator Performance," *IEEE Trans. Microwave Theory Techniques*, MTT-18, 327–331 (June 1970).

27. D. L. Scharfetter, W. J. Evans, and R. L. Johnston, "Double-Drift-Region ($P^{+}PNN^{+}$) Avalanche Diode Oscillators," to be published, *Proc. IEEE*, 58, 1131–1133 (July 1970).

28. H. J. Prager, K. K. N. Chang, and S. Weisbrod, "High-Power, High-Efficiency Silicon Avalanche Diodes at Ultra-High Frequencies," *Proc. IEEE*, 55, 586–587 (April 1967).

29. R. L. Johnston, D. L. Scharfetter, and D. J. Bartelink, "High-Efficiency Oscillations in Germanium Avalanche Diodes below the Transit-Time Frequency," *Proc. IEEE*, 56, 1611–1613 (September 1968).

30. D. L. Scharfetter, D. J. Bartelink, H. K. Gummel, and R. L. Johnston, "Computer Simulation of Low Frequency High Efficiency Oscillation in Germanium," IEEE Solid State Device Research Conference, June 17–19, 1968.

31. W. J. Evans and D. L. Scharfetter, "Characterization of Avalanche Diode Trapatt Oscillators," *IEEE Trans. Electron Devices*, ED-17, 397–403 (May 1970).

32. A. S. Clorfeine, R. J. Ikola, and L. S. Napoli, "A Theory for the High-Efficiency Mode of Oscillation in Avalanche Diodes," *RCA Rev.*, 30, 397–421 (September 1969).

33. B. C. De Loach, Jr., and D. L. Scharfetter, "Device Physics of Trapatt Oscillators," *IEEE Trans. Electron Devices*, ED-17, 9–21 (January 1970).

34. D. L. Scharfetter, "Power-Frequency Characteristics of the Trapatt Diode Mode of High Efficiency Power Generation in Germanium and Silicon Avalanche Diodes," to be published in *Bell System Tech. J.*, 49, 799–826 (May 1970).

35. W. J. Evans, "Computer Experiments on Trapatt Oscillators," *IEEE Trans. Microwave Theory Techniques*, MTT-8, 862–871 (November 1970).

36. D. J. Bartelink and D. L. Scharfetter, "Avalanche Shock Fronts in *p-n* Junctions," *Appl. Phys. Lett.*, 14, 320–323 (May 15, 1969).

37. W. J. Evans, "Circuits for High Efficiency Avalanche Diode Oscillators," *IEEE Trans. Microwave Theory Techniques*, MTT-17, 1060–1067 (December 1969).

38. W. J. Evans, T. E. Seidel, and D. L. Scharfetter, "A Novel Trapatt Oscillator Design," *Proc. IEEE (Correspondence)*, 58, 1294–1295 (August 1970).

39. S. M. Sze, *Physics of Semiconductor Devices*, John Wiley and Sons, New York, 1969.

40. J. G. Josenhans, private communication.

41. W. A. Edson, "Noise in Oscillators," *Proc. IRE*, 48, 1454–1466 (August 1960).

42. T. Misawa, "Microwave Si Avalanche Diode with Nearly-Abrupt-Type Junction," *IEEE Trans. Electron Devices*, ED-14, 580–584 (September 1967).

43. C. N. Dunn, "Empirical Transit-Time-Frequency—Breakdown Voltage Curves for Impatt Diodes," private communication.

44. D. L. Scharfetter, "Power-Impedance-Frequency Limitations of Impatt Oscillators Calculated from a Scaling Approximation," *IEEE Trans. on Electron Devices*, ED-18, (August 1970).

45. R. Edwards, D. F. Ciccolella, T. Misawa, D. E. Iglesias, and V. Decker, "Millimeter-Wave Silicon Impatt Diodes," International Electron Devices Meeting, Oct. 29–31, 1969.

46. C. B. Swan, "Improved Performance of Silicon Avalanche Oscillator Mounted on Diamond Heat Sinks," *IEEE Proc. (Lett.)*, 55, 1617–1618 (September 1967).

47. L. P. Marinaccio, "Ring-Geometry Impatt Oscillator Diodes," *Proc. IEEE (Lett.)*, 56, 1588–1589 (September 1968).

48. E. D. Garlinger and H. L. Stover, "Thermal Spreading Resistance of Ring-Geometry Diodes," *IEEE Trans. Electron Devices*, ED-17, 482–484 (June 1970).

49. R. H. Haitz, H. L. Stover, and N. J. Tolar, "A Method for Heat Flow Resistance Measurements in Avalanche Diodes," *IEEE Trans. Electron Devices*, ED-16, 438–444 (May 1969).

50. B. C. De Loach, Jr., and R. L. Johnston, "Avalanche Transit-Time Microwave Oscillators and Amplifiers," *IEEE Trans. Electron Devices*, ED-13, 181–186 (January 1966).

51. D. E. Iglesias, "Circuit for Testing High-Efficiency Impatt Diodes," *Proc. IEEE*, 55, 2065–2066 (November 1967).

52. F. M. Magalhaes and K. Kurokawa, "A Single-Tuned Oscillator for Impatt Characterizations," *Proc. IEEE (Correspondence)*, 58, 831–832 (May 1970).

53. T. Misawa, "CW Millimeter-Wave Impatt Diodes with Nearly Abrupt Junctions," *Proc. IEEE*, 56, 234–235 (February 1968).

54. J. F. Reynolds, A. Rosen, B. E. Berson, and H. C. Huang, "A Coupled TEM Bar Circuit for Solid-State Microwave Oscillators," 1970 *ISSCC Digest*, 12–13.

55. J. G. Josenhans, "Diamond as an Insulating Heat Sink for a Series Combination of Impatt Diodes," *Proc. IEEE (Correspondence)*, 56, 762–763 (April 1968).

56. F. M. Magalhaes and W. O. Schlosser, "A Microwave Oscillator Using Series-Connected Impatt Diodes," *Proc. IEEE (Correspondence)*, 56, 865–866 (May 1968).

57. C. T. Rucker, "A Multiple-Diode High-Average Power Avalanche-Diode Oscillator," *IEEE Trans. Microwave Theory Techniques*, MTT-17, 1156–1158 (December 1969).

58. I. Tatsuguchi, "A Frequency-Modulated Phase Locked Impatt Power Combiner," 1970 *ISSCC Digest*, 18–19.

59. L. S. Napoli and R. J. Ikola, "An Avalanching Silicon Diode Microwave Amplifier," *Proc. IEEE*, 53, 1231–1233 (September 1965).

60. E. F. Scherer and M. J. Barrett, "A Broadband Multistage Avalanche Amplifier at *X*-Band," 1969 *ISSCC Digest*, 82–83.

61. M. E. Hines, "Negative-Resistance Diode Power Amplification," *IEEE Trans. Electron Devices*, ED-17, 1–8 (January 1970).

62. J. Noisten, "Avalanche Diode Power Amplifiers for Ku Band," 1970 *ISSCC Digest*, 16–17.

63. H. J. Prager, K. K. N. Chang, and S. Weisbrod, "Anomalous Silicon Avalanche Diodes for Microwave Generation." Proc. Cornell Conf. on High Frequency Generation and Amplification, August 1967, pp. 266–280.

64. W. J. Evans, "CW Trapatt Amplification," *IEEE Trans. on Microwave Theory Techniques*, MTT-18, 986–987 (November 1970).

65. J. F. Dienst, R. V. D'Aiello, and E. E. Thomas, "Power Amplification Using Avalanche Diodes," *Electronics Lett.*, 5, 308–309 (July 1969).

66. T. P. Lee and R. D. Standley, "Frequency Modulation of a Millimeter Wave Impatt Diode Oscillator and Related Harmonic Generation Effects," *Bell System Tech. J.*, 48, 143–161 (January 1969).

67. W. J. Clemetson, N. D. Kenyon, K. Kurokawa, W. O. Schlosser, and B. Owen, "An Experimental Millimeter Wave High-Speed Path Length Modulator," 1970 *ISSCC Digest*, 20–21.

68. A. S. Clorfeine, "Self-Pumped Parametric Amplification with an Avalanching Diode," *Proc. IEEE (Correspondence)*, 54, 1956–1957 (December 1966).

69. M. I. Grace, "Down Conversion and Sideband Translation Using Avalanche Transit-Time Oscillators," *Proc. IEEE*, 54, 1570–1571 (November 1966).

70. W. J. Evans, D. L. Scharfetter, R. L. Johnston, and P. L. Key, "Tuning Initiated Failure in Trapatt Diodes," Avalanche Diode Workshop, New York, Dec. 10–11, 1969.

71. S. G. Liu, J. J. Risko, and K. K. N. Chang, "High-Power K-Band Silicon Avalanche Diode Oscillators," *Proc. IEEE (Correspondence)*, 58, 919–920 (June 1970).

72. T. Misawa and L. P. Marinaccio, "100 GHz Si Impatt Diodes for CW Operation," *Proc. of the Symposium on Submillimeter Waves*, Polytechnic Press, N. Y. 1970.

73. C. B. Swan, T. Misawa, and C. H. Bricker, "Continuous Oscillations at Millimeter Wavelengths with Silicon Avalanche Diodes," *Proc. IEEE*, 55, 1747–1749 (October 1967).

74. L. D. Armstrong, "GaAs Impatt Diodes, Oscillators and Amplifiers," 1970 International Microwave Symposium, Newport Beach, Calif., May 11–14, 1970.

75. R. S. Ying, R. G. Mankarious, and D. L. English, "High-Efficiency Anomalous Mode Oscillation from Silicon Impatt Diodes at 6 GHz," 1969 *ISSCC Digest*, 86–87.

76. G. Gibbons and M. I. Grace, "High-Efficiency Avalanche-Diode Oscillators and Amplifiers in X-Band," *Proc. IEEE (Lett.)*, 58, 512–513 (March 1970).

77. W. J. Evans and D. E. Iglesias, "CW Silicon Trapatt Operation," *Proc. IEEE*, 58, 285–286 (February 1970).

78. D. E. Iglesias and W. J. Evans, "High Efficiency CW Impatt Operation," *Proc. IEEE (Lett.)*, 56, 1610 (September 1968).

79. W. J. Evans and R. L. Johnston, "Improved Performance of CW Silicon Trapatt Oscillators," *Proc. IEEE (Lett.)*, 58, 845–846 (May 1970).

80. S. G. Liu and J. J. Risko, "Stacked Kiolwatt Avalanche-Diode Microwave Oscillators," International Electron Devices Meeting, Washington, D. C., Oct. 29–31, 1969.

81. T. E. Seidel and D. L. Scharfetter, "High-Power Millimeter Wave Impatt Oscillators with Both Hole and Electron Drift Spaces Made by Ion Implantation," *Proc. IEEE*, 58, 1135–1136 (July 1970).

82. T. E. Seidel, R. E. Davis, and D. E. Iglesias, "Double-Drift Ion Implanted (p^+pnp^+) Millimeter Wave Impatt Diodes," *Proc. IEEE*, 59, (August 1970).

83. J. C. Irvin, D. J. Coleman, W. A. Johnson, I. Tatsuguchi, D. R. Decker, and C. N. Dunn, "Fabrication and Noise Measure of High Power GaAs Impatt Diodes," International Electron Devices Meeting, Washington, D. C., Oct. 28–30, 1970.

Chapter 4

HIGH-FREQUENCY ULTRASONIC DEVICES

Richard M. White

College of Engineering
Department of Electrical Engineering and Computer Sciences
University of California, Berkeley, California

The inclusion of a chapter on ultrasonic devices in a book about modern electronics might be puzzling to those who thought the subject of acoustics had been in decline since the days of Lord Rayleigh. But, in fact, the subject of high-frequency ultrasonics has received increasing attention since the late 1950s, and today it appears that many new and interesting electronic devices can be made employing ultrasonic waves. *

There are three reasons for this belief. First, in the late 1950s means were developed for generating and detecting ultrasonic elastic waves in solids at very high frequencies, well above the previous practical limit at about 100 MHz. As a result, elastic waves at hundreds of megahertz and at gigahertz frequencies can now be studied and used. Evidence of the progress made is the 1966 report[11] of the generation and detection of coherent elastic waves in quartz at 114 GHz. The second reason is the discovery in the early 1960s that elastic waves could be amplified directly in piezoelectric semiconductors. This discovery stimulated thought about active ultrasonic devices, and encouraged the study and characterization of materials suitable for ultrasonic work at high frequencies. Finally, attention has turned recently to *surface*

---

*The reader may wish to consult some of the following review articles on high-frequency ultrasonics: microwave bulk elastic waves, Refs. 1–4; surface waves and their applications, Refs. 5–10. Many articles on surface waves are included in the *IEEE Trans. Microwave Theory and Techniques*, 17, No. 11 (November 1969), as well as the *Microwave J.* issue of March 1970.

elastic waves which propagate on the exterior surfaces of solids instead of inside them. It now appears that surface wave devices, made
economically by techniques developed originally for integrated circuit
manufacture, can perform many signal-processing functions which
formerly required much more complex circuits and components.

Here we shall first describe the types of elastic waves that can
exist in solids, and then consider the ways by which one can generate
and detect these waves and use them in passive systems. Finally, we
consider the amplification of elastic waves.

4.1 ELASTIC WAVES IN SOLIDS

Figure 1 summarizes many of the kinds of elastic waves that can
propagate in solids. If the solid is unbounded, the only plane elastic
waves that can exist are the bulk waves shown in Figs. 1a and 1b. The
bulk longitudinal wave (Fig. 1a) is like the familiar pressure wave in a
gas, where the particles of the medium move in the direction of propagation of the wave. The transverse bulk wave (Fig. 1b) is analogous
to a wave on a stretched string, with particles moving in planes normal
to the direction of propagation. One can derive the properties of such
one-dimensional waves in linear elastic media by combining Newton's
force equation,

$$\rho \frac{\partial^2 u}{\partial t^2} = \frac{\partial T}{\partial x} \tag{1}$$

with Hooke's law

$$T = cS \tag{2}$$

which expresses the proportionality between the stress, T (force per
unit area), of the wave and the (dimensionless) strain, S, associated
with the wave. In these equations ρ is the density of the solid, c is an
appropriate elastic stiffness constant, u is the displacement of a particle from its equilibrium position in the solid, and t is time. (In this
one-dimensional case, $S = \partial u / \partial x$, where x is the position coordinate
in the direction of propagation.) From these equations one obtains an
elastic wave equation and identifies the phase velocity of the wave as

$$v_p = \sqrt{c/\rho} \tag{3}$$

This velocity, v_p, is independent of frequency from very low frequencies through the microwave range. Thus the bulk elastic waves
are nondispersive. Elastic wave velocities range from 10^3 m/sec to
12×10^3 m/sec. Since these velocities are about 10^5 times smaller than

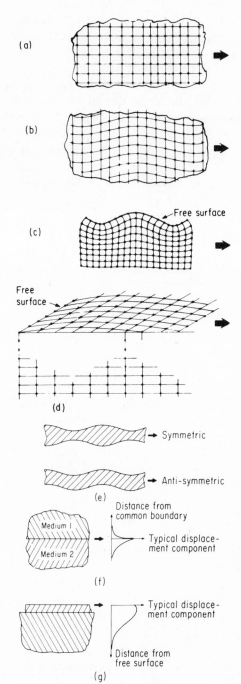

Fig. 1. Elastic waves that propagate in solids. (a) Bulk longitudinal wave, showing sections through an unbounded solid; dots represent instantaneous positions of elementary volumes of solid. The wave propagates in the direction of the arrow. (b) Bulk transverse wave, unbounded solid. (c) Surface (Rayleigh) wave (After I. A. Viktorov, *Rayleigh and Lamb Waves*, Plenum Press, New York, 1967, by permission.) Note that the amplitude of the particle motion decreases with the distance from the surface; particles move both in the direction of propagation and normal to it. (d) Electroacoustic wave; particles move parallel to the free surface and normal to the propagation vector. (e) Plate (Lamb) waves. (f) "Stoneley" wave at the welded boundary of two dissimilar solids. (g) Layered solid. Modes shown are similar to plate modes; Love modes having particle motion parallel to the surface and normal to the propagation vector may also propagate.

electromagnetic wave velocities, transmission-line devices made for elastic waves are about 100,000 times smaller than their electromagnetic counterparts at the same operating frequency. The low velocities mean that time delay devices having a given delay will be correspondingly smaller than electromagnetic delay lines made with coaxial cables or wave guides. The low velocity of elastic waves is probably the characteristic that is most responsible for their use in electronics.

If a solid is bounded, additional elastic waves may exist. The surface wave illustrated in Fig. 1c is the Rayleigh wave that has long been of interest to seismologists. The particle motion, which is elliptical at the free surface, decreases rapidly with depth, becoming very small at a depth of a wavelength or so. Analysis shows that the surface wave motion can be expressed as a sum of several exponential functions of the depth or distance from the free surface. The surface wave velocity has the form of Eq. 3; thus the surface wave velocity is also independent of frequency and the waves are nondispersive. Surface wave velocities lie mostly in the lower third of the range of wave velocities given above, being lower (usually) than the bulk longitudinal and transverse waves in the same solid.

One might wonder why the wave energy remains near the surface. The answer can be found by considering the velocity expression. At the surface the restoring forces on any small volume of the solid are less than in the interior of the solid because of the free boundary. Therefore the surface wave velocity is lower than the bulk velocities, and the energy cannot radiate away from the surface into the interior as a bulk wave.

The other waves illustrated are the "electroacoustic" surface wave existing in certain piezoelectrics and other crystals (Fig. 1d), waves on plates whose thickness is comparable with or smaller than the wavelength (Fig. 1e), the so-called Stoneley waves (Fig. 1f) which can propagate along the common boundary of two semiinfinite solids (provided certain very restrictive conditions on relative densities and elastic constants are satisfied), and a wave on a layered solid (Fig. 1g) in which the wave energy also remains near the free surface.

Studies of the attenuation of bulk and surface elastic waves have shown that the losses in solids at room temperature generally increase as the square of the frequency of the wave. But, fortunately, in many single crystals at room temperature the loss per wavelength for elastic waves is much less than that for waves in electromagnetic transmission lines, even at frequencies in the low gigahertz range. Thus the losses are low enough to permit making practical delay lines and other transmission line devices employing elastic waves.

Measured losses of longitudinal bulk waves in several piezoelectric semiconductors at room temperature are plotted in Fig. 2. The measured attenuation of surface waves in crystalline quartz, which is a piezoelectric insulator, is shown in Fig. 3 as a function of temperature. Losses of either surface or bulk waves decrease significantly at low temperature because cooling reduces the thermal vibrations of the lattice with which the ultrasonic waves interact. At room temperature, even though the loss in dB/cm shown in Fig. 3 is not negligible, at high frequencies the wavelengths are so small that the loss per wavelength is small. (In quartz the wavelength of surface waves at 1 GHz is approximately 3μ, corresponding to an attenuation of only 0.003 dB per wavelength.)

4.2 THE PIEZOELECTRIC EFFECT—BULK AND SURFACE-WAVE TRANSDUCERS

In most modern high-frequency elastic wave devices, the piezoelectric effect is used to convert energy from electrical to elastic form, and vice versa. The piezoelectric effect occurs in many single-

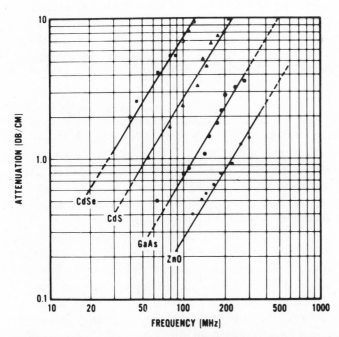

Fig. 2. Attenuation of bulk longitudinal waves in piezoelectric semiconductors as a function of frequency. (Courtesy of F. Hickernell, Motorola Corporation.)

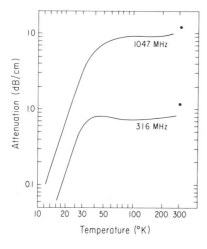

Fig. 3. Variation of attenuation with temperature for surface waves propagated along the X axis on Y-normal quartz. Solid curves show the temperature-dependent part of attenuation (measured electrically); the dots at room temperature are total attenuation (measured optically). (After Salzmann, Pleininger, and Dransfeld.[30])

crystal insulators, the most familiar example being quartz; it also occurs in certain semiconductors (such as cadmium sulfide, zinc oxide, and gallium arsenide) and in ferroelectric ceramics such as lead-zirconate-titanate. When crystals exhibiting the piezoelectric effect are deformed mechanically, the relative locations of the ions of the crystal change and as a result an electric field appears in the crystal. Conversely, when such a crystal is placed in an electric field, it deforms mechanically.

The piezoelectric effect has long been employed in electronics, in temperature-stabilized, high-Q ultrasonic crystal oscillators and filters. A typical oscillator consists of a slab of quartz that is one-half an elastic wavelength thick. Electrodes on the two major surfaces of the slab are connected to an alternating electric voltage source. Such acoustic oscillator plates are very thin and fragile if made for frequencies much above a few tens of megahertz, but two fairly recent developments have made it possible to circumvent this fabrication limit and reach frequencies of above 100 MHz with relative ease.

The first was the development of thin-film piezoelectric transducers made by the vacuum evaporation of piezoelectric material onto a massive substrate. In this way the piezoelectric transducer can be thin enough for efficient high-frequency operation, and yet be fairly rugged. Another approach is illustrated in Fig. 4 in connection with a high-frequency delay line. This is the so-called depletion layer transducer, made from a film of piezoelectric material that is also semiconducting. The semiconductor is rather highly conductive, but a thin layer of it near the left surface is doped or compensated so as to have a high resistivity. The electric field produced by the a-c source shown

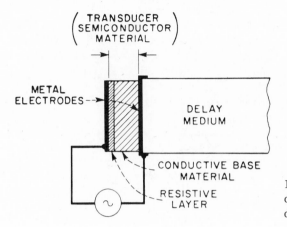

Fig. 4. Depletion layer transducer as it might be used on a
delay line. (From White.[31])

appears mostly in the resistive layer, which thus acts as a piezoelectric source of elastic waves. The depletion layer transducer can be
made in a deposited film, as shown in Fig. 4, or in the delay medium
itself if that is a piezoelectric semiconductor. An example of the latter
is the zinc oxide delay line of Fig. 5, made at Motorola Corporation.
The insertion loss of a typical microwave ultrasonic delay line is
plotted in Fig. 6. The line provides a delay of about 1 μsec with a

Fig. 5. A 700-nsec delay line fabricated for operation at 600 MHz. The microstrip circuitry on the alumina substrate provides a 50-Ω resistive match at the
center frequency. The center band insertion loss was 15 dB, with a 3-dB bandwidth of 120 MHz. (Courtesy of F. Hickernell, Motorola Corporation.)

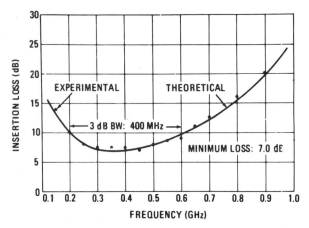

Fig. 6. Theoretical and experimental insertion loss versus frequency for ZnO delay line (matched at each frequency). Delay time: 1 μsec. (Courtesy of F. Hickernell, Motorola Corporation.)

minimum insertion loss of only 7 dB at the center frequency, 400 MHz. In contrast to such very compact microwave delay lines, 700 to 1000 ft of coaxial cable having much higher insertion loss would be required to achieve the same time delay with electromagnetic waves.

The second approach to piezoelectric transduction involves using only one free surface of a piezoelectric medium, as shown in Fig. 7. Here rf electric fields in the microwave cavity at the left produce

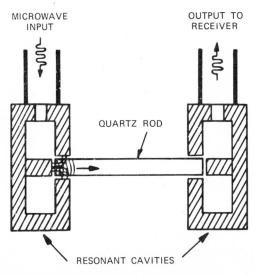

Fig. 7. Excitation and reception of high-frequency bulk ultrasonic waves by piezoelectric coupling in microwave cavities.

stresses and strains in the quartz rod, generating elastic waves that can be detected at the cavity on the right. Such arrangements have been used to measure the elastic wave propagation characteristics of many crystals.

When elastic waves propagate in a piezoelectric medium, electric fields are produced wherever the lattice is strained. Thus, when a bulk elastic wave propagates through a piezoelectric semiconductor, a traveling rf electric field propagates along with the elastic wave. If, in addition, a d-c potential difference is impressed on the semiconductor, the charge carriers in the semiconductor will drift and interact with the traveling rf electric fields associated with the elastic wave. This effect is the basis of the ultrasonic amplifier, which will be discussed at length below.

If a surface elastic wave propagates in a piezoelectric medium, a travelling rf electric field is also set up, both inside the crystal and just outside it (Fig. 8). The field inside the crystal results from the strains in the crystal, and the field decays with depth at the same rate as do the strains. Outside the crystal the field is caused by the charges at the crystal surface produced by the wave. The electric field is essentially electrostatic in nature because of the low phase velocity of the elastic wave; hence the field varies with the distance, z, from the surface as $\exp(-kx)$, where $k = \omega/v_p$ is the propagation constant for the elastic wave of frequency ω. This external electric field can be used in several surface wave devices, including the surface wave ultrasonic amplifiers that are discussed below.

Surface elastic waves may be generated on a piezoelectric solid easily by connecting an a-c voltage source to properly shaped elec-

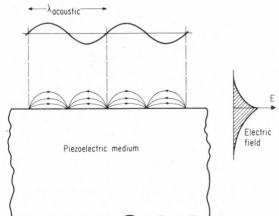

Fig. 8. Electric fields accompanying propagation of surface elastic wave in piezoelectric crystal. (Courtesy of C. W. Turner, University of California, Berkeley.)

trodes deposited on the solid.   The time-varying electric fields near
the electrodes produce stresses in the piezoelectric and thus generate surface, as well as bulk, elastic waves.   The connection between the electric field and the stress of a wave in a piezoelectric
is given by the equations of state for the medium including piezoelec-
tricity,

$$T = cS - eE \qquad \text{(4a)}$$

$$D = eS + \epsilon E \qquad \text{(4b)}$$

where e is the piezoelectric constant for the solid which expresses the
coupling of the electric field, E, to the stress, T, and also the coupling of the strain, S, to the electric displacement, D.   In Eq. 4b, the
dielectric permittivity is ϵ.

A single pair of electrodes will act as a broadband elastic wave
source, but increased efficiency of the coupling results if a number of
electrode pairs are used, as in Fig. 9a.   This interdigital electrode
array acts as a source of surface waves if driven at or near frequencies
for which the periodic length L is equal to the surface wavelength,
$\lambda = v_b/f$.   Waves radiate both to the right and left from such a transducer array (but unidirectional arrays can be made[12]).   Since the piezoelectric effect is reversible, such an electrode array will serve as a
receiver if it is connected to a load impedance instead of a voltage
source.

Two of the many variations on the basic transducer are shown in
Fig. 9c.   The spacing between electrodes changes from one end to
another in the array so as to broaden the frequency response of the array.   Low-frequency waves are generated or detected at the right end

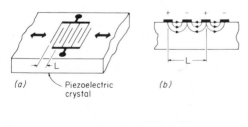

(a)   Piezoelectric   (b)
           crystal

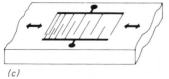

(c)

Fig. 9. Interdigital electrode array
transducers. (a) Array of constant
period, L, on a piezoelectric crystal; (b) sketch showing the approximate distribution of electric fields
at one instant; (c) array having
variable pitch for a broadband operation, and heavier weighting of mid-
band frequency components provided
by varying lengths of opposing electrodes.

of the array, and high frequencies at the left. The lengths of the opposing fingers have been varied so as to provide amplitude shading of the
array which is in many respects like an end-fire electromagnetic antenna. Thus the response for intermediate frequencies has been
enhanced in the example shown.

A high-frequency, surface wave delay line, developed at Litton
Industries, is shown in Fig. 10. The line has a 20% bandwidth about
the 500-MHz center frequency, with a 13-dB insertion loss and a delay between 1 and 2 μsec. The delay crystal is lithium niobate, a
strongly piezoelectric insulator. The delay line itself, in the center
of the block, has an input transducer near its center, and two output
transducers whose output powers are combined in the electromagnetic
stripline circuitry and the small thin-film resistor (dark block between
meander striplines). Another interesting experimental surface wave
device is the i.f. filter designed by workers at Zenith for the i.f.
stage of a color television receiver. The filter characteristic (Fig.
11a) is obtained with several cascaded filters, consisting of interdigital
transducer arrays on an inexpensive lead-zirconate-titanate ceramic
(Fig. 11b).

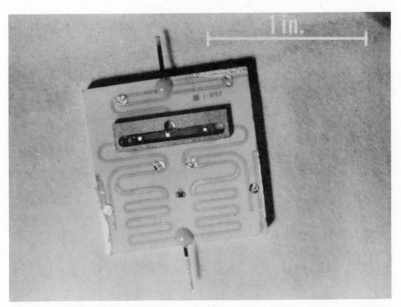

Fig. 10. Surface wave delay line and associated circuitry. A lithium niobate
delay line is set in an oval slot in a small block on a base plate. The input
transducer is the bright spot near the center of the bar; an output transducer
is seen at each end of the bar. The transducers have been emphasized by re-
touching for clarity. (Courtesy of D. Armstrong, Litton Industries.)

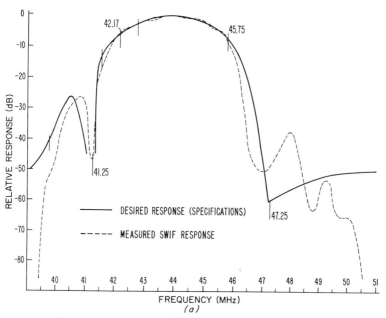

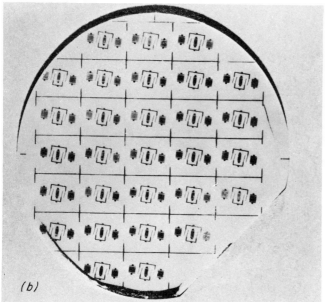

Fig. 11. Surface wave i.f. filters for color television. (a) Desired and measured filter response. (b) Photo of 1.5-in.-diameter lead-zirconate-titanate disk on which about 30 filter elements have been fabricated. Each filter has a single interdigital transducer input in the middle of the element, and two transducers connected together for output. Slanted electrodes provide shielding between the input and output. (Courtesy of A. DeVries, Zenith Radio Corporation.)

198

In situations in which the use of a piezoelectric solid is not feasible for technical or economic reasons, one may generate and receive surface waves in a piezoelectric film or layer that is evaporated or grown on a nonpiezoelectric solid substrate. The epitaxy of piezoelectric materials themselves, and of semiconductors on piezoelectrics, is not yet well developed, but impressive results have already been reported. For example, workers at Autonetics have found only a 15-dB insertion loss at 180 MHz on a delay line made with interdigital transducers on an epitaxially grown zinc oxide film on a sapphire substrate.

The dimensions of the electrode arrays depend upon the frequency, becoming smaller as the frequency increases. Photolithographic processes can be used to make arrays up to about 1 GHz, at which the wavelength for surface waves in most substances approaches a few microns (i. e., line widths in such arrays must be a micron or less). Transducers for higher-frequency use can be made by exposing special resist or masking materials such as poly-methyl-methacrylate with a finely focused electron beam in an electron microscope. As an example, Broers and co-workers[13] have reported fabrication of a 2. 5-GHz delay line, having 12-pair electrode transducers on lithium niobate, made of 800-A thick lines whose widths were approximately 0. 4 μ. This delay line provided 1. 5 μsec delay with a total insertion loss of only 29 dB.

4.3 PASSIVE SURFACE-WAVE DEVICES

To realize the wide range of functions that can be performed with just passive surface wave devices, we should take note of several additional features of surface waves:

1. *Surface elastic waves can be guided.* By treating the surfaces of the solids on which the waves propagate, one can alter the direction of propagation of the waves in the surface plane itself. Several ways of doing this are shown in Fig. 12. The basic idea is the same as that used in electromagnetic striplines: the phase velocity is made lower in the vicinity of the guide than it is in the surrounding region. Thus wave energy remains in the guide region, since it cannot radiate out away from the guide as a free surface wave. A particularly simple way of lowering the velocity is to deposit on the surface a stripe of material that is more dense (and possibly also less stiff) than the substrate; the phase velocity in the region under the stripe will be lowered as the phase velocity formula (Eq. 3) suggests. Alternatively,

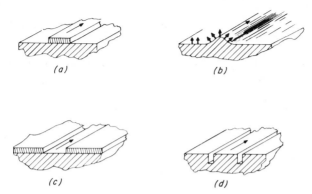

Fig. 12. Surface wave guides. (a) "Slow-on-fast" stripe guide. Stripe may be massive film on substrate, such as gold on fused silica. (b) Topographic guiding with low-height ridge. The ridge is more free of constraint than regions beside it; hence phase velocity is lower near the ridge. (c) "Fast-on-slow" slot guide. (d) Channel guide.

one might lower the velocity simply by having a raised ridge in the surface (the material at the ridge is less constrained than elsewhere, as suggested by the arrows in Fig. 12b), or by putting a film of stiffening material outside the guide region (Fig. 12c), so that the velocity outside is increased. Typical materials for the so-called "slow-on-fast" guide of Fig. 12a are a gold stripe on fused silica; a "fast-on-slow" guide might be made with an aluminum film on T-40 glass, a special glass used for delay lines. Another useful guiding structure that has been examined is the channel formed by cutting side slots (Fig. 12d), between which the waves are continually reflected.

   With such surface wave guides one can make acoustic analogs of electromagnetic transmission line components such as power dividers, directional couplers, reflective filters, and even logic devices (Fig. 13). These components may be useful at low megahertz frequencies because of their very small size. The components may also be parts of entire surface-wave systems where a number of signal processing functions are performed.

   2. *Light beams interact with sound waves.* Because the passage of a sound wave in a solid causes local compression and rarefaction of the solid, and hence causes the index of refraction of the solid to change by small amounts, light passing through a transparent solid will be diffracted and reflected from both bulk and surface elastic waves. In addition, because of the rippling of the surface by the surface wave, light can be diffracted and frequency shifted on reflection from even an

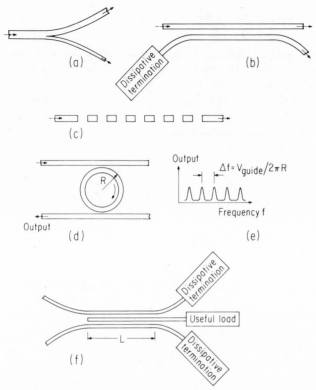

Fig. 13. Surface wave guide components. Arrows show the directions of power flow. (*a*) Power divider. (*b*) Directional coupler. (*c*) Filter employing periodic variations in phase velocity, and hence, wave impedance. (*d*) Frequency-selective coupling. Power couples from the input line to the circular loop; resonances occur at frequencies for which the circumference of the loop is an integral number of guide wavelengths. (*e*) Output from filter in (*d*). (*f*) Logic device. The appearance of the output at useful load depends on the coupling length L and the presence or absence of power on each input line. (From Seidel and White.[33])

opaque solid, where a surface wave is propagating. These effects have been used extensively to study the sound waves themselves but they can be applied in deflecting and modulating light beams. The particle motions associated with elastic waves are usually only a few angstroms or less, and the changes in index of refraction may be one part in a million or smaller, but even these minute changes are large enough to permit realization of useful effects.

    3. *Surface waves can be reflected.* Discontinuities in the propagation medium cause partial reflection of surface waves. The analysis of the reflection process is quite difficult; often conversion of wave

energy from surface to bulk elastic forms occurs along with reflection. Experimental observations have been made of the reflection of surface waves from the ends of the crystals used in delay lines, from single stripes of metal deposited on the surface, and from the electrode array transducers used to generate and detect the waves. Since the reflective properties of the arrays depend upon the load impedance connected to the array terminals, the reflection can be controlled externally, and so reflection can be maximized or minimized at certain desired frequencies with adjustable reactive terminations. The use of active terminations permits the realization of reflection gain, and hence the amplification of surface waves.[14] With controlled reflection one can realize reflection filters, resonant elements, and the like.

4. *Nonlinear elastic effects may occur*. The surface wave energy is concentrated near the free surface, and even a moderate total power input may result in very large power densities near the surface, and so result in the elastic medium's departing from the simple linear behavior described by Hooke's law (Eq. 2). These nonlinear effects can be used for frequency multiplication and mixing, limiting, logic operations, and for parametric amplification.

These four effects, together with the properties mentioned earlier, permit one to imagine (and build) surface wave devices that will perform a host of signal generating and processing functions. For example, one can envision a receiver front end involving a surface wave input transducer, an rf amplifier, a local oscillator, a mixer, and an i.f. amplifier, and finally an output transducer to convert the signal back to the electrical form (amplifiers will be discussed below). The entire processing from rf input to i.f. output could thus be performed with surface waves on a single-crystal plate having suitable overlaying films. One can also combine surface wave components with regular semiconductor devices as hybrid devices interconnected by soldering, lead wires, or beam lead wires, or in monolithic structures. A very promising combined surface-wave–semiconductor system is that employing field effect transistors as wave receivers in a matched filter structure.[15]

Before leaving the passive surface wave devices, we should note a particularly important class of device, the transversal filter (Fig. 14). This structure, which is realizable with surface waves if one simply provides a number of electrode taps on a delay line, can be used to make devices such as pulse compression filters, analog spectrum analyzers, and generators of frequency modulated codes. A very simple form of transversal filter is the interdigital transducer shown in Fig. 9c. The weighting networks are simply the electrode

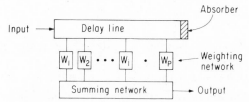

Fig. 14. Transversal filter. This tapped delay line configuration is represen-
tative of many surface wave transducers such as the simple interdigital array
(Fig. 9) and the coded transducer (Fig. 10). (After Squire, Whitehouse, and
Alsup[24], by permission of *IEEE Transactions on Microwave Theory and Techniques*.)

pairs having different gaps, and hence different frequency responses;
while the summing network is simply the common bus bars or feed
leads of the array.   An example of wave form generation and of matched
filtering with such an electrode structure is shown in Fig. 15.   In addition to these uses, surface waves have been used in experimental
scanning and display devices.

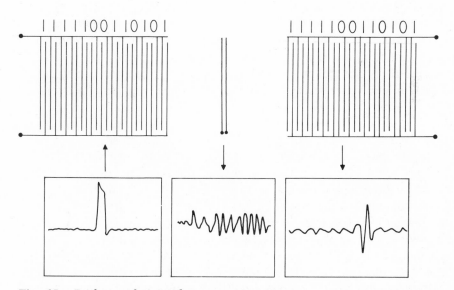

Fig. 15. Barker-code transducers and waveforms. An input of 50-nsec duration to the left transducer produces a complex acoustic waveform of approximately 1-μsec duration, illustrated in the center as detected by a broadband
electrode pair. The waveform is matched to the right transducer which produces the waveform shown. (From White,[9] by permission of *IEEE Transactions on
Electron Devices*.)

4.4 ULTRASONIC AMPLIFICATION

As mentioned in the introduction, it was found in the early 1960s that *bulk* elastic waves could be amplified if they propagated in a piezoelectric semiconductor; for amplification to take place the drift velocity of the carriers must be greater than the phase velocity of the elastic wave. Amplification occurs because the elastic wave creates (by the piezoelectric effect) local electric fields with which the drifting charges can interact to give up some of their energy. Not surprisingly, it is also possible to amplify *surface* elastic waves in structures containing piezoelectric and semiconducting material.

There are two main areas in which ultrasonic amplification might be useful. The first is simply in any system in which electrical signals must be amplified: the signals can be converted to elastic waves, amplified, and then reconverted to electrical form. Here one is concerned with the end product—gain—and not with the process that produces it. The ultrasonic amplifier must compete with other excellent electronic amplifiers not only in terms of gain but also in bandwidth, efficiency, noise figure, dynamic range, stability with respect to ambient conditions, cost, and power supply requirements. The second area of application is in devices or systems in which one already has elastic waves and some gain is needed. An example is a delay line storing signals for hundreds or thousands of microseconds, where an amplifier would help overcome the attenuation of the delay medium. With amplifier structures one can also obtain electronically controlled dispersion and phase shift, attenuation (when the amplifier is operated in the reverse direction or with low-drift voltages), the mixing of signals when the amplifier operates near its saturation region, and oscillation if either electrical or acoustical feedback is present.

An arrangement used for amplifying bulk elastic waves is schematically shown in Fig. 16. Transducers at the input and output convert the signals from electrical to elastic forms and back again; a drift supply connected to ohmic contacts supplies the drift field for the semiconductor. The transformers shown provide d-c isolation. In the bulk wave amplifier the semiconductor might be one of the piezoelectric II-VI compounds such as cadmium sulfide or cadmium selenide which are photoconductive as well and have moderate piezoelectric coupling; amplification has also been observed in the III-V compounds gallium arsenide and indium antimonide, as will be discussed below.

In the bulk wave amplifiers, the solid in which the waves and drifting carriers are present simultaneously is both piezoelectric and semiconducting. Surface wave amplifier structures may have an analogous

ULTRASONIC TRAVELING WAVE AMPLIFIER CONFIGURATION

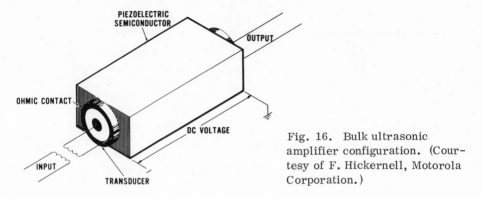

Fig. 16. Bulk ultrasonic amplifier configuration. (Courtesy of F. Hickernell, Motorola Corporation.)

monolithic form (Fig. 17a), or the piezoelectric and the semiconductor may be two different materials. In the latter case, either might occur as a thin film on a more massive substrate (Fig. 17b) or the two media might be separated by a small gap through which traveling electric fields penetrate (Fig. 17c). Actual operating characteristics of these and other ultrasonic amplifiers will be given after the analysis of amplification is summarized.

4.5 AMPLIFIER ANALYSIS

One of the interesting aspects of ultrasonic amplification is that the phenomenon can be looked at and analyzed in so many different ways. The "classical" approach[16] is to begin with the piezoelectric equations of state (Eqs. 4a and 4b) which relate elastic and electric field quantities, put in a current of drifting charges allowing for both drift and diffusion, and then obtain an expression for gain or loss as an imaginary part arises in the effective elastic stiffness constant for the material. One finds that gain of the waves occurs if the carrier drift velocity exceeds the phase velocity of the wave; otherwise loss occurs, as energy is given to the electrons by the waves and the energy is ultimately given up as resistive heating. The gain per unit length increases as the piezoelectric coupling becomes stronger. The gain is dependent upon the signal frequency, conductivity, and the diffusion constant.

In this approach one has for the current density, J, of majority charges (assumed electrons) taking account of drift and diffusion:

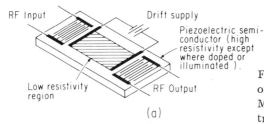

RF Input

Drift supply

Piezoelectric semi-conductor (high resistivity except where doped or illuminated).

Low resistivity region

RF Output

(a)

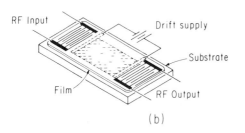

RF Input

Drift supply

Substrate

Film

RF Output

(b)

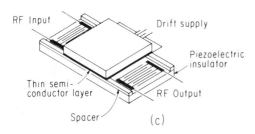

RF Input

Drift supply

Piezoelectric insulator

Thin semi-conductor layer

RF Output

Spacer

(c)

Fig. 17. Schematic representation of surface wave amplifiers. (a) Monolithic amplifier. Interdigital transducers are used on high-resistivity piezoelectric semiconductor block. Between transducers, gain occurs, owing to carrier drift in the shallow doped or illuminated low-resistivity region. (b) Film amplifier. The film and substrate are different materials, one a piezoelectric insulator and the other a semiconductor (in this illustration, the film is piezoelectric). (c) Separated-medium amplifier. Drifting charges in the thin semiconductor layer on the supporting block interact with the electric fields as waves propagate in the piezoelectric crystal. Spacers keep the members separated by a very small distance (the height of the spacers is greatly exaggerated for clarity).

$$J = q(n_0 + n)\mu E + q\mathfrak{D}\frac{\partial n}{\partial x} \qquad (5)$$

In this expression q is the electronic charge, n_0 the equilibrium charge concentration, n the a-c charge concentration, μ the mobility, E the electric field, and $\mathfrak{D}$ the diffusion constant. One then uses the current continuity equation,

$$\frac{\partial J}{\partial x} = q\frac{\partial n}{\partial t} \qquad (6)$$

and Poisson's equation,

$$\frac{\partial D}{\partial x} = -qn \qquad (7)$$

and Newton's force equation

$$\frac{\partial T}{\partial x} = \rho \frac{\partial^2 u}{\partial t^2} \tag{8}$$

in order to derive a wave equation that includes the effects of the carriers and their drift. Upon linearizing the resultant equation and assuming single-frequency waves and fields, one ultimately obtains a modified wave equation

$$\rho \frac{\partial^2 u}{\partial t^2} = c' \frac{\partial^2 u}{\partial x^2} \tag{9}$$

in which the effective elastic stiffness constant c' is a complex quantity that is substituted into the velocity expression (Eq. 3) to yield a complex velocity. Thus one finds both propagation, and gain or loss, of the waves in this system. The resultant attenuation constant may be written

$$\alpha = \omega \left(\frac{\rho}{c}\right)^{1/2} \frac{K^2}{2} \frac{(\omega_c/\gamma\omega)}{1 + (\omega_c/\gamma\omega)^2[1 + (\omega^2/\omega_c\omega_D)]^2} \tag{10}$$

where some abbreviations have been used: $K^2 = e^2/c\epsilon$ is the electro-mechanical coupling coefficient characteristic of the solid (K^2 must lie between 0 and 1; for a strong piezoelectric it might be 0.1 or greater, and K is related to the fraction of the wave energy that is in electrical, rather than purely mechanical, form); $\omega_c = \sigma/\epsilon$ is the dielectric relaxation frequency, reciprocal of the dielectric relaxation time, $\tau_R = \epsilon/\sigma$, where σ is the conductivity; ω_D is the so-called diffusion frequency ($\omega_D = v_0^2/\mathfrak{D}$); and $\gamma = 1 - v_{drift}/v_0$ is a term that accounts for the drift of the carriers relative to the propagating elastic wave (v_{drift} is the drift velocity, v_0 the wave velocity).

A plot of this attenuation expression appears in Fig. 18. One notes that gain occurs if the carriers drift in the same direction the waves travel but at a higher velocity. The gain is greater the larger the electromechanical coupling coefficient, K. Furthermore, study of the gain expression shows immediately that the maximum gain per unit length (if we neglect diffusion) occurs when the signal frequency ω equals the dielectric relaxation frequency ω_c. Good agreement with this theoretical performance is obtained in carefully compensated material, as shown by the results of Fig. 19 for bulk transverse wave amplification at 84 MHz in CdS. Detailed study shows that, at high frequencies, important deleterious effects result from carrier diffusion, which tends to debunch the electrons that have been bunched by the traveling electric fields, and by carrier trapping which tends to immobilize the carriers.

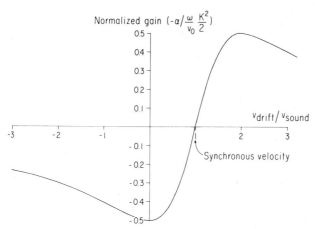

Fig. 18.   Plot of electronic gain expression of Eq. 10 for an ultrasonic amplifier versus ratio of drift to sound velocities, with $\omega_c/\omega = 1$.

The physical meanings of the salient features of the gain expression have been pointed out by Adler,[18] who compares the bunching process in the ultrasonic amplifier with the induction of charge in a conducting material by the presence of a charged comb situated near it (Fig. 20). If such a comb is brought close to a metallic solid, a pattern of induced charge forms in the solid at a rate that depends upon the dielectric relaxation time for the solid: if the solid is a perfect conductor ($\omega_c \rightarrow \infty$), the pattern forms instantaneously; if it is somewhat resistive (ω_c finite), the pattern takes some time to form. Now assume that the comb represents the source of the piezoelectrically produced electric fields, due to the acoustic wave; if the comb is moving to simulate the effect of the travelling wave, the bunches of induced charge may follow precisely only if the conductivity is infinite.

Now consider charge bunching in a resistive material to which a drift electric field is applied. Charge bunching will occur at a rate that depends on the dielectric relaxation time and also upon the relative velocities of the comb (i.e., the acoustic wave) and the charge. If the comb and the charges move at the same velocity, the induced charge pattern can form and ultimately produce a field in the direction of propagation which exactly cancels that produced by the charged comb itself; this case corresponds to the synchronous velocity ($v_{drift}/v_0 = 1$) in Fig. 18, at which there is neither gain nor loss for the wave. If the comb and the charges move with different velocities, the degree of bunching and the phase relations between the comb (which causes bunching) and the bunches formed will depend on the relative velocities and on the

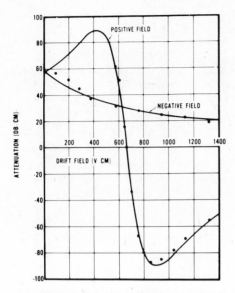

Fig. 19. Comparison of experimental with theoretical electronic gain and attenuation of 84-MHz transverse ultrasonic waves in a pulsed CdS amplifier. The
curve for the negative field (drift opposite to the wave) has been "folded over"
onto the positive plot. Material characteristics: resistivity 8×10^3 Ω-cm;
mobility 270 cm^2/V-sec; amplifier length 0.5 cm. (Courtesy of F. Hickernell,
Motorola Corporation.)

dielectric relaxation time required to form a bunch. Full analysis of
this model shows that the effective bunching depends upon the relative
frequency, $\gamma\omega$, with which the comb fingers pass the partially formed
bunches. The optimum bunching occurs when this effective frequency
equals the dielectric relaxation frequency, i.e., $\omega_c/\gamma\omega = 1$. One finds,
by considering the effect of the bunching on the effective elastic constant for the material, c', that this point corresponds to the gain peak
seen in Fig. 18. At velocities to the right of the synchronous point,
the charge bunches lead the wave and the wave is amplified; to the left
of the synchronous velocity the bunches lag and extract energy from
the wave. The full expression derived from this physical approach involving charge bunching and the effect of the wave velocity is precisely
the same as that of Eq. 10, when diffusion is neglected.

This very physical approach can also be applied to the various
forms of surface wave amplifier. For example, in the case of a film-
type amplifier, in which a semiconductor film is situated on a piezoelectric substrate, the results suggest that, if highly conductive semiconductor films are to be used, they must be very thin, so that the effec-

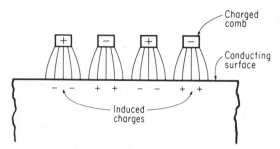

Fig. 20. A charged comb brought near the conducting surface summons an induced charge in the surface in a manner similar to the redistribution of a
charge in a piezoelectric semiconductor caused by an elastic wave. (Courtesy
of R. Adler, Zenith Radio Corporation.)

tive dielectric relaxation frequency is comparable with the effective
signal frequency. A very high relaxation frequency means that a high
effective signal frequency will be required for appreciable gain; hence
a very large drift electric field will be required which will lead to a
large power dissipation in the film and very inefficient operation.

Some factors that have not been included here are the ordinary
acoustic attenuation in the medium (which usually increases as the
square of the signal frequency and which must be subtracted from the
electronic gain given by Eq. 10); the detailed effects of diffusion (which
are more severe in surface wave amplifiers than in bulk amplifiers);
the effects of carrier trapping and low-surface mobility; and the effect of the air gap in the separated-medium surface wave amplifier
(which must be made very small if there is to be an appreciable electric field in the semiconductor to cause carrier bunching). Some of
the amplifier results obtained experimentally will illustrate these
points.

4.6 AMPLIFIER RESULTS

Experimental results on a number of bulk elastic wave amplifiers
are summarized in Table 1, from Hickernell.[19] Some developmental
bulk amplifiers appear in Fig. 21. In general, results on bulk ultrasonic amplifiers can be summarized as follows:

1. Substantial electronic and net gains are obtainable, particularly with pulsed drift fields. (Some values are given in Table 1 and the
caption of Fig. 21.)

TABLE 1.   RESULTS ON BULK ULTRASONIC AMPLIFIERS REPORTED BY
HICKERNELL[a]

| Material | Frequency, MHz | Acoustic Gain, dB/cm | Gain Factor, dB/cm-MHz | |
			Observed	Theoretical
CdS	60	150	2.5	2.9
	550	425	0.8	2.9
	1000	300	0.3	2.9
GaAs	30	3	0.1	0.23
CdSe	45	15	0.3	1.7
ZnO	30	52	1.7	3.4

[a]Zinc oxide data are for longitudinal waves, from work done at Minnesota Mining
and Manufacturing Co.   All other data are for transverse waves.

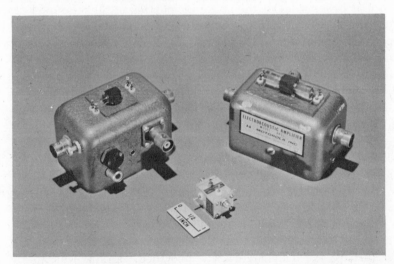

Fig. 21.   Evolution of ultrasonic bulk amplifiers at Motorola Corporation.   The
amplifier at the right has quartz transducers, fused quartz buffer rods for isolation, and CdS as the amplifier medium.   The device at the left has quartz
transducers bonded directly on the centimeter-long CdS crystal; the amplifier
provided a 40-dB net gain at 60 MHz.   The amplifier at the bottom includes heat
sinking for operation with a duty cycle up to 0.9.   (Courtesy of F. Hickernell,
Motorola Corporation.)

2. Removal of heat from the interior of a bulk amplifier is usually difficult and is a serious consideration in the selection of materials. May[20] has shown that, at frequencies above 500 MHz for operation with maximum gain and reasonable power dissipation densities, one should employ transverse waves in zinc oxide. The use of a static magnetic field, discussed below, helps considerably to reduce the power dissipation in high-mobility semiconductors.

3. Noise figures of bulk amplifiers tend to be high largely because of transducer loss. Spontaneous oscillations resulting from amplification of thermal vibrations or acoustic transients in the lattice may occur (spontaneous acoustic oscillations are discussed near the end of this chapter). Amplifier efficiency may be a few percent.

Turning from bulk to surface wave amplifiers, one finds that substantial electronic and net gains have been obtained over large bandwidths, that heat removal from the amplifier is easier than with bulk amplifiers provided that the semiconductor is in the form of a thin film or epitaxial layer on a relatively good conductor of heat, and that little is known at present about the noise properties of these amplifiers.

Results have been reported for monolithic surface wave amplifiers and for separated-medium amplifiers; to date, film amplifier results are few because of the difficulty of fabricating high-quality semiconductor films on piezoelectric substrates.

Monolithic amplifiers made of CdS in pulsed operation, with the hexagonal axis perpendicular to the free surface, have shown electronic gains of 21 dB/cm at 8 MHz for ordinary surface waves.[21] Electronic gain also has been observed for electroacoustic surface waves (Fig. 1d) at 23 MHz, propagating on a CdS plate having its hexagonal axis lying in the plane of the plate and normal to the propagation direction.[22]

At low frequencies, Fischler and Yando[23,24] have observed gains up to 15 dB/cm at 2 MHz in a lead-zirconate-titanate ceramic plate adjacent to a silicon wafer. They observed oscillation as well, when either the acoustic waves reflected at the ends of the plate or electrical feedback through the transducers was provided. Mixing was also observed, when an additional signal was introduced, either acoustically or electrically.

In experiments at 108 MHz, Lakin and co-workers[25] have observed more than 50-dB net terminal gain in pulsed, separated-medium amplifiers having lithium niobate as the piezoelectric and either a silicon bar or a silicon-on-sapphire epitaxial film* as the semiconductor.

*The silicon-on-sapphire was fabricated at Autonetics; the amplifier measurements were made at Stanford University.

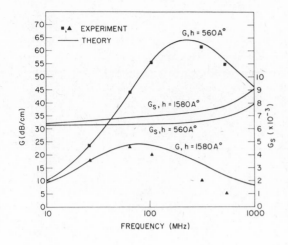

Fig. 22. Measured and theoretical electronic gain (G) for a separated-medium, silicon-on-sapphire amplifier at 108 MHz for two different gap sizes (h). (G_S is a semiconductor gain function defined in Lakin and Shaw.[25])

Resistivities in the range of 10 to 100 Ω-cm were used. The gain agreed well with the theoretical predictions, as shown in Fig. 22. It should be noted that the two media must be very close together if appreciable electric fields are to be present in the semiconductor; the exponential decay mentioned earlier, $\exp(-kz)$, dictates spacings of only about 500 A for 100-MHz operation and values near 100 A for 1-GHz operation. Electronic gain has been observed on such a separated-medium 100 MHz amplifier for frequencies up to about 1 GHz, even though the 560-A gap was large for the highest operating frequencies. The use of an epitaxial film semiconductor permits effective cooling of the semiconductor for operation with a continuous drift field. Terminal gains of up to 30 dB were obtained at 108 MHz under continuous wave conditions. The total insertion loss of the array transducers and the delay medium itself amounted to about 9 dB at 108 MHz in these devices.

In all the surface wave amplifiers, it is possible to break the drift region into small segments so as to operate with modest drift voltages, in spite of the high-drift electric fields required to provide a drift velocity in excess of the sound velocity. The semiconductor film in either the film amplifier or the separated-medium amplifier may be segmented (Fig. 23), and all the segments connected in parallel to a low-voltage supply. Alternatively, one may provide a drift field that alternates in sign by using drift electrodes in the interaction region[26] to achieve the same result. Interaction over lengths of a centimeter with drift fields of 1 kV/cm have been obtained with a drift voltage of only 50 V. A complete experimental epitaxial, separated-medium amplifier is shown in Fig. 24.

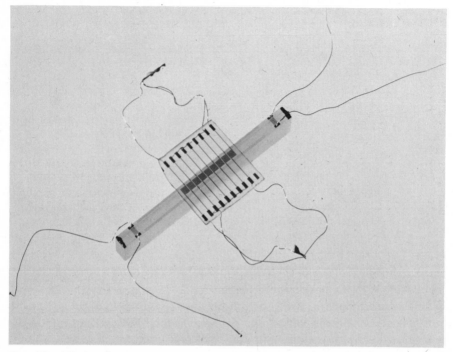

Fig. 23. Photo of epitaxial silicon for use in a separated-medium, surface wave amplifier. Silicon is segmented so that low drift voltage can be used. (From Lakin and Shaw.[25])

Finally, it should be noted that promising initial results on the amplification of surface waves with actively terminated transducer reflectors and parametric amplification have been reported.[14,27]

4.7 ULTRASONIC AMPLIFICATION IN A STATIC MAGNETIC FIELD

The addition of a static magnetic field at right angles to the drift electric field can greatly improve the operation of bulk acoustic amplifiers made with high-mobility semiconductors. Two structures that can be used are sketched in Fig. 25. The lower structure is nearly as effective as the upper, since only in the end regions near the drift contacts does the carrier drift motion depart appreciably from the longitudinal.

The full analysis of amplification in a "crossed" magnetic field shows that the classical result for the gain applies, provided that the following modified parameters are substituted:

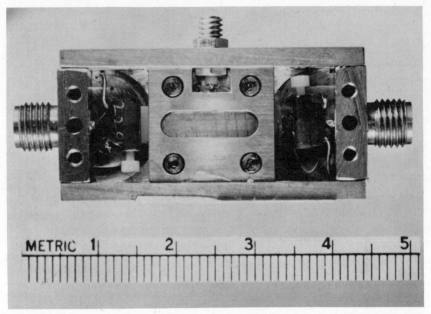

Fig. 24. Separated-medium, surface wave amplifier. The amplifier is located in a brass box, whose top cover has been removed. Silicon-on-sapphire is seen through the lithium niobate bar; the semiconductor has been divided into nine segments, permitting operation from the 180-V supply. The amplifier has a net terminal gain of 25 dB at 200 MHz, with a 0.5-W power input (d-c terminal is a top of the box). The 3-dB bandwidth is 40 MHz; matching inductors are used to give a 50-Ω input impedance. (Courtesy of P. J. Hagon, Autonetics.)

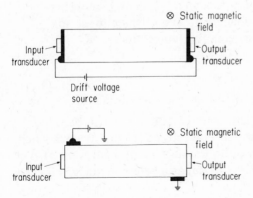

Fig. 25. Two possible arrangements for crossed-field ultrasonic amplifier having static magnetic field at right angles to the drift electric field. (Courtesy of C. W. Turner, King's College, London.)

1. ω_c (dielectric relaxation frequency) becomes

$$\omega'_c = \frac{\omega_c}{[1 + (\omega_M \tau)^2]} \tag{11}$$

2. ω_D (diffusion frequency) becomes

$$\omega'_D = \omega_D [1 + (\omega_M \tau)^2] \tag{12}$$

Here

$$\omega_M \tau = \mu B = \left(\frac{q\tau}{m}\right) B \tag{13}$$

where ω_M is the cyclotron frequency in the static magnetic field, B, and τ the mean time to collision of the carriers. The frequency of the maximum gain per unit length in the amplifier without the magnetic field is $\omega_{max} = (\omega_c \omega_D)^{1/2}$, and this is clearly unchanged by application of the magnetic field. The significant effect is that the optimum drift velocity for the maximum gain which is

$$v_{opt} = v_0 \left[1 + \frac{2(\omega_c / \omega_D)}{(1 + (\omega_M \tau)^2)}\right] \tag{14}$$

is greatly reduced by the application of the magnetic field, provided that the product $\omega_M \tau$ is large. The dramatic effect is shown in Fig. 26; clearly, the use of the magnetic field enables one to operate the amplifier with much less power dissipation in the semiconductor.

The reduction in the drift field for the maximum gain is observed to agree with theory of Route and Kino[28] for an InSb amplifier operated at liquid nitrogen temperature (77 °K). Results obtained with a crossed-field GaAs amplifier at room temperature in operation at 1 GHz are shown in Fig. 27. The use of the static magnetic field makes possible amplification at high frequencies in high-mobility materials; substantial gains are observed, even though these materials have low electromechanical coupling coefficients.

4.8 ACOUSTOELECTRIC EFFECT

Finally, we consider briefly an acoustic effect that can be either useful or harmful. This is the acoustoelectric effect that occurs when elastic waves propagate through a solid having mobile charges. If the solid is piezoelectric,* the sound wave produces a relatively large electric

---

*In nonpiezoelectric solids, deformation potential coupling will produce carrier motion also, but the effect is usually much smaller.

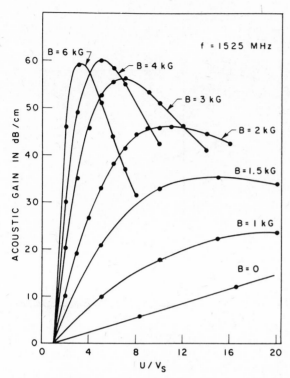

Fig. 26. Shift of drift velocity for maximum gain to lower values, resulting from the use of the static magnetic field (B) with high-mobility piezoelectric semiconductor, InSb. (From Route and Kino.[28]) Drift velocity is U, sound velocity is V_S.

field which acts on the carriers, say electrons, and drags them along. In other words, momentum is transferred between the sound waves and the electrons. The so-called acoustoelectric current density produced, J_{AE}, is

$$J_{AE} = \frac{-2I\mu\alpha}{v_{sound}}$$ (15)

where I = the intensity of the sound wave
  μ = the mobility of the carriers
  α = the attenuation constant of the medium.
The acoustoelectric current can be observed in unbiased piezoelectric semiconductors but it becomes most significant if a sufficient drift field is present in an acoustic amplifier. If one measures the I-V characteristic of such a material (e.g., CdS), one finds results[37] as sketched in Fig. 28. At low fields the acoustic waves are attenuated

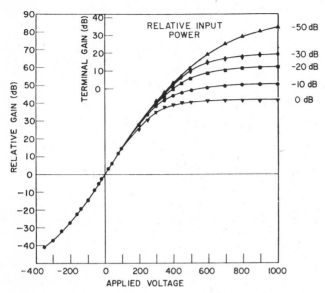

Fig. 27. Gain of room temperature GaAs, crossed–field, bulk transverse wave, 1–GHz amplifier, n-type GaAs, 14.7 Ω-cm. $\mu = 4200$ cm^2/V-sec, length = 0.88 cm, $f = 1015$ MHz.[38]

($\alpha > 0$), so that $J_{AE} < 0$; this current density adds to the negative (for electrons) drift current density. When the drift field exceeds the synchronous value, the sign of α changes ($\alpha < 0$), the acoustic intensity I increases, the sign of J_{AE} changes ($J_{AE} > 0$), and the acoustoelectric current density opposes the drift value. Hence the I-V characteristic exhibits a sharp break or kink.

If the gain is large enough, the random acoustic flux present in a long bar of piezoelectric amplifier material, as a result of thermal vibrations, may be strongly amplified and cause a large drop in the current through the leads to the bar. The result is a current oscilla-

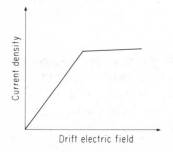

Fig. 28. Current density versus drift electric field for a piezoelectric semiconductor. The change of slope away from the low–field ohmic value occurs when the field is high enough for ultrasonic amplification.

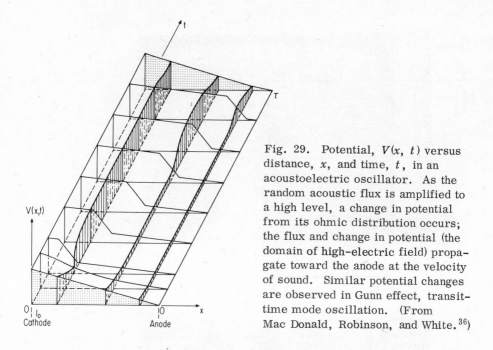

Fig. 29. Potential, $V(x, t)$ versus distance, x, and time, t, in an acoustoelectric oscillator. As the random acoustic flux is amplified to a high level, a change in potential from its ohmic distribution occurs; the flux and change in potential (the domain of high-electric field) propagate toward the anode at the velocity of sound. Similar potential changes are observed in Gunn effect, transit-time mode oscillation. (From Mac Donald, Robinson, and White.[36])

tion, very reminiscent of the Gunn oscillations in the transit-time mode.* Thermal acoustic flux present in the structure is amplified; the current through the diode is sharply reduced; and a domain of high flux travels through the sample at the velocity of sound. The potential distribution in the sample changes as the flux travels (Fig. 29) and current oscillations are observed in the external circuit. These effects have been observed in CdS, GaSb, GaAs, InSb, and in tellurium. The current wave form and observations of the potential changes as a domain of high flux passes through a CdS acoustoelectric oscillator are shown in Fig. 30.

These oscillatory effects have been identified as being responsible for some of the emission of microwave power ("Larrabee emission") in InSb in a magnetic field.[29] They have also been used to make self-excited rf current sources. Such an oscillator operating at 1 MHz, producing 430 mW power at about 2% efficiency, is shown in Fig. 31. Paige and others at the Royal Radar Establishment in England have made very thin CdS platelets with plane parallel faces oscillate at multiples of the one-way transit frequency (fundamental frequencies

---

*In some cases, a nonoscillatory current saturation is observed.

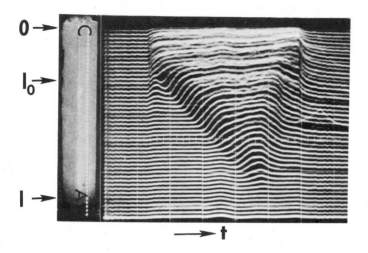

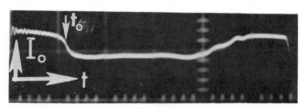

Fig. 30. Acoustoelectric oscillations in CdS bar. Top: scanning electron microscope picture of the bar with cathode, C, and anode, A, identified. Propagation of the domain corresponds to a change in "height" on the contour presentation at the right. Bottom: terminal current versus time for oscillator. (From MacDonald, Robinson, and White.[36])

for 30 to 40 μ thick platelets are in the range 25 to 30 MHz; current oscillations at third, sixth, and ninth harmonics are observed, with outputs of a fraction of a milliwatt with a few volts drive). These oscillators may be useful for short-range telemetry and other purposes.

4.9 CONCLUSIONS

A number of ultrasonic effects and devices have been described in order to show how elastic waves may be useful in high-frequency electronics. These devices offer advantages of small size because of the

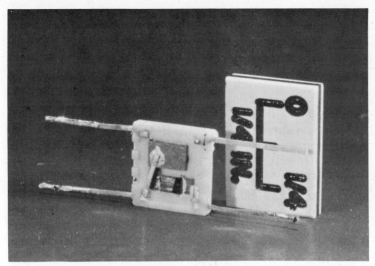

Fig. 31. CdS acoustoelectric oscillator, providing a 430-mW, 1-MHz output signal with an efficiency of about 2%. The oscillator is a CdS crystal 2 mm long, mounted on a beryllium oxide substrate for removal of heat. (Courtesy of F. Hickernell, Motorola Corporation.)

low velocity of elastic wave propagation. The use of surface elastic waves permits devices to be fabricated by photolithographic processing techniques which have been highly developed for the manufacture of integrated circuits. Active ultrasonic devices such as amplifiers and oscillators have been developed for various specialized purposes; but, at present, they appear to have their greatest potential for use in connection with relatively complex surface wave devices, in which information storage or delay for long times is needed or a number of signal processing operations are accomplished. The passive ultrasonic effects, particularly those involving surface waves, permit the economical realization of many functional devices such as pulse compression filters and matched filters. Although there are a number of inherent problems such as the fabrication problems associated with high-frequency transduction and undesirable nonlinear effects, one may anticipate the appearance of an increasing number of ultrasonic devices in the megahertz- and gigahertz-frequency ranges.

REFERENCES

1. K. Dransfeld, "Kilomegacycle Ultrasonics," *Scientific American*, 208, No. 4, 60–68 (April 1963).
2. P. M. Rowell, "Microwave Ultrasonics," *Brit. J. Appl. Phys.*, 14, 60–68 (1963).
3. P. E. Tannenwald, "Microwave Ultrasonics," *Microwave J.*, 6, 61–64 (1963).
4. S. W. Tehon and S. Wanagua, "Microwave Acoustics," *Proc. IEEE*, 52, No. 10, 1113–1127 (October 1964).
5. J. H. Collins and P. J. Hagon, "Applying Surface Wave Acoustics," *Electronics*, 42, No. 23, 97–103 (Nov. 10, 1969).
6. J. H. Collins and P. J. Hagon, "Amplifying Acoustic Surface Waves," *Electronics*, 42, No. 25, 102–111 (Dec. 8, 1969).
7. J. H. Collins and P. J. Hagon, "Surface Wave Delay Lines Promise Filters for Radar, Flat Tubes for Television, and Faster Computers," *Electronics*, 43, No. 2, 110–122 (Jan. 19, 1970).
8. I. A. Viktorov, *Rayleigh and Lamb Waves*, Plenum Press, New York, 1967.
9. R. M. White, "Surface Elastic-Wave Propagation and Amplification," *IEEE Trans. Electron Devices*, 14, No. 4, 181–189 (April 1967).
10. R. M. White, "Surface Elastic Waves," *PROC. IEEE*, 58, No. 8, 1238–1276 (August 1970).
11. J. Ilukor and E. H. Jacobsen, "Generation and Detection of Coherent Elastic Waves at 114,000 Mc/sec.," *Science (USA)*, 153, 1113–1114 (Sept. 2, 1966).
12. W. R. Smith, H. M. Gerard, J. H. Collins, T. M. Reeder, and H. J. Shaw, "Design of Surface Wave Delay Lines with Interdigital Transducers," *IEEE Trans. Microwave Theory Techniques*, 17, No. 11, 865–873 (November 1969).
13. A. N. Broers, E. G. Lean, and M. Hatzakis, "1.75 GHz Acoustic-Surface-Wave Transducer Fabrication by an Electron Beam," *Appl. Phys. Lett.* 15, No. 3, 98–101 (Aug. 1, 1969).
14. A. J. Bahr, "Electrically Controlled Reflection of Acoustic Surface Waves," Post-deadline paper, IEEE Ultrasonics Symposium, St. Louis, Mo., Sept. 24–26, 1969.
15. L. T. Claiborne, C. S. Hartmann, and W. S. Jones, "Annual Technical Report for Surface Wave Devices: 15 January 1969–15 January 1970," ONR Contract No. N00014-69-C-0173, Texas Instruments Company, Dallas, Tex., March 1970.
16. D. L. White, "Amplification of Ultrasonic Waves in Piezoelectric Semiconductors," *J. Appl. Phys.*, 33, No. 8, 2547–2554 (August 1962).
17. J. H. McFee, "Transmission and Amplification of Acoustic Waves in Piezoelectric Semiconductors," from *Physical Acoustics*, Vol. 4, Part A, W. P. Mason, Ed., Academic Press, New York, 1966.
18. R. Adler, "A Simple Theory of Acoustic Amplification," Paper A-5, IEEE Ultrasonics Symposium, St. Louis, Mo., Sept. 24–26, 1969.
19. F. Hickernell, private communication.
20. J. E. May, Jr., "Ultrasonic Traveling-Wave Devices for Communications," *IEEE Spectrum*, 2, No. 10, 73–85 (October 1965).

21. R. M. White and F. W. Voltmer, "Ultrasonic Surface-Wave-Amplification in Cadmium Sulfide," *Appl. Phys. Lett.*, <u>8</u>, No. 2, 40–42 (Jan. 15, 1966).

22. V. R. Learned and C. W. Turner, private communication.

23. C. Fischler and S. Yando, "Amplification of Acoustic Waves in a Coupled Semiconductor-Piezoelectric Ceramic System," *Appl. Phys. Lett.*, <u>13</u>, No. 10, 351–353 (Nov. 15, 1968).

24. C. Fischler and S. Yando, "Amplification of Guided Elastic Waves in Piezoelectric Plates through Electrical Coupling to a Semiconductor," *Appl. Phys. Lett.*, <u>15</u>, No. 11, 366–368 (Dec. 1, 1969).

25. K. M. Lakin and H. J. Shaw, "Surface Wave Delay Line Amplifiers," *IEEE Trans. Microwave Theory and Techniques*, <u>17</u>, No. 11, 912–920 (November 1969).

26. C. W. Turner, A. Shomon, and R. M. White, "Reduced-Voltage Operation of a Surface-Elastic-Wave Amplifier," *Electronics Lett.*, <u>5</u>, No. 11, (May 29, 1969).

27. G. Chao, "Parametric Amplification of Surface Acoustic Waves," *Appl. Phys. Lett.*, <u>16</u>, No. 10, 399–401 (May 15, 1970).

28. R. K. Route and G. S. Kino, "Sound Wave Amplification in Indium Antimonide," *Appl. Phys. Lett.*, <u>14</u>, No. 3, 97–99 (Feb. 1, 1969).

29. C. W. Turner, "The Role of Acoustic Wave Amplification in the Emission of Microwave Noise from InSb," *IBM J. Res. Develop.*, <u>13</u>, No. 5, 611–615 (September 1969).

30. E. Salzmann, T. Pleininger, and K. Dransfeld, "Attenuation of Elastic Surface Waves in Quartz at Frequencies of 316 MHz and 1047 MHz," *Appl. Phys. Lett.*, <u>13</u>, No. 1, 14–15 (July 1, 1968).

31. D. L. White, "Depletion Layer Transducer—A New High Frequency Ultrasonic Transducer," *IEEE Trans. Ultrasonic Eng.*, <u>UE-9</u>, No. 1, 21–27 (July 1962).

32. P. M. Rowell, "Microwave Ultrasonics," *Brit. J. Appl. Phys.*, <u>14</u>, 60–68 (1963).

33. H. Seidel and D. L. White, "Ultrasonic Surface Waveguides," U.S. Patent 3,406,358, Oct. 15, 1968.

34. A. H. Meitzler and H. F. Tiersten, "Elastic Surface Waveguide," U.S. Patent 3,409,848, Nov. 5, 1968.

35. W. D. Squire, H. J. Whitehouse, and J. M. Alsup, "Linear-Signal Processing and Ultrasonic Transversal Filters," *IEEE Trans. Microwave Theory Techniques*, <u>17</u>, No. 11, 1020–1040 (November 1969).

36. N. D. MacDonald, G. Y. Robinson, and R. M. White, "Time-resolved Scanning Electron Microscopy and Its Application to Bulk-Effect Oscillators," *J. Appl. Phys.*, <u>40</u>, No. 11, 4516–4528 (October 1969).

37. W. H. Haydl, K. Harker, and C. F. Quate, "Current Oscillations in Piezoelectric Semiconductors," *J. Appl. Phys.*, <u>38</u>, No. 11, 4295–4309 (October 1967).

38. "Microwave Acoustic and Bulk Device Technique Studies," Technical Report RADC-TR-222, Stanford University, Stanford, Calif., May 1969, pp. 53–56.

Chapter 5

SUPERCONDUCTIVITY

J. E. Mercereau

Department of Physics
California Institute of Technology
Pasadena, California

5.1 INTRODUCTION

Superconductivity brings to mind a state of zero electrical resistance—at least below some critical transition temperature and magnetic field. Discovered in 1911, this phenomenon has now been found in more than 30% of the elements and over 1000 alloys, with transition temperatures ranging to above 20 °K and critical magnetic fields of hundreds of kilogauss. The purpose of this chapter is to emphasize the unique quantum nature of this phenomenon and to outline the application of superconductivity to electronic devices.

5.2 PHENOMENOLOGY OF SUPERCONDUCTIVITY

Electrons in metals, as elsewhere, are essentially quantum mechanical in behavior, and any description in classical terms is at best incomplete. Nevertheless, it is sometimes useful to have even an imperfect model as an aid to identifying relationships in a complex process. In this spirit the following pseudodescription of electronic conduction may help to focus on the unique interactions necessary for superconductivity and define some of the common terminology.

In metals, conduction electrons are free to move within the ionic lattice. Actually, since the lattice is not perfectly rigid, attractive Coulomb forces between the electrons and ions distort the lattice near each electron and create a slight local positive charge cloud around it which tends to cancel the electric field of the electron at large distances. Thus, to first order, each conduction electron is effectively isolated from direct Coulomb interaction with other electrons far away. The conduction electrons can be thought of as a negatively charged but noninteracting gas, constrained to move through a background lattice of positive ionic charge. But since the lattice distortion around each electron must accompany the electron as it moves, electrons are said to be "coupled" to the lattice. This coordinated motion of lattice and electron is usually accounted for in the dynamics of the electron by assigning it an effective mass. In most metals this coupling to the lattice also causes electrical resistance, since the momentum and energy transferred to the lattice distortion by the moving electron is not necessarily returned coherently to electronic motion and eventually ends up as heat.

However, this is not quite the end of the story because, in some metals, electrons are able to influence each other indirectly and create a unique electronic state called superconductivity. Fundamentally, this electronic state is unique because its elemental components, the electrons, are not allowed to act independently but are forced by quantum regulations to move in precise cooperation with all other superelectrons. This comes about in the following way. The only electrons free to interact are at the edge of the Fermi distribution and move very quickly (at the Fermi velocity, v_F). But because of their larger mass, ions in the lattice respond rather more slowly (at the Debye frequency ω_D, at best). Thus the lattice distortion (or positive cloud) which we have been discussing cannot instantly follow the motion of an electron and is not exactly "around" it but strings out behind as sort of a positively charged tail along the electron trajectory, lasting about ω_D^{-1} sec. If a second electron passes through this region of induced positive charge before it dies, its energy will be reduced by the net attraction of the induced charge. Thus a "pairing" interaction can occur via the intermediary of the lattice—two electrons interacting together with a "pair potential," V. The duration of the interaction is longest, and thus the total interaction largest if the second electron travels the whole length of the tail of the first. Therefore, pairs of electrons will interact most strongly if they have exactly opposing momenta. (A similar coupling between electrons of precisely parallel momenta might be expected but is not allowed because of the Exclusion Principle.)

If there is some center of mass motion of the pair, the resultant momentum of each of the electrons in the pair can be written as $p_+ = p_f + p_{cm}$ or $p_- = -p_f + p_{cm}$, where p_f is the Fermi momentum and p_{cm} is caused by the center of mass motion. Quantum mechanically, each electron can be described by the amplitude and a phase of a wave function $\Psi = ae^{i\gamma}$, where for a plane wave the spacial dependence of the phase is $\gamma = 1/\hbar(p \cdot x)$. Thus each pair could be described as a single quantum state, $\Psi_1\Psi_2$, characterized by an amplitude A, and a common center of mass phase γ_{cm}, as

$$\Psi = \Psi_1\Psi_2 = A\exp\left(\frac{2p_{cm}x}{\hbar}\right) = Ae^{i\gamma_{cm}}$$

Note that this pairing of electrons occurs via momentum and does not imply a permanent localized grouping of two electrons in space.

This induced attractive interaction or pairing between electrons would certainly be expected to reduce the energy of the electronic system. However, the unusual aspect of the phenomena is that no matter how weak the interaction, it results in an entirely new electronic state,[1,2] separated from the noninteracting state by a *discretely* lower energy. The difference in energy (per electron) between these states is known as the "energy gap," Δ.

The reason for this *discrete* difference in energy comes about because of the large number of electrons (all those within $\hbar\omega_D$ of the Fermi energy) which can participate in the pairing interaction. And even this normally large number is enhanced by a rearrangement of the electronic momentum distribution, which is also due to the pairing. By analogy this effect might be thought of as the equivalent of surface tension in fluids—but in this case acting in momentum space at the surface of the Fermi sphere where an "interface" is formed, because of the pairing interaction, separating different phases of the electronic system. This state becomes possible when the temperature is lowered sufficiently so that thermal fluctuation effects no longer destroy the state (or $kT \lesssim \Delta$).

But for our purposes this is still not quite a complete description of "superconductivity." Superconductivity requires, in addition, that the center of mass phase must be identically the same for *all* pairs. The spatial extension of a single pair (or the distance over which the electron phase is correlated by the pairing interaction) is called the coherence distance, ξ, and arises from the uncertainty principal $\Delta E \Delta t \geq \hbar$. Since the energy of interaction is Δ and the electrons involved move at the Fermi velocity, the distance covered by an electron during the time required to specify the binding energy is $\xi \sim (\hbar v_F/\Delta)$. For a typical superconductor, $v_F \sim 10^8$ cm sec^{-1}, and $\Delta \sim 10^{-15}$ erg, so that the pairs are rather large, approximately 10^{-4} cm. The size and number of pairs can become so large, in fact, that the pairs overlap in the metal; any single electron may find itself simultaneously "paired" with several others, each of the separate couples bound by energy 2Δ to a common phase γ.

Since the same situation can occur for any electron—once the pairs overlap—all electrons must then coalesce into a common phase state. And long-range phase order is thus created out of overlapping short-range order. This creation of long-range order by short-range forces is a relatively common phenomenon in configuration space, as may be witnessed in the formation of a single crystal or a ferromagnet. However, the occurrence of a single-*phase* state, or macroscopic wave function, $\Psi = \sqrt{\rho}\, e^{i\gamma}$, common to large numbers (density ρ) of electrons sharing a common center of mass phase, is somewhat more unique and is what has become known as the phenomenon of superconductivity.

5.3 MACROSCOPIC WAVE FUNCTION

Usually, the quantum phase is a relatively unimportant parameter in macroscopic experiments; however, in the case of superconductivity, with so many particles sharing the same quantum phase, the phase does become important and, in fact, may be considered a real physical variable. The most familiar characteristics of superconductivity (zero resistance and zero internal magnetic field) follow directly as a consequence of this fact. The quantum mechanical equations of motion for a wave function are as follows:

$$i\hbar\frac{\partial\Psi}{\partial t} = \mu\Psi$$

$$-i\hbar\nabla\Psi = p\Psi$$

where μ and p are, respectively, the energy and momentum of the particle. If we apply these relationships to an electron pair and then ask for the average value of the energy $\langle\mu\rangle$ and momentum $\langle p\rangle$, we find that for the pair:

$$\hbar\frac{\partial\gamma}{\partial t} = \langle\mu\rangle \tag{1a}$$

$$\hbar\nabla\gamma = \langle p\rangle = 2\langle mv_{cm} + eA\rangle \tag{1b}$$

where v_{cm} is the center of mass velocity and A is the vector potential ($\nabla\times A = B$). But this pair phase is also shared by all other pairs and is also the phase for the macroscopic system. Since all pairs also have the identical v_{cm}, we can identify the electric current density j as $j = \rho 2ev_{cm}$ and obtain macroscopic equations of motion for the quantum phase of the system. By recalling that the electric field E is defined in terms of potentials as $E = \nabla\Phi - \dot{A}$, these phase relationships can be written in terms of fields and currents as follows. Taking the gradient of equation (1a) leads to:

$$\frac{\partial}{\partial t}(\nabla\gamma) = \hbar^{-1}\nabla\mu = \left(\frac{2e}{\hbar}\right)\nabla\Phi$$

By substituting equation (1b) for $\nabla\gamma$ and defining the ratio of constants, $(m/\mu_0\rho e^2)^{-1/2}$ as a length λ we get:

$$m\dot{v}_{cm} = e\{\nabla\Phi - \dot{A}\} = eE$$

or

$$\mu_0\lambda^2\frac{\partial}{\partial t}(j) = E$$

This result is just the same as Newton's law applied to free electrons— a force causes an acceleration. But note that this implies no dissipation, or zero resistance.

Similarly, by taking the curl of equation (1b)—and since the curl of a gradient is always zero:

$$\nabla \times (\lambda^2 j) = -H$$

here we have used the fact that the magnetic field H is defined as $\mu_0 H = \nabla \times A$. When the above relation is combined with the Maxwell equation $\nabla \times H = j$ we can get a single equation representing the behavior of the magnetic field in a superconductor,

$$\nabla^2 H = \lambda^{-2}H$$

This result expresses the fact that an externally applied magnetic field is excluded from a superconductor except for a "penetration depth," λ, at the surface. For most materials $\lambda \simeq 10^{-5}$ cm.

These relationships, of course, express the usual phenomena of superconductivity (resistance and magnetic field are zero) and come directly from the concept of a macroscopic quantum state. But you should also note that in the mathematical manipulations we have used to get this far, we have lost $(\hbar)$, the relationship to quantum mechanics.

5.4 JOSEPHSON EFFECT

In 1962 Josephson[3] considered what would happen if two superconductors were sufficiently close together so that current could flow from one to the other by some sort of tunneling process. He derived the following expressions for current (density, j_J) flowing between two superconductors:

$$j_J = j_0 \sin \left\{ \gamma_1 - \gamma_2 - \frac{2e}{\hbar} \int_1^2 A \cdot dx \right\} \tag{2a}$$

$$\frac{\partial}{\partial t} (\gamma_1 - \gamma_2) = \frac{2e}{\hbar} V \tag{2b}$$

where $\gamma_1 - \gamma_2$ is the relative quantum phase between superconductors, V is the potential difference, A is the magnetic vector potential. The amplitude of the current j_0 depends on the strength of coupling between the superconductors and on the energy gap and may be considered as a temperature-dependent parameter of the system; whereas the control of this current lies in the quantum phase and its dependence on the potentials. Note especially that now the current depends on the sine of the phase difference between the two superconductors, rather than on the gradient of the phase at a point that we found true for a single superconductor. This difference is the essential factor in the development of superconducting quantum electronics.

The usual experimental configuration for observing the Josephson effect is essentially two superconductors separated by a very thin oxide layer—fundamentally a capacitor with a very thin dielectric. However, here the dielectric is so thin that the supercurrent (see Eq. 2) can be carried through the dielectric by electron tunneling. At low frequency this conduction mechanism dominates over the usual displacement currents of the capacitor; and to a good approximation total current through such a device, called a Josephson junction is gotten by integrating Eq. 2 over the contacting areas. The result is just[5]:

$$I = I_0 \left[\sin \left(\frac{e}{\hbar} \Phi_J \right) \Big/ \left(\frac{e}{\hbar} \Phi_J \right) \right] \sin \delta\gamma \tag{3a}$$

$$\delta\gamma = \frac{2e}{\hbar} Vt + \alpha \tag{3b}$$

Thus the Josephson junction is a device that will carry a zero-voltage current of maximum amplitude

$$I_0 \left\{ \sin \left(\frac{e}{\hbar} \Phi_J \right) \right\} \Big/ \left(\frac{e}{\hbar} \Phi_J \right) ,$$

while at the finite voltage, V, it supports an *oscillating* current of the above magnitude at a frequency $\omega = (2e/\hbar)V$. The maximum current is a function of the magnetic flux Φ_J and has a diffractionlike modulation with a period given by $h/2e$. This magnetic flux, Φ_J, is defined to be the product of the field applied in the plane of the junction by the area that the field penetrates between the superconductors. So far in this discussion we have been assuming that the junctions are small enough and the frequency low enough so that variations of the field within the junction area may be neglected. In this case the dependence of the phase, $\delta\gamma$, on voltage is simply given as above, where at zero voltage the constant α is a boundary condition determined by the current source and, at finite voltage, the instantaneous frequency is $2eV/\hbar$. The maximum zero-voltage current occurs when $\alpha = \pi/2$, but this maximum current is itself a function of the magnetic flux, as we have previously described. At high frequency, or for large junctions, the spatial variation of the fields must also be taken into account[6,7]; which sometimes gives rise to complicating resonance effects, but the fundamental response to the voltage and field is still as given above.

A typical junction of this type will have a resistance in the normal state of from 1 to 10^{-2} Ω and a maximum Josephson current from 1 to 10^2 mA. As you might expect, there is a fundamental relation between the current, the resistance, and the energy gap; it can be shown[3] that this relation is given by $I_J = (\pi/2)(\Delta/R)$. Also, the maximum current is related to the total energy in the Josephson process which is just $I_0(\hbar/2e)$, or, for a one mA junction, only 10 eV for the *entire* device. Thus the phenomenon is rather delicate and sensitive to external influences. Since room temperature thermal energy is approximately 10^{-2} eV per degree of freedom, considerable care is required to shield the device in such a way that these intrinsic quantum phenomena become visible. However, when this is done, both effects expressed in Eq. 3 can easily be observed.[5,8]

Figure 1 shows the diffraction effect anticipated in Eq. 3a out to the fifth side lobe. These data were taken by recording the maximum zero-voltage current that can pass through the junction as a function of the applied field. This junction was about $\frac{1}{4}$-mm wide and had a total field penetration depth of about 1200 A. Thus the indicated field periodicity of about 0.7 G converts to a flux period of approximately 2×10^{-7} G/cm^2. This flux periodicity is a universal phenomenon of superconductivity—depending only on the ratio of fundamental constants, $h/2e$ $= 2.07\times10^{-7}$ g/cm^2—and not at all on the particular superconductor or its material properties.

Figure 2 shows the radio frequency output of a junction biased at about 7×10^{-8} V. Near 35 MHz, there is a maximum in the rf response, corresponding to the supercurrent oscillations expected in Eq. 3. The

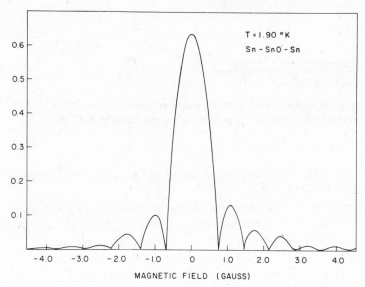

Fig. 1. Maximum zero-voltage Josephson current as a function of the magnetic field. (From J. E. Mercereau, "Quantum Engineering," in M. H. Cohen, Ed. *Superconductivity in Science and Technology*, Chicago University Press, Chicago, 1968, p. 63, by permission.)

width of this oscillation is a measure of the noise in the superconducting process and will be discussed later. Again, the ratio of the voltage to the frequency is a universal property of superconductivity and depends on the same ratio of fundamental constants, $h/2e$. Oscillation has been observed over a wide range of frequencies, down to less than 1 Mc, where it is limited by noise, to above the gap frequency 10^{12} Hz, where the limits are still being explored.

5.5 QUANTUM INTERFERENCE

These Josephson junctions are essentially the key to the development of superconducting quantum electronics. Through the action of the junction the supercurrent can respond to electric and magnetic fields in a fundamental way determined by the direct intervention of quantum mechanics. The junction, used in conjunction with other superconducting elements, gives rise to circuits called quantum interference devices[9] which directly demonstrate the interaction of the quantum aspects of a superconductor with electric and magnetic fields. The simplest of these devices is essentially a Josephson junction contained in a super-

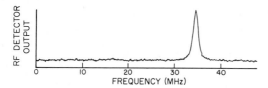

Fig. 2. Radio frequency response of a Josephson device biased at about 7×10^{-8} V. Oscillation near 35 MHz represents power of approximately 10^{-13} W.

conducting loop, or stated differently, a junction "shorted" by a superconducting inductor (see Fig. 3a).

In this simple circuit, the quantum phase is controlled by the electric and magnetic fields in a particularly simple way. This phase, in turn, acts to control the Josephson current through the junction, or therefore around the circuit. Thus the current in this circuit is just the Josephson current $I = I_0' \sin \delta\gamma$, where we have rewritten Eq. 3a to contain all the magnetic diffraction effects in I_0'; or $I_0' = \{\sin(e/\hbar \Phi_J)\}/ (e/\hbar \Phi_J)$. Combining this with Eq. 3b, we get the following expression for the Josephson current in the circuit:

$$I = I_0' \sin \frac{2e}{\hbar} \{ \int V\, dt + \alpha \}$$

where V is the voltage across the junction and α is the phase angle across the junction in the absence of any voltage. But this voltage and phase angle must, by continuity, also be the voltage and phase angle around the superconducting loop. Thus the phase angle $[(2e/\hbar)\alpha]$ is also the phase difference around the superconducting loop in absence of any voltage, or $(2e/\hbar)\, \alpha = \int_{\text{loop}} (\nabla\gamma) dx$, and V is the voltage around the loop. But in a superconductor we have seen that the

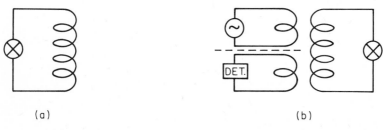

(a)   (b)

Fig. 3. (a) Schematic representation of a superconducting interferometer, essentially a series circuit containing a Josephson device, $\otimes$, and a superconducting inductor. (b) Schematic representation of the circuit in Fig. 3(a) with drive and detector circuits.

phase gradient is related to currents and fields as in Eq. 1b, so that

$$\alpha = \frac{\hbar}{2e} \oint \frac{2}{\hbar} \{mv_{cm} + eA\} dx = \frac{m}{e} \oint v_{cm} dx + \oint A dx$$

where we write $\oint$ to indicate integration around the superconducting loop. Note that $\oint 2mv_{cm} dx$ is the mechanical angular momentum (l) for the electron pair around the loop and $\oint A dx$ is the magnetic flux enclosed by the loop, Φ_T.

Consequently, the electrical response of this macroscopic circuit is dominated by quantum effects in much the same way as though it were a giant Bohr atom. In this circuit the Josephson current oscillates in time at an instantaneous rate determined by the potential ($\omega = 2eV/\hbar$), and the phase of this oscillation is modulated by both the magnetic flux in the loop and the angular momentum of the electrons around the loop.

Usually, the superconducting ring is driven by an alternating voltage $v \sin \omega t$; in this case the voltage integral is itself a simple function, so that the induced current in the circuit becomes

$$I = I_0' \sin \frac{2e}{\hbar} \left\{ \frac{v}{\omega} \cos \omega t + \Phi_T + \frac{1}{2e} l \right\} \tag{4}$$

But the current is now highly nonlinear. However, a usual arrangement to observe these effects is to arrange a circuit such as in Fig. 3b, where the superconducting circuit is driven inductively and the resultant current is also detected inductively at the drive frequency (ω). At the drive frequency the amplitude of the current in the superconducting circuit (Eq. 4) reduces to

$$I = I_0' J_1 \left(\frac{2ev}{\hbar\omega} \right) \cos \left(\frac{2e}{\hbar} \right) \left(\Phi_T + \frac{1}{2e} l \right)$$

where J_1 is a first-order Bessel function.

In most situations, the angular momentum contribution is not important and the amplitude of the supercurrent at frequency ω simplifies to

$$I = I_0' J_1 \left(\frac{2ev}{\hbar\omega} \right) \cos \left(\frac{2e}{\hbar} \Phi_T \right). \tag{5}$$

Equation 5 represents the current induced in a superconducting circuit such as we have described. Without the junction, the current in the loop would simply follow Faraday's induction law, increasing or decreasing to compensate for flux changes impressed on the circuit, and would faithfully follow the amplitude and frequency of the applied voltage. With the junction however, the effects of quantization (or interference) become immediately apparent. The current now becomes periodic in magnetic flux, Φ_T, even reversing direction relative to the drive voltage for certain flux values. This periodicity is also evident as a function of drive voltage, where again the amplitude and sign of the current are also quasiperiodic in this parameter.

Typical results from an interferometer circuit are shown in Fig. 4. In the upper curve the drive voltage was held fixed and the induced current measured as a function of the magnetic flux. In the lower curve the magnetic flux was kept constant and the induced current measured as a function of the drive voltage. In both cases the expected periodic response represented in Eq. 5 is clearly evident, indicating the true quantum nature of the device.

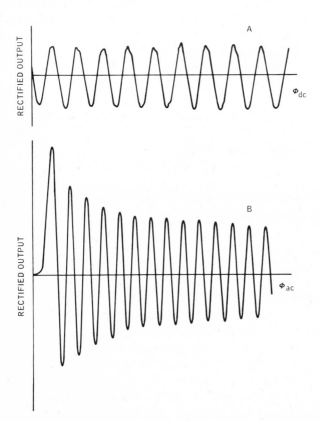

Fig. 4. Response of a superconducting interferometer such as that represented in Fig. 3(b). Upper curve: drive voltage constant shows periodic response to magnetic flux, φ_T. Lower curve: magnetic flux constant shows periodic response to rf voltage, $v = \omega\varphi_{ac}$. Flux period is $h/2e$, voltage period $\hbar\omega/2e$. (From Ref. 18, by permission of *Revue de Physique Appliquée*.)

Thus the main result of this kind of circuitry is a quantum device that responds periodically to magnetic field, with a flux period given by $h/2e = 2.07$ g/cm^2; which also responds quasiperiodically to voltage at a frequency $\nu = 2eV/h$, $= 484$ MHz/μV.

5.6 FREQUENCY MODULATION, MIXING, AND DETECTING

As we noted previously, the Josephson relation (Eq. 2) can be highly nonlinear. This aspect of the phenomenon has been used for signal mixing and the detection of radiation. The self-oscillation of the device (see Fig. 2) can be frequency-modulated by a radiation field and produces many useful results.

If the device is biased to a voltage V_0 and, in addition, has impressed an alternating voltage $v \cos \omega t$, the resulting supercurrent is

$$I = I_0' \sin \frac{2e}{\hbar} \left[V_0 t + \frac{v}{\omega} \sin \omega t + \alpha \right]$$

This expression simply and explicitly relates the frequency modulation effects we have stated above. If we expand this expression in harmonics, we get

$$I = I_0' \sum_{-\infty}^{+\infty} J_n \left(\frac{2ev}{\hbar \omega} \right) \sin \left[\left(n\omega + \frac{2eV_0}{\hbar} \right) t + \alpha \right]$$

Thus the Josephson current has components or side bands at infinitely many frequencies but with amplitudes dependent on the high-order Bessel function, J_n.

In particular, there may be side bands at zero frequency which means a d-c current in response to a-c stimulation, whenever $V_0 = n\hbar\omega/2e$. These zero-frequency currents correspond to the appearance of the Josephson d-c current at finite voltage. The result of impressing an alternating voltage on a Josephson junction thus alters the structure of the d-c current-voltage characteristic; current "steps" appear at voltages related precisely to the impressed frequency as above. Figure 5 shows the result of impressing a 35-GHz signal on a junction. The zero-voltage current is suppressed and steps occur at voltages $V = n(\hbar\omega/2e) \simeq n(7 \times 10^{-5})$ V. Experimental confirmation of this behavior was first obtained by Shapiro,[8] who found the expected structure and confirmed the expected voltage-frequency relationship. His experiments were done at a high frequency, at which our simple model does not directly apply. At the higher frequencies the spatial variation of both the quantum phase and the electric field must be taken

into account to determine the proper coupling. Except for this point
the physical consequences are the same, especially the expectation of
current steps at voltages of exactly $n(\hbar\omega/2e)$.

In addition to the effects at finite voltage, the zero-voltage supercurrent is also affected by radiation, as we have already seen. At
zero voltage the above expression reduces to

$$I = I_0' J_0 \left(\frac{2ev}{\hbar\omega}\right)$$

Thus the current at zero voltage responds at low-radiation power
levels by decreasing proportional to v^2, while the first step increases
as v. However, since the current at zero voltage responds to all frequencies, this response can be used as a wide band detector; in the
same sense, effects at finite voltage correspond to narrow band detection. These effects have been proposed for use as detectors, especially in the millimeter wavelength region, where sensitivities of 10^{-13} to
10^{-14} W have been reported.[12]

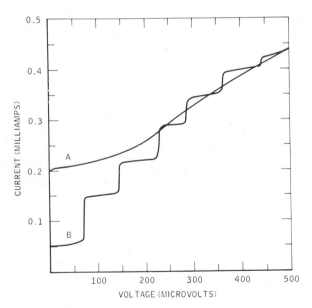

Fig. 5. Current-voltage (I-V) characteristic of a Josephson device. Upper
curve is without microwave stimulation; lower curve is the result of applying a
35-GHz signal to the device.

5.7 WEAK SUPERCONDUCTIVITY

The primary difficulty that has been encountered in using these circuits for instrumentation purposes has been the relatively fragile nature of the Josephson junction. This fact has led to considerable effort to replace the Josephson junction by a more stable configuration. It has been found that a circumstance now called weak superconductivity[14] exhibits many of the characteristics of the Josephson junction. Superconductivity of this form exists at finite voltages and is a time-dependent thermodynamic phase involving time dependence of both the phase and amplitude of the wave function, rather than simply the time-dependent quantum phase we discussed previously. This kind of superconductivity usually involves inhomogeneities in the superconductor, either geometric inhomogeneities such as the point contact[15] or Dayem[16] bridges or actual material inhomogeneities[17] implanted in the superconductor itself. The detailed understanding of this kind of superconductivity is somewhat fragmentary at present, but the experimental facts are that this kind of superconductivity seems to be quite similar to the behavior of the Josephson junction and can be obtained in structures sufficiently stable to be useful for instrumentation purposes.

5.8 INSTRUMENTATION

These superconducting elements have been built into superconducting circuits of various kinds and used as the basis for a novel kind of instrumentation.[18] At low frequencies the principal emphasis has been on the magnetic field sensitivity of the devices, and several kinds of magnetometers, ammeters, voltmeters, etc., have been fabricated. All of these devices essentially depend on the magnetic field sensitivity of a multiply connected superconducting circuit. For example, a circuit with an aperture of 1 cm^2 produces a signal modulation with a magnetic field periodicity of 2.07×10^{-7} G; see Fig. 4a. With conventional electronic techniques this fundamental period has been further subdivided by as much as a factor of 10^4 (using a 1-sec time constant), leading to magnetic field sensitivity for devices such as this of about 10^{-10} to 10^{-11} G. These magnetometers have also been used as voltmeters by simply converting a potential to a known current and thus to a magnetic field. Voltmeters with a sensitivity of 10^{-14} to 10^{-15} V, have already been achieved. At high frequencies, the main application of these devices so far has been as mixers or as detectors of radiation. Incoming radiation tends to frequency modulate the oscillation of the device and, in particular, can give side bands at zero frequencies.

These side bands then show up as steps in the d-c voltage-current characteristic[8]; see Fig. 5. The appearance of these steps has been used as a basis for a detection process,[19] where sensitivities of about 10^{-14} W have been achieved and the mixing process[20,21] has been observed to frequencies as high as approximately 10^{12} Hz. This step structure has also been used to determine precise values for the ratio of the fundamental constants, h/e, by a simultaneous measurement of the frequency and voltage.[22]

These devices have also been used as oscillators[23]; however, output power is severely limited and the present level of emitted power is only about 10^{-11} W at frequencies of 10^{10} cycles/sec. The devices commonly operate at currents ranging from 10^{-5} to 10^{-2} A and at potentials ranging from 10^{-9} to 10^{-3} V. Thus the power dissipated by the superconducting circuit ranges from about 10^{-14} to 10^{-5} W, depending mostly on the frequency of operation of the device: the higher the frequency, the higher the voltage. Aside from their intrinsically low-power characteristics, the main problem here has been the difficulty in coupling to the low-impedance devices, either to extract energy or, as detectors, to couple energy in.

5.9 NOISE

Fundamental limits on frequency or sensitivity for these devices have not yet been determined. So far the main limitation has been from the noise in the conventional electronics associated with the use of these devices. Even so, noise in the devices has been measured[24] and so far can be attributed to essentially conventional noise sources. Since the devices operate primarily at finite voltages, Johnson noise is to be expected from the finite dynamic resistance of the device. This has been seen and, in fact, is being exploited for use as a noise thermometer[25] at low temperatures. Another noise source is quasi-particle[26] or shot noise arising from electrons thermally excited across the tunnel barrier. This noise depends on the square of the dynamic resistance and also on the current through the device. These noise currents show up as phase noise or line width in the Josephson oscillation frequency; see Fig. 2. However, these noise effects can be reduced to any appropriate level of operation simply by reducing the resistance or temperature of the device.

5.10 PROGNOSIS

These devices are now beginning to be used in the scientific community as ultrasensitive sensors of many types, ranging from use in radio astronomy through materials research to biomedical studies.[27] In addition, there is a real possibility that the phenomenon will become a basic technique for establishing the values of fundamental constants.[28] All in all, it is likely that we are only at the beginning of a transition that will take superconductivity from a purely scientific study into a useful electronic technology.

REFERENCES

1. L. N. Cooper, *Phys. Rev.*, $\underline{104}$, 1189 (1956).
2. J. Bardeen, L. N. Cooper, and J. R. Schreiffer, *Phys. Rev.*, $\underline{106}$, 162 (1957); $\underline{108}$, 1175 (1957).
3. B. D. Josephson, *Phys. Lett.*, $\underline{1}$, 251 (1962).
4. B. D. Josephson, *Advan. Phys.*, $\underline{14}$, 419 (1965).
5. P. W. Anderson and J. M. Rowell, *Phys. Rev. Lett.*, $\underline{10}$, 230 (1963).
6. R. E. Eck, D. J. Scalapino and B. N. Taylor, *Phys. Rev. Lett.*, $\underline{13}$, 15 (1963).
7. Milan D. Fiske, *Rev. Mod. Phys.*, $\underline{36}$, 221 (1964).
8. S. Shapiro, *Phys. Rev. Lett.*, $\underline{11}$, 80 (1963).
9. R. C. Jaklevic, J. Lambe, A. H. Silver, and J. E. Mercereau, *Phys. Rev. Lett.*, $\underline{12}$, 159 (1964).
10. P. W. Anderson, *Progress in Low Temperature Physics*, Vol. 5, North Holland Publishing Co., 1967.
11. J. E. Mercereau, *Superconductivity*, Vol. 1, Marcel Dekker, 1969.
12. A. H. Dayem and C. C. Grimes, *Appl. Phys. Lett.*, $\underline{9}$, 47 (1966).
13. C. C. Grimes, P. L. Richards, and S. Shapiro, *Phys. Rev. Lett.*, $\underline{17}$, 431 (1966).
14. P. W. Anderson, *Rev. Mod. Phys.*, $\underline{38}$, 298 (1966).
15. J. E. Zimmerman and A. H. Silver, *Phys. Lett.*, $\underline{10}$, 47 (1964).
16. P. W. Anderson and A. H. Dayme, *Phys. Rev. Lett.*, $\underline{13}$, 195 (1964).
17. H. A. Notarys and J. E. Mercereau, Conference on Superconductivity, Stanford University, Stanford, Calif., to be published in *Physica*.
18. J. E. Mercereau, *J. Physique*, $\underline{5}$, 13 (1970).
19. C. C. Grimes, P. L. Richards, and S. Shapiro, *J. Appl. Phys.*, $\underline{39}$, 3905 (1968).
20. C. C. Grimes and S. Shapiro, *Phys. Rev.*, $\underline{169}$, 397 (1968).
21. D. G. McDonald *et al.* to be published.
22. W. H. Parker, B. N. Taylor, and D. N. Langenberg, *Phys. Rev. Lett.*, $\underline{18}$, 287 (1967).
23. D. N. Langenberg, D. J. Scalapino, B. N. Taylor, and R. E. Eck, *Phys. Rev. Lett.*, $\underline{15}$, 294 (1965).

24. W. H. Parker, Conference on Fluctuations in Superconductors, Asilomar, Calif. 1968.

25. R. A. Kamper, Symposium on Physics of Superconducting Devices, University of Virginia, 1967.

26. A. J. Dahm, A. Denenstein, D. N. Langenberg, W. H. Parker, D. Rogouin, and D. J. Scalapino, *Phys. Rev. Lett.*, 22, 1416 (1969).

27. D. Cohen, E. A. Edelsack, and J. E. Zimmerman, *Appl. Phys. Lett.*, 16, 278 (1970).

28. B. N. Taylor, W. H. Parker, and D. N. Langenberg, *The Fundamental Constants and Quantum Electrodynamics*, Academic Press, New York, 1969.

Chapter 6

SOLID STATE LIGHT SOURCES: LASERS

W. V. Smith

IBM Research Laboratory
8803 *Ruschlikon, ZH*
Switzerland

6.1 CHARACTERISTIC PROPERTIES OF LASERS AND COHERENT RADIATION

Maiman's historical achievement in 1960 of a pulsed ruby laser[1,2] extended the techniques of generating coherent electromagnetic radiation from the millimeter to the visible region of the spectrum in a single giant step. During the rest of the decade the gap between these frequencies has been filled by a multitude of sources, first with fixed frequencies and more recently with tunable frequencies. The upper end of the frequency range has also been extended well into the ultraviolet, with speculation of further possible extensions into the X-ray domain. The first laser was a solid state device, as are many of the other important types of lasers, whose properties we shall develop in this chapter. It is a consequence of the fact that the devices are physically large compared to a wavelength that the coherence properties of the radiation emphasize spatial, as well as temporal, features. The achievement of ultrashort picosecond (10^{-12} sec), ultraintense terawatt (10^{12} W) pulses of radiation by time compression techniques is also a manifestation of the coherence of the light. The dramatic rate of increase in available power output from lasers—a factor of 100 every 3 years, both in pulsed and continuous output—is summarized in Fig. 1.

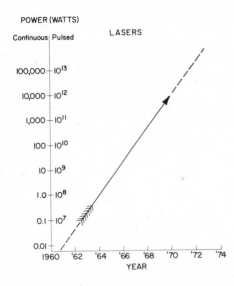

Fig. 1. Maximum laser output power by year of achievement.

6.1.1 Stimulated Emission and Amplification

Most lasers are oscillators, combining resonant feedback circuits (mirrors) with an amplifying medium. The novel and crucial feature of the device, however, is the process by which amplification is achieved, *stimulated emission of radiation*. This is simply the inverse of the familiar process of *absorption* of resonance radiation. Consider an assemblage of atomic or atomlike oscillators such as impurity ions in a solid state lattice (Cr^{3+} ions in Al_2O_3; i.e., ruby). These atoms can exist in a number of different possible energy levels, illustrated in Fig. 2. External forces acting on the atoms cause them to change from one energy level to another with the absorption or emission of quanta, $h\nu$, of energy. The energy-conserving relation is

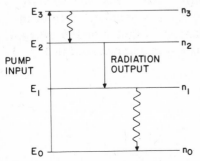

Fig. 2. Atomic energy levels, populations and pumping cycle. Pump input raises atoms from E_0 to E_3, whence they decay to E_2, with population n_2. If $n_2 > n_1$, a stimulated radiation output results, transferring atoms from E_2 to E_1. The cycle is completed by further energy decay to E_0. Normally, alternative pumping and decay branches also exist, the details of which determine the efficiency of the process. (From Smith,[38] by permission of Artech House, Dedham, Mass.)

$$hv_{ij} = E_i - E_j \tag{1}$$

where h is Planck's constant and i and j identify the initial and final energies of the transition.

A particularly important influence causing these transitions is the electromagnetic energy flux I_ν incident on the atoms. From arguments dating back to Albert Einstein in 1917,[3] it can be shown that the net fractional absorption of energy flux per unit path, $k(\nu)$, with the convention $E_i > E_j$, is given by

$$k(\nu) = -\frac{dI_\nu/dz}{I_\nu} = -\frac{\lambda^2 Nf(\nu - \nu_{ij})}{8\pi n^2 t_s \Delta\nu} \tag{2}$$

where $N = N_i - N_j(g_i/g_j)$. N_i and N_j are the populations, and g_i and g_j are the statistical weights of the levels concerned, with a ratio usually near unity. Also, t_s is the time constant for the spontaneous emission of radiation, $f(\nu - \nu_{ij})$ a function expressing the spectral line shape of the transition involved, n the refractive index of light of the wavelength $\lambda = c/\nu$, and $\Delta\nu$ the full width at half intensity of the spectral transition. For a monochromatic incident energy flux matching the spectral transition, $\nu = \nu_{ij}$, $f(0)$ has the approximate value of unity.

Written as Eq. 2, it is of course transparently obvious that, since the absorption depends on the *difference* in populations or *inversion* N, any process that makes this factor positive will make the absorption negative, resulting in a net gain or *amplification* of the input radiation:

$$g(\nu) = -k(\nu)$$

$$I_\nu(z) = I_\nu(0)\exp[g(\nu)z] \tag{3}$$

(Note that, in thermal equilibrium, N is negative; hence the term *inversion* for positive N.)

The principle of the LASER (Light Amplification by Stimulated Emission of Radiation) is to achieve a population inversion of an excited state over a lower lying state in a system of atomic energy levels by pumping energy into the system in some selective fashion favoring populating state i. The selective energy input may either be by optically exciting the atoms from the ground state E_0 directly to E_i or to a higher state whence they decay to E_i (Section 6.2), by injecting minority charge carriers (electrons or holes) across a p-n junction in a semiconductor (Section 6.3), or by any of a number of other mechanisms. Lasing action has been achieved in all three states of matter: gas, liquid, and solid. This chapter will concentrate on solid state lasers, the original variety achieved and a continuing practically important one.

We have written Eqs. 2 and 3 in the form most suited to describe the laser. In their original 1917 formulation the practical implications were less obvious. It took the World War II association of physicists and engineers in developing microwave radar to orient the work of a few imaginative investigators to recognize and achieve the amplification possible, first in the microwave MASER (1954)[4] and then the laser (1960).[1,2] For these achievements three scientists—C. H. Townes, in the United States; and N. G. Basov and A. M. Prokorov, in the Soviet Union—achieved the Nobel Prize in Physics in 1964.

The numerical values of the various parameters of Eq. 2 vary widely among different practical and impractical laser materials, as we shall discuss in subsequent sections. As an example of a particularly easily energized laser, the ion neodymium (Nd^{3+}) in the host material yttrium aluminum garnet (YAG) requires an inversion of only 1.1×10^{16} cm^{-3} to achieve a gain of 1%/cm at its lasing wavelength of 1.065 μm.[5]

The minimum theoretical rate of energy input that can sustain an inversion, with its stored energy $W_s = N(E_i - E_j)$, is

$$P_s = \frac{\Gamma W_s}{t_s} = \Gamma N_i \frac{h\nu_{ij}}{t_s} \approx \Gamma N \frac{h\nu_{ij}}{t_s} \tag{4}$$

where the right-hand form of Eq. 4 assumes $N_j(g_i/g_j) \ll N_i \cdot P_s$ is the power per cm^3 radiated spontaneously, multiplied by a factor Γ representing the added energy above $h\nu_{ij}$ required to raise an atom to the appropriate excited state. For Nd^{3+}, $\Gamma_{min} = 1.2$ and $t_s = 550$ μs[5], so that in principle the required 1.1×10^{16} cm^{-3} inversion could be maintained continuously with an input power of only 1.4 W/cm^3. In practice the required input power is substantially above this but still a practically attainable figure.

6.1.2 Oscillation and Coherence of Radiation

A material with an inverted population has frequency selective characteristics in its gain, with a maximum at ν_{ij} and a band pass of approximately $\Delta\nu$, as expressed by $f(\nu - \nu_{ij})$ of Eq. 2; see also Sections 6.2.1 and 6.3.2. It also has geometrically selective characteristics from the shape of the material, i.e., a rod has one preferred long direction of maximum total gain. Consequently, the optical noise or amplified spontaneous emission from the rod also has substantial coherence, peaking spectrally at ν_{ij}, with a band pass less than $\Delta\nu$ and a directional maximum along the axis of the rod. Indeed, in some lasers this amplified spontaneous emission is sufficiently coherent to

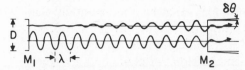

Fig. 3.   Laser Fabry-Perot. Interferometer reflection at mirrors and growth of waves in amplifying medium is shown, as well as diffraction by aperture. From Smith,[38] by permission of Artech House, Dedham, Mass.)

make a useful laser source.* Usually, however, additional frequency selective and directional elements are employed in the form of mirrors at each end of the laser, one of which is made partially transmitting to emit the beam. The feedback from the mirrors through the amplifier also makes the device resemble more conventional lower frequency oscillators and, indeed, the mirror combination can be viewed as a multiresonant circuit. In the simple case of a laser rod directly terminated by plane mirrors, the fundamental longitudinal resonances of this circuit are given by

$$\frac{q\lambda}{n} = 2l \tag{5}$$

where l is the length of the rod of refractive index n and q is an integer.

The constructive phase additions of successive reflections at the mirrors are illustrated in Fig. 3, which also shows the increased directivity of the output beam. For the ideal case of purely longitudinal modes of oscillation that are coherent across the entire cross section of the laser, the beam directivity is determined essentially by diffraction at the mirror disks that terminate the rod. The light is emitted in a pattern of concentric cones with intensity nodes between the cones. Over 80% of the light is emitted in the central cone and about 50% of the radiation is emitted within a solid angle

$$\delta\Omega = k\frac{\lambda^2}{D^2} \tag{6}$$

where D is the rod diameter, $k \approx 1$, and the effect of the refractive index has been ignored. Actually, none of these postulates is completely realistic for solid state lasers. Inhomogeneities or anisotropic properties of the material excite other modes of oscillation or restrict the oscillation to filamentary paths, tending to increase k and $\delta\Omega$. These

---

*This is the case, for instance, with the ultraviolet N_2 laser.[6]

effects will be discussed in Section 6.2.2, and the planar rather than cylindrical geometry appropriate to injection lasers in Section 6.3.2. For some purposes laser rods are incorporated into external interferometer cavities, with various combinations of plane or spherical mirrors and limiting apertures to define other mode configurations. These and other variants of the fundamental Fabry-Perot interferometer of Fig. 2 will be discussed as needed.

Spatial Coherence; *Radiance*. The *brightness* or *radiance* B of a light source is measured in watts per cm^2 per steradian. Combining this definition with Eq. 6, and approximating the mirror area A by D^2, we find that

$$B = \frac{P}{A\delta\Omega} \approx \frac{P}{k\lambda^2} \tag{7}$$

Thus the brightness of lasers of comparable power increases quadratically with the frequency. The process of energy conversion in the laser tremendously increases the brightness of any pumping source; however, after leaving the laser, the light can be transformed in divergence and directionality by conventional lenses, light pipes, etc., which preserve the brightness of the beam. Figure 4 illustrates a "black box" concept of the conversion of input pumping energy to a coherent directional output focused to a spot by an appropriate lens-mirror combination. Figure 5 shows the brightness preserving tradeoffs between collimated light beams and focusing to a spot. A diffraction limited ($k = 1$) 10-W continuous wave NdYAG laser (commercially available) has a radiance of 10^9 W cm^{-2} sr^{-1}, which may be compared with a radiance of 500 for a 10-W, high-pressure mercury arc; or 2300 for the sun. The millionfold advantage of the laser explains its usefulness in any application emphasizing directivity or point focusing. Surveying, optical radar, and materials processing such as spot welding or drilling holes in refractory materials are obvious and presently practical applications. Multikilovolt electron beams have brightnesses comparable to lasers; hence electron beams are competitive for some materials processing applications.

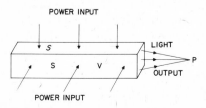

POWER INPUT

LIGHT OUTPUT

POWER INPUT

Fig. 4. Black box concept of a laser. Power incident through one or more of the surfaces S bounding a volume V is partially converted throughout the volume into light, all of which can be focused through the vicinity of a point P. (From Smith,[38] by permission of Artech House, Dedham, Mass.)

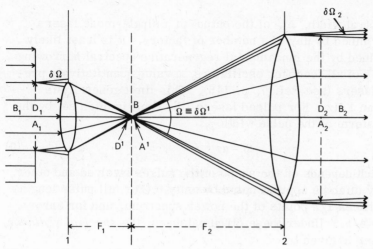

Fig. 5. Lens transformation preserving the brightness in a light beam, and including focusing to a spot. (From Smith,[38] by permission of Artech House, Dedham, Mass.)

Time Coherence; Spectral Radiance, Time Compression. The spacing between the longitudinal modes of Eq. 5 is often much less than the line width of the lasing spectral transition. For instance, the line width of a Nd glass laser is about 200 cm^{-1}, that of NdYAG[5] 5.2 cm^{-1}, whereas the longitudinal mode spacing,

$$\delta\nu = \frac{c}{2ln} \qquad (8)$$

for $n = 1.84$ and $l = 5$ cm, is 1.6×10^9 Hz or 0.054 cm^{-1}.

The initial emphasis in the evolution of lasers was to achieve single-mode operation in order to attain maximum monochromaticity, or time coherence, of the laser radiation. This remains a desirable and often attainable objective for continuous wave and long pulse operation of lasers. It is an objective more easily attained in gas lasers than in solid state ones, both because the initial line widths are narrower and because refractive index variations and other thermal effects are smaller. Hence the maximum time coherence achievable with solid state lasers is attained by oscillator-amplifier combinations, where the oscillator is a gas laser and the amplifier a solid state one, whose band pass overlaps the oscillation frequency of the gas (Section 6.2.4). More practical, perhaps, is the combination of NdYAG oscillator and Nd glass amplifier for narrowing the spectral output of high-power glass lasers.

The spectral width, Δf, of the output of a single-mode laser may be determined by any of a number of factors. It is least likely to be determined by the fundamental regenerative spectral narrowing characteristic of all feedback oscillators, a value quantitatively predictable for lasers (see Ref. 7, p. 115). This limit generally is much less than 1 Hz. For pulsed lasers, the limit is often set by the Fourier transform of the pulse width τ,

$$\Delta f = a\tau^{-1} \tag{9}$$

where a, which depends on the shape of the pulses involved and other criteria, may often be approximated by unity. (For full pulse lengths measured between $1/e$ points of the power spectrum, and for error functions, $a = 4/\pi$.) Under these circumstances, the *spectral radiance* $B/\Delta f$ of a laser is given by

$$\frac{B}{\Delta f} = \frac{P\tau}{ak\lambda^2} = \frac{E}{ak\lambda^2} \tag{10}$$

where E is the energy in the laser pulse. Thus the spectral radiance, under the above assumptions, is *independent of the laser pulse length for constant energy in the pulse*. The Q-switching and mode-locking techniques discussed in Section 6.2.3 result in nanosecond or picosecond pulses, respectively, for ruby and Nd lasers. Using Eq. 9, the corresponding band widths are 10^9 and 10^{12} Hz. The first figure is comparable to the mode spacing of typical lasers and hence may still be viewed as single-mode operation. The second figure encompasses many modes, however, which must have appropriate phases with respect to each other. Hence picosecond pulse lasers are also described as "mode locked." The techniques of achieving such short pulses also involve a *time compression* of the input pumping radiation; therefore, time compression may be considered as another manifestation of laser coherence.

An energy output of 50 J in a pulse width of 5 psec has recently been reported from a Nd glass oscillator-amplifier combination.[8,9] This represents a peak power in the pulse of 10 terawatts. The spectral radiance of this source, if diffraction limited, would be 5×10^9 J cm^{-2} sr^{-1}. The same 50-J output in a 5-nsec pulse would have the same spectral radiance but "only" 10 gigawatts of peak power.

Threshold Inversion and Pumping Power. In a laser oscillator, the round trip gain given by Eq. 3 for $z = l$ must balance the nonresonant volume losses due to absorption and scattering in the material, k_{NR}, and the end losses at the mirrors which are usually determined by the reflectivities R_1 and R_2. The relation is

$$R_1 R_2 \exp(g - k_{NR})2l = 1 \tag{11}$$

or

$$g = k_{NR} - \frac{1}{2l} \ln(R_1 R_2) \approx k_{NR} + \frac{1}{2l}(1 - R_1 R_2)$$

where $g = -k(\nu)$ is given by Eq. 2 and the far right-hand approximation for g holds for highly reflecting mirrors, $R \approx 1$. As with amplifiers, the parameters entering Eq. 11 vary greatly with the type of laser. For the Nd-YAG example k_{NR} may be neglected, and Eqs. 2, 3, and 11, solved for the threshold inversion N_{th} as

$$N_{th} = \frac{4\pi n^2 t_s \Delta\nu}{l\lambda^2}[-\ln(R_1 R_2) \approx (1 - R_1 R_2)] \tag{12}$$

where $f(\nu - \nu_{ij})$ has been approximated as unity. Typical values of 5 for l and 0.1 for $(1 - R_1 R_2)$ require a 1%/cm gain from Eq. 11 and a threshold inversion of 6.5×10^{15} cm^{-3} from Eq. 12, using the already-quoted parameters of n is nearly equal to 1.84, $\Delta\nu = 5.2$ cm^{-1}, and t_s = 550 μs. This is half the more carefully calculated inversion quoted for 1%/cm gain in Section 6.1.1. For a typical rod diameter of 2 to 3 mm, resulting in a volume of 0.5 cm^3, the theoretical required threshold input power from Eq. 4 is less than 1 W. A more practical figure to consider, however, is the approximately 2% operating efficiency of these lasers, a factor that includes the inefficiency of conversion of electrical input power to pumping light, and the further inefficiency of absorption of this light by the laser.

For pulsed operation, a knowledge of t_s is relevant (see Section 6.2). However, particularly when continuous laser operation is involved, it is useful to eliminate t_s between Eqs. 12 and 4 with P_s = the threshold power, P_{th}, with the result

$$P_{th} = \frac{7.5 n^2 \Delta\nu (1 - R_1 R_2)\Gamma}{l\lambda^3} \tag{13}$$

W/cm^3 for $\Delta\nu$ in cm^{-1}, l in centimeters, and λ in microns. This expression also eliminates N_{th} (but *not* a concern for optimum levels of doping. See Section 6.2.1).

6.2 OPTICALLY PUMPED SOLID STATE LASERS

6.2.1 Materials

Efficient photoluminescence in solids—the first prerequisite of an optically pumped laser material—invariably requires the combination

of optically active impurities in a primarily optically passive host.
The passive host must be transparent to the pumping radiation, as
well as to the lasing radiation; it must also accept the desired dopant
and be growable in useful sizes with excellent optical quality. We
shall restrict the present chapter to hosts transparent to visible radiation. Laser host materials opaque to visible radiation are semiconductors and will be treated in Section 6.3.

We may view a transparent solid as a material with an energy band
structure like that of Fig. 2, but with the energy levels broadened into
bands by the interatomic forces. The lower-lying bands are full of
electrons and the upper bands empty; there is an energy gap E_g between
the top of the highest filled band and the bottom of the lowest empty
band. (Others of the bands may or may not overlap each other. See
Section 6.3.1 and Fig. 9 for energy band discussions particularly appropriate to semiconductors.) If the frequency ν in Eq. 1, with E_g replacing $E_i - E_j$, lies in the ultraviolet, the material is optically transparent and an insulator. For E_g expressed in electron volts and λ_g in
microns, the relation becomes

$$\lambda_g E_g = 1.24 \tag{14}$$

Sapphire (Al_2O_3), yttrium aluminum garnet (YAG, $Y_3Al_5O_{12}$), and glass
are typical laser host materials. These and other optical materials
actually are transparent only to radiation lying within a pass band,
where λ_g is the shortest wavelength passed and the long wavelength ir
limit is determined by the lattice vibrations (reststrahlen) of the material. For sapphire, this pass band extends from 0.14 to 6.5 μm,
with the limits defined as the points at which transmission through a
2-mm sample drops to 10%.

The most suitable optically active impurities for wide band gap
laser hosts are ions with radii comparable to those of the positive ions
in the host lattice (e.g., Y^{3+} and Al^{3+}) and with the same valency. Furthermore, their electronic transitions must be well shielded from interaction with the host lattice, which means that they are restricted to
the incomplete inner d and f electron shells of transition and rare
earth ions such as Cr^{3+} and Nd^{3+}, respectively. Host lattices with
divalent ions, CaF_2, are suitable for incorporation of divalent impurities such as Dy^{2+}; however, there is a smaller selection of divalent
ion-divalent host lattice combinations with suitable chemical and physical properties from which to choose than is the case for trivalent
ions. Somewhat similar considerations hold for trivalent actinide
series ions such as U^{3+}.

The impurity ion also must have a suitable energy level structure.

In particular, for the low threshold it is desirable that the terminal
laser state lie high enough above the ground state so that it is not
appreciably populated at the operating (usually room) temperature.
This limit is set by the Boltzmann relation

$$g_j n_j = g_0 n_0 \frac{\exp(E_j - E_0)}{kT} \tag{15}$$

where k is Boltzmann's constant and kT for room temperature is 0.026
V or 210 cm^{-1}. For $g_0 = g_j$ and $n_j/n_0 = 10^{-2}$, $E_j - E_0 = 970$ cm^{-1}.

There are other subtler properties of the impurity ions and the
host lattice that are important. The value of the spontaneous emission
time constant t_s (Eq. 2) determines the minimum doping level N_0 required to make available the requisite inversion N for the desired gain,
since $N < N_0$. The inner shell rare earth and transition ions are examples of electric dipole forbidden transitions for which t_s typically is
about 1 msec; whereas, for semiconductor lasers, t_s corresponds to
an allowed electric dipole transition and is about 1 nsec. For pulsed
operation, energy can be stored in an inverted population for a time
up to approximately t_s, and then emitted in a single burst of radiation.
Hence appreciable energy storage is possible only in long time constant optically pumped lasers, not in semiconductor lasers.

For a given gain, the minimum inversion and doping level also is
proportional to the line width $\Delta \nu$ (Eq. 2). Consequently, so are the
threshold pump input (Eq. 13) and the stored energy W_{th}. Furthermore, the optical transitions from the ground state to higher levels,
from which the initial laser state is populated, must have certain
properties that differ somewhat from those of the lasing transitions.
The former line widths preferably should be broad to accommodate
the broad bandwidth generally associated with convenient efficient
pumping radiation, and the absorption coefficients should be such
that the radiation is efficiently but fairly uniformly absorbed throughout the bulk of the material. The trade-offs involved include the incorporation of sensitizer ions to enhance the absorption of pump energy and to transfer the energy to the laser ions, and the deleterious
effects associated with trying to incorporate particularly high doping
levels (see Ref. 7, Chaps. 3 and 6). Finally, we must note that, while
the general principles of using wide-band gap host lattices incorporating shielded inner shell impurity ions of matching radii and valence,
minimize host-ion interactions, they do not eliminate them. The energy level structure of the free ion is modified, reduced somewhat in
energy scale, with a splitting of the energy levels by these interactions
which depends on the symmetry of the host lattice and the electronic
state of the ion (see Ref. 7, Chap 6). Furthermore, the lattice vibra-

tions of the host interact with the ion and tend to induce nonradiative transitions. This is not entirely a negative feature for, although such transitions from the initial laser state decrease laser efficiency, they are also responsible for populating the initial state from higher energy levels populated by the pumping radiation, and for depopulating the terminal laser state. Hence a nice balance must be achieved between the phonon spectrum of the host lattice and the energy level structure of the active ion. Optimally, the ion energy levels should be fairly closely spaced, say at 1000 cm^{-1} intervals or less, except for the desired laser transition (10, 000 to 20, 000 cm^{-1}), so that nonradiative transitions involving only a few acoustical phonons (of energies ~ 300 cm^{-1}) can take place between the other levels and establish paths for population and depopulation of the laser states (see Ref. 7, Chap. 6; Ref. 10).

The above considerations have resulted in a distillation from over 50 ion-lattice combinations[11] of a relatively few useful ones. Table 1 lists these (essentially rows 1, 2, and 4), plus a few other representative combinations that emphasize the superiority of the first few, particularly for room temperature operation. The combination Nd^{3+} in $CaWO_4$ illustrates the point that equal valency of impurity and displaced lattice ion (Ca^{2+} in this case) are not absolutely necessary, although the lower local symmetry generally caused by charge compensation usually results in a more complicated spectrum, as evident here. Similarly, Nd^{3+} in glass has several transitions that, moreover, are broadened by the local variations in the crystal field. The uniquely large and favorable energy separation of the Nd^{3+} terminal state from the ground state is also evident. The lower separation of the other trivalent rare earth ions partially accounts for their poorer performance, and the divalent rare earths are seen to be under even more of a handicap in this regard. On the other hand, the continuing usefulness of the ruby laser, with its identity of terminal and ground state, illustrates that the four-level operation of Fig. 2 is not an absolute necessity; when the ground state is also the terminal state, inversion can be obtained by transferring over half the ground state population to the initial laser state. The broad pumping bands of ruby, combined with its narrow fluorescence width, help outweigh the disadvantage of three-level operation. Competition between radiative and nonradiative transitions (fluorescence efficiency) is not shown in Table 1. It is this factor, as much as the low-lying terminal state, that prohibits high-temperature operation of the divalent rare earth lasers. In contrast, the fluorescence efficiency of Nd^{3+} in YAG is 99. 5% even at room temperature.[29]

The evident superiority of ruby, Nd in YAG, and Nd in glass over the other ion-lattice combinations of Table 1 warrant our concentration on these three combinations for the balance of this section. As noted earlier, sensitizer ions are sometimes added to the material to increase the absorption of pump energy, with subsequent transfer of this pump energy to the lasing ion. For example, a glass rod codoped with Yb^{3+} and Er^{3+}, and clad with a glass sheath codoped with Nd^{3+} and Yb^{3+}, transfers a substantial fraction of the energy absorbed in the pumping bands of Nd^{3+} and Yb^{3+} to Er^{3+} which lases at 1.543 or 1.536 μm (room temperature, and in a different host lattice than Table 1), although less efficiently than a simple Nd glass laser.[30] There is military interest in developing lasers in the atmospheric window around 1.5 μm to minimize eye damage, as the eye is opaque to this radiation.

6.2.2 Continuous Wave and Long Pulse Operation

Although all three of the useful solid state lasers have been operated continuously, by far the best operation has been achieved with the Nd-YAG laser that combines the three favorable factors of four-level operation, narrow laser line width, and good thermal conductivity of the host lattice. Increasingly, specific laser types are being optimized for their favorable properties, and it seems unlikely that any other solid state laser will exceed Nd-YAG for continuous wave performance in the reasonably foreseeable future.* Furthermore, many uses of laser radiation do not depend very critically on the operating wavelength; when other wavelengths are required, they can be generated by nonlinear optical techniques such as frequency doubling from efficient primary radiation sources (Chapter 7). Frequency doubling extends the usefulness of the Nd-YAG laser to the peak of the visible at 0.532 μm. For the above reasons we shall describe the continuous wave operation of this laser only. The 1971 advertising literature describes Nd-YAG lasers with overall efficiencies over 2% and up to 200-W continuous output in multimode operation or 10 W in single-transverse-mode, diffraction-limited operation, achieved by inserting a 1-mm aperture in an external mirror configuration. A similar structure used with a ruby laser is shown in Fig. 6. A substantial contribution to the improved performance characteristics of these lasers lies in the development of krypton arc lamp

---

*However, Yalo, yttrium orthoaluminate, is also a promising host lattice,[31] and in the form of clad fibers, glass lasers have attractive features.

TABLE 1. OPTICALLY PUMPED SOLID STATE LASER MATERIALS

Ion	Host Material	Wavelength, μm	Fluorescence Linewidth, cm⁻¹	Fluorescence Lifetime t_s, msec	Energy of Terminal Level, cm⁻¹	Absorption Bands, μm	Temperature, °K	Mode of Operation	Threshold	References
Cr^{3+}	Al_2O_3	0.6943	10	3	0	0.32–0.42 0.5–0.6	300	P CW	>1000 J 840 W	12
Nd^{3+}	$Y_3Al_5O_{12}$	1.0648	5.2	0.55	2001	0.57–0.6	300	P CW	1 J 350 W	5, 13
Nd^{3+}	$CaWO_4$	1.0652 1.3392	15 —		~2000 ~4000	0.57–0.6	295	P P	3 J 3.6 J	14, 15
Nd^{3+}	Glass K–Ba–Si	1.06	200	0.56	1950		300	P	50 J	16
Nd^{3+}	Glass (Barium crown)	1.06		0.4	~2000		300	CW	900 W	17
Nd^{3+}	Glass (La–Ba–Tb–B)	1.37			4070		300	P	460 J	18
Nd^{3+}	Glass (Na–Ca–Si)	0.92			470		77	P	700 J	19

TABLE I. (Continued)

Ion	Host Material	Wavelength, μm	Fluorescence Line-width, cm⁻¹	Fluorescence Lifetime t_s, msec	Energy of Terminal Level, cm⁻¹	Absorption Bands, μm	Temperature, °K	Mode of Operation	Threshold	References
Eu^{3+}	Y_2O_3	0.6113	10.6 at 77°K 24 at 300°K	0.87	859	0.2–0.28 0.46–0.47 0.52–0.54	220	P	85 J	20
Ho^{3+}	$Y_3Al_5O_{12}$	2.0975 2.0914 2.1223			462 — 518		77	P	44 J 1760 J 410 J	21
Er^{3+}	$Y_3Al_5O_{12}$	1.6602 1.6452			525 525		77	P	80 J 470 J	21
Yb^{3+}	$Y_3Al_5O_{12}$	1.0296	~10	~1	600	0.9–1	77	P	325 J	21
Dy^{2+}	CaF_2	2.36	<0.1	12	28.8	0.8–1	77 7	P CW	20 J 600 W	22–25
Sm^{2+}	CaF_2	0.7085	~1	0.002	263	0.4–0.45 0.58–0.68	20	P	<0.1 J	26, 27
Sm^{2+}	SrF_2	0.6967	~1	14	263		20	P	~4 J	28

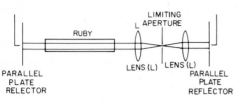

Fig. 6. Optical system in a diffraction-limited ruby laser.
(From Skinner and Geusic,[87] by
permission of Columbia University
Press, New York).

pumping sources with a relatively good match to the Nd pumping bands.
As noted in Section 6.1, this 10-W diffraction-limited, 1.06-μm laser
source has a radiance of 10^9 cm^{-2} sr^{-1}. It is practically competitive
with continuous wave gas laser sources of comparable radiance: 2.5-
W argon lasers (0.5 μm) and 1000-watt CO_2 lasers (10.6 μm). The
comparable radiances of these lasers at widely different power levels
arise from the wavelength dependence of Eq. 7. There are, of course,
some differences in applications associated with the different spectral
regions and with the different power levels involved.

For long (millisecond) pulsed operation, the lasers to compare
are ruby and Nd glass. Both are commercially available in types with
outputs ranging from 1 to 500 J. On this time scale, the output power
of the laser roughly follows that of the pumping pulse but is severely
and irregularly modulated by the more rapid pulsations (spiking) described in the next section. Heating of the laser during the pulse
causes a shift of operating wavelength, in discrete steps, as the peak
in the fluorescence gain moves over successive resonance modes of
the cavity, which are also shifting as a result of variations in the
refractive index and length of the laser rod (see Ref. 7, Chaps. 3
and 6). For these and other reasons, such as the importance of high
radiance for nonlinear optical applications, and the interest in short-
time-scale phenomena, there is more interest in nsec and psec pulsed
lasers than in long pulse lasers. However, we should note the record
of over 5000 J of output in a 3-msec pulse obtained from a Nd glass
laser.[30, 32] The 3% Nd doped, 30-mm diameter by 1-m-long laser rod
was clad with a 38-mm-diameter glass sheath of a composition that
absorbed the laser wavelength. This cladding also had the unusual
property of a higher refractive index than the core, removing off-
axis laser light and confining the output to a 10-milliradian cone
angle.

6.2.3 Nanosecond and Picosecond Pulse Operation

The really interesting type of pulsed laser operation compresses
the input pumping pulse into one or more output pulses, each orders

of magnitude shorter than the input pulse. The pulses so generated fall into two major categories: nsec Q-switched and psec mode-locked giant pulses.

Nanosecond Q Spoiled Lasers. An intense laser pulse in the nsec time domain is obtained by suddenly changing the reflectivity of a laser cavity mirror from a low value to a high value after the pumping input pulse has been on for a time preferably of the order of magnitude of the spontaneous emission time of the material. By the proper adjustment of the pumping pulse, the inverted population of the laser at that time will have reached a high quasiequilibrium value N_{LR}, just below the threshold for laser oscillation with this low reflectivity. This high N_{LR} will also provide a high gain g_{LR}. When a high reflectivity is now suddenly established, the inversion will greatly exceed the new threshold N_{HR}; because of the high gain, the radiation density W of a standing wave pattern in the laser cavity will build up rapidly, with an initial exponential rate,

$$\frac{dW}{dt} = \frac{W_{th}}{t_s} \exp\left(\frac{t}{t'}\right)$$

$$(t')^{-1} = g_{LR} \frac{c}{n}$$

$$W_{th} = N_{LR} h\nu \tag{16}$$

where W_{th} represents the stored energy density in the inverted population at the initial high threshold, and W_{th}/t_s the spontaneous emission rate of conversion of this energy to radiation. Although Eq. 16 is only valid for $W \ll W_{th}$, its integration to $W = W_{th}$ yields a lower limit t_L to the emitted radiation pulse:

$$t_L = t' \ln\left(1 + \frac{t_s}{t'}\right) \tag{17}$$

For a Nd glass laser, Eq. 12, using $\ln(1/R_1 R_2)$, the low reflectivity case, and using the parameters of row 4, Table 1, yields for $R_1 = 1$, $R_2 = 0.1$, $l = 5$ cm, $V = 0.5$ cm^3

$$N_{LR} = 5.3 \times 10^{19} \text{ cm}^{-3}$$

$$W_{th} = 10 \text{ J/cm}^3 \qquad\qquad W_{th} V = 5 \text{ J}$$

$$g_{LR} = 2.3 \text{ cm}^{-1} \qquad\qquad t' = 2.7 \times 10^{-11} \text{ sec} \tag{18}$$

$$t_L = 0.5 \text{ nsec} \qquad\qquad P = W_{th} V/t_L = 10 \text{ gigawatts}$$

More accurate calculations (see Ref. 7, Chap. 4) give times of several nanoseconds, in essential agreement with observations.

This important type of giant pulse laser operation was proposed by Hellwarth[33] and demonstrated by McClung and Hellwarth,[34] using ruby as the laser. A variety of methods has been proposed and used for varying the reflectivity of the laser mirror. Among them are (see Ref. 7, Chap. 4) an electrooptic Kerr cell, a rotating prism, and a bleachable dye inserted between the laser and its mirror. The dye is opaque to low levels of radiation at the laser frequency but transparent to intense radiation which equalizes the populations of the ground and excited states of the dye transition. Hence, from Eq. 2, the dye becomes transparent, allowing a giant laser pulse to emerge. Of these control methods the first can be accurately timed for a single pulse and the second for a sequence of pulses, but the timing of the third method is determined by an interaction between the dye, the laser, and the pumping pulse intensity. Figure 7 is a schematic diagram comparison of rotating prism and bleachable dye Q-switch lasers.

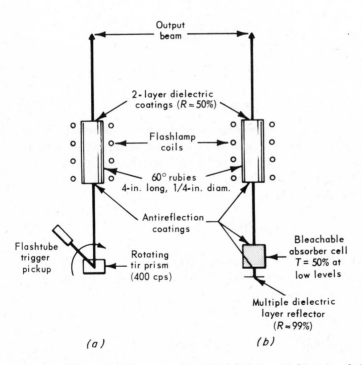

Fig. 7. Schematic diagram of methods of Q-switching involving (a) an externally controlled shutter, and (b) a bleachable absorber. (From Smith and Sorokin,[7] by permission of McGraw-Hill, New York.)

Picosecond Mode-Locked Lasers. Equation 16 applies to the relatively slowly and smoothly varying envelope of a standing wave pattern in the laser cavity. The simple one-dimensional structure of a
Fabry-Perot cavity (Fig. 2) allows decomposition of the standing
waves into two traveling waves, one to the right and one to the left.
An alternative picture of the development of a laser pulse proceeds
from a consideration of one of these traveling waves. Suppose, con-
ceptually, that it were possible to rotate the prism in Fig. 7 at a speed
such that the period of light modulation corresponded to the round
trip period of light through the laser to the top mirror and back. The
laser output would then develop as a series of pulses spaced at a
period,

$$T = \frac{2(ln + l')}{c} = \frac{2L}{c} \tag{19}$$

where l is the length of the amplifying section of the refractive index
n, l' the distance between the laser and the rotating prism, and L
$= ln + l'$ the effective length of the cavity. Equation 19 is a generalized
reciprocal of Eq. 8. The pulse length τ of the individual pulses
would be a small fraction of T, and a spectrum analysis of the resulting wave train would reveal it to be a carrier with amplitude-modu-
lated sidebands.

A more practical, actually achieved, device is to use a fixed mirror and insert an electrooptic or acoustooptic modulating device in the
cavity with the same modulation period T. The result is that a number
m of modes spaced in frequency by the reciprocal of Eq. 19 lock together to form a spectrum[35,36] (in the approximation $n = 1$)

$$E(x, t) = \frac{1}{2} \sum_{a=-m/2}^{+m/2} \cos\left[(q+a)\frac{\pi}{L}(x-ct) - \cos(q+a)\frac{\pi}{L}(x+ct)\right] \tag{20}$$

where the modes are assumed to have equal amplitudes. This analysis predicts a pulse width, to the first zero in the oscillation envelope
of the m locked modes, of

$$\tau = \frac{T}{m} \tag{21}$$

with the peak field amplitude equal to m times each individual mode
amplitude and the peak power $P_p m$ times the average power P_{av}.

$$P_p = mP_{av}. \tag{22}$$

The question of what determines the number m is a current subject of intense research. The most obvious assumption is to assume

that the number of modes that can oscillate is limited by the band-
width $\Delta\nu$ of the laser transition, i.e., Δf (Eq. 9) $= \Delta\nu$, in which case,

$$\tau \approx \Delta\nu^{-1} \tag{23}$$

This criterion predicts, for instance, $\tau \approx 0.2$ psec for the Nd glass
laser. As a rule, pulses one or two orders of magnitude longer than
this are observed, the increase being caused by the dispersion in the
glass. These pulses, which are frequency chirped, can be time compressed to the expected limiting value by introducing a compensating
dispersive delay line external to the laser.[37]

Equation 20 is essentially a consequence of the rate equation analysis of laser phenomena. This approach becomes invalid for time constants shorter than those of Eq. 23, but it does not follow that Eq. 23
is therefore the physical limit for short pulses. Rather, further sharp-
ening, if possible, can only be analyzed by an approach that considers
interactions among the individual oscillating atoms coupled by the radiation field. For a qualitative discussion of this approach, with references to the original literature, see Smith.[38]

Mode-locking can be achieved by a passive bleachable organic dye
in the configuration of Fig. 7b, as well as by externally imposed modulation, provided that the time constant of the bleachable dye can follow
the round trip period of the light pulse. In this case, the traveling
pulse can be viewed as building up from noise peaks at the laser frequency which bleach the dye, hence are less attenuated than valleys in
the noise, both being linearly amplified during their passage through
laser. (See Ref. 38 for a more complete discussion.)

While ruby and Nd glass lasers are operated in mode-locked pulse
trains developed during millisecond pumping pulses, Nd-YAG lasers
can be operated with a continuous pumping source, and a resulting
continuous train of picosecond pulses. The nanosecond giant pulse
and picosecond mode-locked pulse characteristics of these several
lasers are summarized in Table 2 which also includes some oscillator amplifier combinations.

6.2.4 Amplifiers

Higher and more coherent output powers are available from oscil-
lator-power amplifier combinations than from amplifiers alone. The
applicability of this general principle to lasers is particularly clear.
In pulsed lasers more energy can be stored in a chain of amplifier
sections separated by isolator sections than in a single oscillator, even
using Q-switching techniques. Furthermore, the stored energy in this

TABLE 2. CHARACTERISTICS OF REPRESENTATIVE OPTICALLY PUMPED SOLID STATE LASERS

Laser Type	Pulse Length, sec	Output Pulse Energy, J	Output Peak Power, W	Radiance W/cm² sr	Comments	Reference
CW Nd YAG Single Mode			10	10^9		42
Multi-Mode			200			43[c]
Q-Switch Nd YAG	3×10^{-8}	9×10^{-4}	3×10^4		CW pumped av. power 0.4 W, rep. rate 200 sec^{-1}	44
Mode-Locked Nd YAG	3×10^{-11}	3×10^{-9}	100		CW pumped av. power 1.2 W, rep. rate (2.6 nsec)$^{-1}$	45
Long Pulse Nd Glass	3×10^{-3}	5000	1.7×10^6	2.3×10^9	3% efficiency above threshold	30, 32
Long Pulse Ruby	10^{-3}	400	4×10^5			d
Q-Switched[a] Nd Glass	2.2×10^{-8}	90	4×10^9	2×10^{17}	$R/\Delta f = 3 \times 10^5$ ($\Delta f = 20$ Å)[b]	39[e]
Q-Switched Ruby	2×10^{-8}	4×10^{-2}	2×10^6	4.2×10^{13}	Single transverse mode	46
Mode-Locked Nd Glass	5×10^{-12}	50	10^{13}			8, 9
Mode-Locked Ruby	3×10^{-12}	1.5×10^{-2}	5×10^9			47

[a]These are oscillator-amplifier chains.

[b]Some line structure was observed in the output, and the measured coherence length of 20 mm greatly exceeded the value of 0.5 mm one would deduce from the line width.

[c]100 nsec Q-switching at kilohertz rates, with kilowatt peak powers, are also achieved.

[d]Typical commercial literature.

[e]These results are for the combination of a pinhole mode selected oscillator, followed by two amplifier sections. No optical isolation was used between the sections.

chain can be dumped by the pulse of energy from a single-mode, diffraction-limited oscillator, in principle amplifying this mode without
distortion. In practice, of course, imperfections in the amplifier do
introduce distortions that are particularly serious at high powers,
where thermal effects cause inhomogeneities in the refractive index.
Finally, the highest spectral radiance can also be achieved with oscillator amplifier combinations. For most purposes, the combination of
a Nd-YAG oscillator with a Nd glass amplifier gives adequate spectral
purity. When greater monochromaticity is desired, however, an ap-
propriately chosen gas laser can be used as the starting point.[40, 41]

Stored Energy and Isolation. One limit on the stored energy WV
in an amplifier section is given by the threshold inversion (Eq. 12)
multiplied by $h\nu lA$, where A is the area of the laser rod.

$$W_{th}V = \frac{-7.5An^2\,\Delta\nu t_s}{\lambda^3}\ln(R_1R_2) \tag{24}$$

joules for $\Delta\nu$ in cm^{-1} and λ in μm. This expression is independent of
the length of the rod, i.e., one can in principle achieve the same total
amplification in a short rod as in a long rod by increasing the inversion of the short one; it is the total amplification that is balanced by
the length-independent losses of the mirrors. In practice maximum
practical doping levels, efforts to minimize thermal distortion, etc.,
dictate unit amplifier lengths of about a meter for high brightness Nd
glass lasers but allow shorter lengths for ruby lasers.

While very low reflectivities can be attained by suitable optical
coatings, they are susceptible to damage. Also, the logarithmic form
of Eq. 24 and the importance of the neglected background absorption
k_{NR} for low reflectivities in Eq. 11 make reducing reflectivity below,
say, 1%, a relatively unprofitable way to increase the stored energy.
Our frequently used parameters for Nd glass, together with a 1% reflectivity on each end and a 3-cm diameter, result in a stored energy of
67 J per amplifier section. Adding the appropriate wavelength, linewidth, and t_s terms from Table 1, we see that the corresponding values
for ruby and Nd-YAG are, respectively, 65 and 3.3 J. (The favorable
wavelength and t_s factors for ruby almost exactly balance the unfavor-
able wavelength factor.) However, it is more difficult to obtain large
diameter, optically homogeneous ruby boules than glass rods, and the
disadvantage of the low thermal conductivity of glass can be compensated for by slicing the glass into thin disks, so that more practical
comparisons of the systems would decrease the comparative areas of
ruby and YAG lasers to Nd glass by a factor of 10, reducing the stored
energy to 6.5 J in ruby and 0.33 J in YAG.

There is another limit to the stored energy per amplifier section than that of Eq. 24. The spontaneous emission $(\pi D^2/4t_s)W_s dz$ from a segment dz of the rod is amplified in its passage through the rod in the same manner that an input signal is amplified. To a first crude approximation, only that fraction $(D/2l)^2$ (the solid angle subtended by the far end of the laser) of the spontaneous emission is subject to this amplification. The integration of Eq. 3, multiplied by this geometrical factor, for the distance l corresponding to one pass through the amplifier, yields an energy flux P_{AVAL}

$$P_{AVAL} = \frac{\pi D^4 W_s}{16 l^2 t_s g}[\exp(gl) - 1] \tag{25}$$

through the output mirror of the laser. If this flux exceeds the isotropic spontaneous emission rate $W_s V/t_s$, it obviously distorts the previously derived threshold and stored energy relations. This inequality can be put in the form

$$\frac{\exp(gl) - 1}{gl} > \frac{16 l^2}{D^2} \tag{26}$$

as a condition to be avoided if effects of photon avalanches are to be minimized. For an aspect ratio of $l/D = 30$, the critical value of gl is about 12, which would just begin to become important in the example of Eq. 18, a 5-cm-long Nd glass rod with a threshold inversion of 5.3×10^{19} cm^{-3} and a gain of 2.3 cm^{-1} if the aspect ratio were indeed 30 instead of the actual value of 14. In the example of Eq. 24, $N = 5.1 \times 10^{17}$ cm^{-3} and $g = 0.022$ cm^{-1}, with $gl = 2.3$, well below the critical value. For a 10-times higher doping level (and gain), however, amplified spontaneous emission would become of primary importance.

Isolation between amplifier sections is achieved either by use of a Faraday rotator or a nonlinear absorber (see Ref. 7, Chap. 4). In the high radiance Q-switched laser of Table 2, however, no isolation elements were used. That is, the reflectivity of the final amplifier termination was kept low enough (below 0.5%) to prevent oscillations with the 17% output reflector of the oscillator.[39] The losses through the system also include the effects of mode-limiting apertures and mode-matching telescopes. This loss was 8.5 dB and the small signal gain 30 dB. In practice, still lower reflectivity, 0.1%, was required to prevent damage to the oscillator by the amplified return signal.

The highest spectral purity, solid state lasers are achieved by combining a gas laser with a solid state laser amplifier chain. The spectral purity of such lasers is also measured by the coherence length

$$l_{ch} = \frac{c}{\Delta f} \qquad (27)$$

at which interference fringes can be seen through a two-beam Michelson interferometer. One such amplifier chain consisting of a 1.0621-μm He-Ne laser followed by an isolator, a single-mode fiber laser amplifier, plus further laser amplifiers (all of composition with a fluorescence maximum at 1.062 μm) had a coherence length in excess of 12 m ($\Delta f = 25$ MHz) with an output power of 3.5 W in 1 msec.[40, 41]

6.3 SEMICONDUCTOR LASERS AND ELECTROLUMINESCENT DIODES

6.3.1 *P-n* Junction Electroluminescence

The success of low-voltage semiconductor technology has provided an impetus for seeking to extend this technology to the efficient generation of light in solids. At present both fields have achieved their greatest success with single-crystal materials. Hence this review will be restricted to single-crystal *p-n* junction electroluminescence, both for its intrinsic interest and as a background for understanding semiconductor lasers, where the demands of optical quality make single-crystal material even more necessary.

Junction electroluminescence inherently makes two demands on materials. As semiconductors, they must be capable of being doped both n type (mobile negative electrons) and p type (mobile positive electron holes), and they must provide dominant recombination paths for these carriers that are radiative, rather than nonradiative. It is desirable but not absolutely necessary that all these properties are combined in a single material. However, it is also possible for one material to be grown epitaxially on another, with which it has well-matched crystal properties. If one material is n type and the other p type, the interface is a heterojunction which is another critical element that must have favorable electrical and optical properties. To date, relatively poor electroluminescence has been achieved with heterojunctions.

The requirement of electrical conductivity has major consequences in determining the types of optical transitions that are useful in semiconductor lasers. There must be an interaction between the charge carriers and the luminescent centers. Consequently, the rare earth and transition metal ions with their shielded f shell and d shell inner electron transitions that are so important in optically pumped lasers

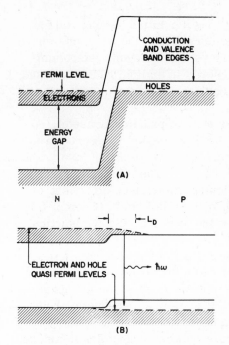

Fig. 8. Diagram of energy versus distance for a p-n junction. In the unbiased case (a) the Fermi level is continuous across the junction. When a forward bias is applied (b), reducing the barrier to the flow of electrons and holes, electrons are injected into the p side and holes into the n side, resulting in a population inversion between those levels whose separation in energy is less than the separation of the quasi-Fermi levels for electron and holes. (From W. P. Dumke,[86] by permission of Marcel Dekker, New York.)

have proven unsuccessful as luminescent centers in semiconductors. One class of successful luminescence centers are those impurities that simultaneously play a role as donors or acceptors for the conduction mechanism of the semiconductor. Usually, these are ions whose normal valency differs by unity from the corresponding ions in the host lattice. Divalent zinc and tellurium replacing trivalent gallium and arsenic, respectively, in GaAs are examples. In the equilibrium $Te^{3-} = Te^{2-} + e^-$, the electron is loosely bound, since Te is normally divalent. Hence Te^{3-} is a donor that furnishes a mobile electron to the conduction band. Similarly, Zn^{3+} is an acceptor that furnishes a mobile hole to the valence band.

The junction between a heavily doped n region of a semiconductor and a heavily doped p region is shown in Fig. 8. The horizontal Fermi level E_F in Fig. 8a continuous across the junction, marks the midpoint in the energy distribution function for electrons, half of which lie above E_F and half below E_F. The heavy doping places this demarcation line in the conduction band of the n region of the semiconductor and in the valence band of the p region.*

*For a more detailed description of injection lasers, see Ref. 7, Chap. 7.

Figure 8*b* shows the effect of applying a positive biasing voltage *V* to the semiconductor. Electrons are drawn into the *p* region, where they recombine with the majority carriers. In efficient electroluminescent materials the recombination energy is emitted radiatively and, as seen from the figure, has a value comparable to the energy gap of the semiconductor.

In Section 6.2 we noted that the number of practically useful, optically pumped solid state lasers was limited; indeed, we concentrated on three materials, ruby, Nd glass, and Nd YAG, emitting laser radiation at two wavelengths, 0.6943 and 1.06 μm. The situation in efficient electroluminescent materials is developing in the same direction, except that the choices among competing materials is not yet completely clear. Table 3 lists some characteristics of representative junction electroluminescent materials.

Column 3 of Table 3 describes the nature of the band gap in the materials. The energies of the electrons in the conduction band and the holes in the valence band are different functions of the momenta of the mobile charges in each band. If the minimum conduction band energy and the maximum valence band energy occur at the same momenta, the band gap, i.e., the difference between conduction minimum and valence maximum, is termed direct, since hole and electron can recombine directly with conservation of momentum. For indirect band gaps the two energy extrema occur at different momenta, so that the recombination of electron and hole require the mediation of a third party such as a lattice phonon to conserve momentum. Figure 9 illustrates these two classes of semiconductors.

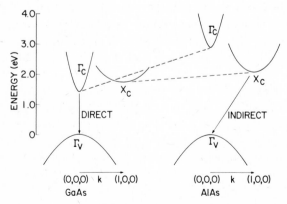

Fig. 9. Schematic representation of the band structure of GaAs and AlAs. The lowest conduction band symmetry of AlAs is assumed. (Reprinted by permission from Proceedings of the Conference on Materials and Characterization, held at Chania, Crete.)

TABLE 3. SOME PROPERTIES OF REPRESENTATIVE JUNCTION ELECTROLUMINESCENT MATERIALS[a]

Material	E_G, eV	d, i[b]	Dopants don.	Dopants acc.	Emission Peak eV	Emission Peak Å	η_{ext}	Probable Recombination Mechanism or Centers
GaN[49]	3.39	d	—	—	—	—	—	—
SiC (6H)	3.02	i	N	Al	2.40	5200	10^{-5}	Al, N pairs
SiC (6H)	3.02	i	N	B	2.12	5850	10^{-3}	B, N pairs
GaP	2.26	i	Te, S, O	Zn	1.8	7000	7×10^{-2}	ZnO assoc. nearest neighbor pairs
GaP	2.26	i	Te, S	Zn	2.2	5700	10^{-3}	$e^- (S, Te) - p^+$
GaP	2.26	i	Te, S (N)	Zn	2.22	5600	6×10^{-3}	Exciton bound to N
GaAs$_{1-x}$P$_x$[64]	1.4–1.9	d	Te	Zn	1.4–1.9	8900–6500	$10^{-2} - 10^{-3}$	$e^- - p^+ (Zn)$
Ga$_{1-x}$Al$_x$As[64]	1.4–1.9	d	Te	Zn	1.4–1.9	8900–6500	$10^{-2} - 10^{-3}$	$e^- - p^+ (Zn)$
GaAs[64]	1.4	d	Sn, Te	Zn	1.4	8900	10^{-1}	$e^- - p^+ (Zn)$
GaAs[65]	1.4	d	Si	Si	1.34	9300	3×10^{-1}	$e^- - p^+ (Si)$
InAs[c]	0.41	d						$e^- - p^+$
PbSe[c]	0.15	d						$e^- - p^+$

[a]Much of the material of this table is taken from Lorenz.[48]
[b]d=direct band gap; i=indirect band gap.
[c]Properties at cryogenic temperatures. See Ref. 7, Chap. 7.

Examination of Table 3 reveals the interesting trend that, with the exception of the newly investigated material GaN,[49] those materials which can be doped both p and n type have direct band gaps for small values of the gap (less than 1.5 to 2 eV), whereas the higher-band-gap materials have indirect gaps. Table 3 omits the broad class of efficient II-VI phosphors such as ZnS because no satisfactory way has been found to dope them both p and n type and, to date, heterojunction techniques employing one p and one n type member of this class of materials have not generated light efficiently.

The external quantum efficiency, η_E, column 8, reveals an interesting maximum at GaAs, the largest extensively studied, direct band gap material in the table. (Here η_E is the ratio of the number of photons emitted by an electroluminescent diode to the number of electrons flowing across the p-n junction of the diode per unit time.) In low-current diodes with typical resistances of a small fraction of an ohm, almost the entire voltage drop occurs across the junction and is approximately equal to the band gap, so that the overall power efficiency is Γ times the quantum efficiency, with Γ defined as in Eq. 4. That is, $\Gamma = 1.24/\lambda_f E_g$ for E_g in eV and the fluorescence maximum λ_f in micrometers. At high currents the ir drop must be taken into account.

The maximum in η_E combines a maximum in η_i, the internal quantum efficiency of generating a photon within the material, which also peaks at GaAs, with an increasing internal absorption of the generated photons by free carriers as one goes to smaller energy gaps and longer wavelengths. The maximum in η_i occurs because the direct nature of the optical transitions makes them more "allowed" by optical selection rules, and hence with a shorter t_s (Eq. 4) than for indirect band gap materials. For GaAs, with doping levels around 10^{18}/cc, t_s is about 1 nsec (see Ref. 7, Chap. 7) compared to 1 msec for the "forbidden" inner shell transitions of the optically pumped lasers of Section 6.2. This short radiative lifetime is necessary for radiative transitions to compete favorably with nonradiative ones in these materials. Nonradiative transitions may be expected to become more competitive for lower energy gap materials, where a cascade of a small number of optical phonons can bridge the energy gap.

In GaAs the nature of the optical transition at high doping levels is generally viewed as one between an injected free electron and a bound hole, i.e., the acceptor also plays the role of a localized recombination center. In the lower band gap materials recombination processes between free electrons and holes predominate.

In the indirect, large band gap materials, the spectroscopic properties of the donor and acceptor centers become more important. Recombination often proceeds in two steps. The material is generally compensated, i.e., the entire crystal is first doped n type (as, also, is the case for GaAs), and then an overriding p doping is diffused into one side of the material. A particularly efficient radiative recombination mechanism in GaP is for a hole and an injected electron to be trapped, i.e., to form a spatially localized exciton on a nearest neighbor ZnO impurity pair replacing a GaP pair.[50, 51] The exciton recombines radiatively, essentially substituting the optical transition probabilities and selection rules of the impurities for those of the host lattice. This process leads to the efficient red electroluminescence in GaP. The green electroluminescence result from excitonic recombination at a neutral N impurity.*[52]

For weakly bound impurities, host lattice and impurity transition probabilities often differ very little (the wave function of the impurity can be described as a sum of contributions of wave functions of energy bands of the host lattice, with the closest energy band predominating unless impossible by symmetry conditions). For deeper impurities, however, other selection rules may apply and, in particular, the localized nature of the sites, both in space and energy, reestablish a direct character to the transition. For weakly bound impurities in indirect band gap materials such as the exciton or electron hole pair bound to nitrogen in GaP, the source of the green emission, the luminescent transition is accompanied or assisted by a phonon, as is the case for indirect band to band recombination.[48]

The p-n junction electroluminescence is directly useful in such applications as indicator lights or alpha numerical displays. In such cases a practical measure of importance is the spectral response that peaks in the green, thus rendering the relatively inefficient green emission competitive with the more efficient red electroluminescence.[48] The economics of these applications also demand inexpensive fabrication technology. Reliability and long life are traditional solid state objectives which an increasing body of data indicate are being achieved by electroluminescent diodes.

6.3.2 P-n Junction Lasers

GaAs Lasers. The step from efficient p-n junction electroluminescence to a p-n junction laser was historically a swift one, with the

---

*N is isoelectronic with P and hence, in substitution for P, is neutral on a covalent bond picture. Alternatively, both N and P are trivalent on the ionic picture we have used in describing the equilibria of trapped and free donors and acceptors.

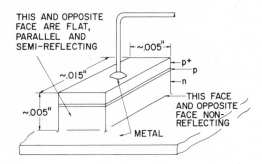

Fig. 10. Typical injection laser structure. The nonreflecting sides suppress all but the Fabry-Perot modes. (From W. P. Dumke[86] by permission of Marcel Dekker, New York.)

first reports of efficient electroluminescence in GaAs in 1962, followed by the achievement of GaAs lasers in the same year (see Ref. 7, Chap. 7; Refs. 53–55). Laser action was initially observed at low temperatures, and in pulsed operation; however, improvements in materials preparation, in understanding of the physics involved, and in engineering, have extended useful pulsed performance to room temperature and continuous oscillation has been achieved at cryogenic temperatures.*

Several key considerations have emerged. The most important of these is the requirement that the lasing transition be a *direct* one, since otherwise the nonresonant loss k_{NR} of the free carriers overwhelms the gain of the lasing transition.[57] This theoretical consideration has been experimentally verified by the observation that, in GaAsP[58] and GaAlAs[59] alloys, where the band gap changes from direct to indirect with increasing P and Al concentrations, lasing can be achieved over most of the direct band gap range of the alloys but not in the indirect range.

Other important conclusions are drawn from a more detailed examination of conditions near the p-n junction. The lasing action takes place in the plane of the junction, on the p doped side, from the population inversion established by the injection of electrons across the junction. Usually, the resonant cavity is formed in the long dimension of the diode by the refractive index GaAs-air discontinuities at cleaved 110 boundaries of the GaAs (see Fig. 10). In the gain-loss balance of Eq. 11, both the background absorption k_{NR} (free carrier absorption) and the reflection loss are large but are balanced by a large gain. An even more important loss mechanism, however, can be the diffraction

*Continuous oscillation has also recently been achieved in a four-layer heterostructure laser in which the width of the active region is less than 1 μ .[56]

loss caused by the very narrow lateral dimension—about 2 μ—to which the population inversion is confined by the spatially inhomogeneous recombination and diffusion properties of the injected electrons and the doping profile of the material. Fortunately, the refractive index discontinuities or gradations at the boundaries of the inverted region tend to trap the radiation in the junction of the plane. A gradual realization and analysis of this feature has led to deliberate incorporation of refractive index inhomogeneities, by an epitaxial GaAs-GaAlAs boundary grown about 2μ from the junction on the p side.[60-63] It is believed that this boundary, in addition to internally reflecting the light,[63] reflects the injected electrons and provides a more uniform and denser inverted population.[61] Further designed discontinuities in the n side of the junction result in further improvement.[56]

The above efforts to engineer the dimensions of the inversion region are only the latest stage in a progression of improvements that have demonstrated the importance of maximizing doping concentration consistent with good crystal perfection, and capable of achieving planar dislocation free junctions. The high doping moves the Fermi levels into the n and p conduction bands and increases the gain in the inversion region (while, unfortunately, simultaneously increasing the free carrier loss). Planar, dislocation free junctions improve the optical characteristics of the cavity. The cumulative effect of these improvements has been to lower the room temperature thresholds of GaAs injection lasers from an initially reported 100,000 A/cm^2 to less than 2,000 A/cm^2,[66] and to increase the temperature of continuous wave operation from 4 °K to over 300 °K.[66]

As evident from Fig. 10, the optical cavity characteristics of an injection laser are rather unique, consisting of a rectangular slab perhaps $2 \times 10 \times 40$ μm, or $2 \times 100 \times 400$ μm. The radiation pattern from the 2 μm by 10 or 100 μm cross section typically consists of a cone of full angular spread to half power points about 20 to 30 degrees perpendicular to the junction plane and 4 to 10 degrees parallel to this plane, approximately corresponding to the diffraction pattern of the junction in the narrow dimension. However, as with other lasers, lasing often occurs in filamentary modes that do not fill the entire junction, thus accounting for the relatively large 4 to 10-degree spread parallel to the junction plane. Generally, the spectrum consists of several oscillating frequencies near the peak of the inversion gain. These characteristics are not ideal for applications that make maximum demands upon the coherence properties of the laser. However, rather simple lenses can provide a collimation of the beam to 1 or 2 degrees in each dimension. For more specialized applications an external resonant cavity can select a single oscillating geometrical mode[67]; this in combination with a grating can also select a single frequency.[68]

Injection lasers are particularly easy to modulate, at low voltages and high speeds. This is a consequence of their small dimensions and short recombination times. Their compactness, high efficiency, and light weight are also attractive properties. Their short recombination times limits their time compression features. However, pulses in the 30 to 100 psec range[69, 70] have been achieved by inverting the population of only a fraction of the length of a laser (by controlling the location of the current flow). This is essentially a form of bleachable dye Q-spoiling, the uninverted portion of the laser playing the role of the bleachable dye. Speculations have been advanced that such pulses can be controlled in a fashion useful for performing computer logic operations; such operations have been achieved but their practicality is dubious. More important, periodic current modulation at gigacycle or near gigacycle frequencies of conventionally constructed lasers results in a train of pulses at the modulation frequency, each about 200 psec long, which can be modulated in forms useful for communications applications.[69, 71-73] The lasers can be operated either in a mode-locked region analogous, but not identical, to the mode-locked operation of optically pumped solid state lasers[72] or, probably more usefully, away from this region, thus making them more accessible to external control.[73] Current modulation pulls the modulation frequency and can be used both in frequency control feedback circuits and pulse-time modulation circuits.[72, 73]

We shall not discuss the range of semiconductor materials preparation that technology has investigated and used in injection lasers. These preparations include crystal pulling, vapor growth, diffusion, and liquid phase epitaxy which are well covered in several review articles.[66, 74] Since the devices are small and utilize high-current densities, heat sinking techniques are also important.[66] For some applications it is useful to fabricate arrays of these lasers.[75] Some characteristics of individual lasers and laser arrays are given in Table 4. Note that collimation techniques for laser arrays are more complicated than for individual lasers.

Other Injection Lasers. All of the direct band gap materials listed in Table 3 have been made to lase, and several others as well (InSb, InP, GaSb, PbTe, and alloys of some of these and some materials of Table 3). The short wavelength limit of these materials is about 0.64 μm in $GaAs_{0.59}P_{0.41}$ at 77 °K, and the long wavelength limit 28 μm in $Pb_{0.73}Sn_{0.27}Te$ lasers at 12 °K.[76] This is a wider wavelength range than has been achieved with optically pumped solid state lasers and, further-

TABLE 4.   CHARACTERISTICS OF GaAs LASERS AND LASER ARRAYS

Mode of Operation	Threshold[a,b] Current Density, A/cm^2	Operating[b] Current Density, A/cm^2	Pulse Length, μsec	Efficiency, %	Peak Output, W	Modulation Frequency, kHz	Beam Spread with Typical Lenses, degrees
CW, 77 °K single laser, λ = 0.84 μm	1,500	3,000	—	20	1	1	1–2
Pulsed 77 °K single laser, λ = 0.84 μm	1,500	5,000	1	40	10–100	1–3	1–2
Pulsed, room temperature single laser, λ = 0.9 μm	10,000	30,000	0.2	10	10–50	1	1–2
Pulsed, arrays 77 °K, room temperature	—	—	0.2–1	—	10^3–10^4	1–3	2–6

[a]For a typical laser length of 400μm. The threshold current density decreases with increasing length, since the reflectivity losses become less important. For example, the threshold at 77° extrapolated to infinite laser length is about 300 A/cm^2.

[b]These illustrations are chosen as typical for high power applications. For lower powers double heterostructure lasers have lower thresholds and lower possible operating currents. See text.

more, has been extended into the ultraviolet by electron beam pumping
(see next subsection). The band gaps, particularly for the long wavelength materials, are quite pressure-sensitive, and substantial pressure tuning of some of these materials has been achieved: from 7. 5 to
22. 5 μm in PbSe.[77,78] The wavelengths and, particularly, the efficiencies of some of these lasers, InSb and InAs, are also magnetic-field-
sensitive.

As already noted in Section 6. 3. 1, the electroluminescent efficiency of these materials peaks at GaAs, and so does their performance
as lasers. For minor amounts of alloying into GaAs, however, the
degradation in performance is also minor, so that a range of wavelengths between 0. 8 and 0. 9 μm are available, particularly from
GaAlAs alloys, with laser characteristics approaching those of GaAs.
In night vision applications, the rapid improvement of image intensifier receivers as the incident wavelength is decreased renders these
alloys of potential importance. For the most part, however, the
types of applications that make use of laser coherence can be satisfied
as well by infrared radiation at the GaAs emission as by another choice
of wavelength. Most potential applications of visible electroluminescence—to displays, for instance—seem better satisfied by nonlasing
electroluminescent diodes than by lasing ones, largely because the
former are so much less expensive and are capable of fabrication into
more complicated arrays.

6. 3. 3 Other Types of Excitation; Electron Beam Pumped Lasers

We have described the use of a heterostructure to reflect light
and electrons injected across a p-n junction, and have mentioned that
the p-n junction itself could also be a heterojunction between easily
p doped material and easily n doped material. Other variants of junction lasers have also been proposed and tried but none has proved as
successful as the simple ones already described. An alternative to injection of minority carriers is the creation of electron hole pairs by
ionization from energetic mobile charges. In an all solid state structure, this process is achieved by avalanche multiplication of carriers
in regions of high-electric field, and has been achieved at high-resistivity interfaces between Mn-doped GaAs sandwiched between two Zn
diffused layers.[79] However, the device is much less efficient, and
with a higher threshold, than a standard GaAs laser. At present, the
most interesting extension of semiconductor lasers appears to be to
pumping these by high-voltage electron beams, thus generating electron hole pairs in the short depth region penetrated by the beams. This
form of excitation was suggested relatively early in the history of the

development of semiconductor lasers and has indeed proved to be an efficient type of excitation. Efficiencies of the order of 25 to 30% have been achieved at liquid N_2 temperatures in both CdS[80] and GaAs[81] electron beam pumped lasers. Less efficient operation in the ultraviolet (0. 33 μm) has also been achieved, with ZnS, [82] and less efficient operation at higher thresholds is also possible at room temperatures.

The work of Bogdankevich *et al.*[81] with GaAs affords a good comparison with injection lasers of the same material. The samples, typically rectangular parallelepipeds 0. 2×0. 5×2. 5 mm in dimensions, are bombarded perpendicularly to the large cross section and lase in the long dimension. Optimum doping is $1-2\times10^{18}$ cm^{-3} Te at 85 °K and $2-4\times10^{18}$ cm^{-3} at room temperature. The increase in threshold with temperature is less severe than with injection lasers, amounting only to a factor of 3 to 5. However, the power threshold in watts per square centimeter at room temperature is about a factor of 10 greater than that for injection lasers. (The localized power input near the junction of the injection laser is greater, however.)

Basov, who heads Soviet research in semiconductor, as well as high-power picosecond, lasers, has also investigated a "mirror" geometry for laser action in the direction of the electron beam,[83, 84] i.e., perpendicular to the platelet. His geometry, shown in Fig. 11, employs a 0. 1-mm-thick GaAs platelet pressed against a 5-mm-long sapphire resonant cavity, terminated by an 80% reflecting metal film mirror, with the beam area defined by a 0. 48-mm-diameter hole in a mask in front of the GaAs platelet. A comparable geometry without the sapphire cavity, with a 0. 21-mm hole, gave 900 W output power at a 25% efficiency for a 190-kV beam voltage. The threshold at liquid N_2 temperature was 13 A/cm^2 and the beam divergence 5 to 7 degrees. With the cavity the threshold was reduced to 3 A/cm^2 and the beam divergence to 45 min. Note the higher voltage, compared to that in Table 5, required for mirror excitation. This is because the beam population must be inverted to a greater depth to achieve lasing action perpendicular to the surface.

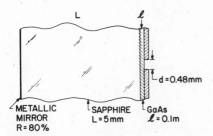

Fig. 11. Electron beam pumped semiconductor laser with sapphire cavity.

TABLE 5. CHARACTERISTICS OF SOME ELECTRON BEAM PUMPED LASERS

Laser Material	Beam Voltage, kV	Threshold Current Density, A/cm^2	Temperature, °K	Overall Power Efficiency	Pulse Length, µsec	Peak Power Output, W	Repetition Rate, kHz	Wavelength, µm
CdS[80]	60	0.05	110	26	0.5	350	3	0.49
GaAs[81]	45	0.6–1 3	85 300	30 11	0.1 0.1	400 200	1 1	0.84
ZnS[82]	40	0.4	77	2		0.6		0.33

Interest in electron beam pumped lasers stems not only from the greater wavelength range encompassed but also from device concepts associated with the possibility of scanning the excitation region with an electron beam (laser beam deflection). Even at this early stage of development, there are some reasonably optimistic forecasts of the feasibility of achieving hundreds of kilowatts of power by electron beam pumping without catastrophic destruction of the material by the heat dissipation.[85] Little is known about degradation problems, however.

REFERENCES

1. T. H. Maiman, *Nature*, 187, 493 (1960).
2. T. H. Maiman, *Brit. Commun. Electron.*, 7, 674 (1960).
3. A. Einstein, *Phys. Z.*, 18, 121 (1917).
4. J. P. Gordon, H. J. Zeiger, and C. H. Townes, *Phys. Rev.*, 95, 282 (1954).
5. T. Kushida, H. M. Marcos, and J. E. Geusic, *Phys. Rev.*, 167, 289 (1968).
6. D. A. Leonard, *Appl. Phys. Lett.*, 7, 4 (1965).
7. W. V. Smith and P. P. Sorokin, *The Laser* McGraw-Hill, New York, 1966.
8. G. Gobeli, *Electronic News*, 14, 72 (Mar. 17, 1969).
9. G. Gobeli, *Laser Focus*, 5, 24 (April 1969).
10. L. A. Riseberg and H. W. Moos, *Phys. Rev. Lett.*, 19, 1423 (1967).
11. B. Di Bartolo, *Optical Interactions in Solids*, John Wiley & Sons, New York, 1968, Appendix 1.
12. T. H. Maiman, *Phys. Rev.*, 123, 1151 (1961).
13. J. E. Geusic, H. M. Marcos, and L. G. Van Uitert, *Appl. Phys. Lett.*, 4, 182 (1964).
14. L. F. Johnson, *J. Appl. Phys.*, 33, 756 (1962).
15. L. F. Johnson and R. A. Thomas, *Phys. Rev.*, 131, 2038 (1963).
16. E. Snitzer, in *Quantum Electronics* 3, P. Grivet and N. Bloembergen, Eds., Columbia University Press, New York, 1964, p. 999.
17. C. G. Young, *Appl. Phys. Lett.*, 2, 151 (1963).
18. P. B. Maurer, *Appl. Optics*, 3, 153 (1964).
19. P. B. Mauer, *Appl. Optics*, 2, 87 (1963).
20. N. C. Chang, *J. Appl. Phys.*, 34, 3500 (1963).
21. L. F. Johnson, J. E. Geusic, and L. G. Van Uitert, *Appl. Phys. Lett.*, 7, 127 (1965).
22. Z. J. Kiss and R. C. Duncan, *Proc. IRE*, 50, 1531 (1962).
23. A. Yariv, *Proc. IRE*, 50, 1699 (1962).
24. L. F. Johnson, *Proc. IRE*, 50, 1691 (1962).
25. Z. J. Kiss, H. R. Lewis, and R. C. Duncan, *Appl. Phys. Lett.*, 2, 93 (1963).
26. P. P. Sorokin and M. J. Stevenson, *IBM J. Res. Develop.* 5, 56 (1961).

27. W. Kaiser, C. G. B. Garrett, and D. L. Wood, *Phys. Rev.*, <u>123</u>, 766 (1961).
28. P. P. Sorokin, M. J. Stevenson, J. R. Lankard, and G. D. Pettit, *Phys. Rev.*, <u>127</u>, 503 (1962).
29. T. Kushida and J. E. Geusic, *Phys. Rev. Lett.*, <u>21</u>, 1172 (1968).
30. C. G. Young, *Proc. IEEE*, <u>57</u>, 1267 (1969).
31. G. A. Massey, *Appl. Phys. Lett.*, <u>17</u>, 213 (1970).
32. C. G. Young, *Laser Focus*, <u>3</u>, 36 (February 1967).
33. R. W. Hellwarth, in *Advances in Quantum Electronics*, J. R. Singer, Ed., Columbia University Press, New York, 1961, p. 334.
34. F. J. McClung and R. W. Hellwarth, *Proc. IRE*, <u>51</u>, 46 (1963).
35. M. DiDomenico, Jr., *J. Appl. Phys.*, <u>33</u>, 2870 (1964).
36. A. Yariv, *J. Appl. Phys.*, <u>36</u>, 388 (1965).
37. E. B. Treacy, *Phys. Lett.*, <u>28A</u>, 34 (1968).
38. W. V. Smith, *Laser Applications*, Artech House, Dedham, Mass., 1969, Chap. 6.
39. W. F. Hagen, *J. Appl. Phys.*, <u>40</u>, 5111 (1969).
40. G. C. Holst, E. Snitzer, and R. Wallace, *J. Opt. Soc. Amer.*, <u>51</u>, 499 (1961).
41. G. C. Holst, E. Snitzer, and R. Wallace, *IEEE J. Quant. Electronics*, <u>QE-5</u>, 342 (1969).
42. *Laser Focus*, <u>5</u>, 17 (April 1969).
43. W. Koechner, *Laser Focus*, <u>5</u>, 29 (September 1969).
44. V. Evtuhov and J. E. Neeland, *IEEE J. Quant. Electronics*, <u>QE-5</u>, 207 (1969).
45. L. M. Osterink and J. D. Foster, *J. Appl. Phys.*, <u>39</u>, 4163 (1968).
46. J. E. Bjorkholm and R. H. Stolen, *J. Appl. Phys.*, <u>39</u>, 4043 (1968).
47. M. E. Mack, *IEEE J. Quant. Electronics*, <u>QE-4</u>, 1015 (1968).
48. M. R. Lorenz, *Trans. Met. Soc. AIME*, <u>245</u>, 539 (1969).
49. J. I. Pankove, H. P. Marushka, and J. E. Berkeyheiser, *Appl. Phys. Lett.*, <u>17</u>, 197 (1970).
50. T. N. Morgan, B. Welber, and R. Bhargava, *Phys. Rev.*, <u>166</u>, 751 (1968).
51. C. H. Henry, P. J. Dean, and J. D. Cuthbert, *Phys. Rev.*, <u>166</u>, 754 (1968).
52. D. G. Thomas and J. J. Hopfield, *Phys. Rev.*, <u>150</u>, 680 (1966).
53. R. N. Hall, G. E. Fenner, J. D. Kingsley, T. J. Soltys, and R. O. Carlson, *Phys. Rev. Lett.*, <u>9</u>, 366 (1962).
54. M. I. Nathan, W. P. Dumke, G. Burns, F. H. Dill, Jr., and G. J. Lasher, *Appl. Phys. Lett.*, <u>1</u>, 62 (1962).
55. T. M. Quist, R. H. Rediker, R. J. Keyes, W. A. Krag, B. Lax, A. L. McWhorter, and J. H. Zeiger, *Appl. Phys. Lett.*, <u>1</u>, 91 (1962).
56. I. Hayashi, M. B. Panish, P. W. Foy, and S. Lumski, *Appl. Phys. Lett.*, <u>17</u>, 109 (1970).
57. W. P. Dumke, *Phys. Rev.*, <u>127</u>, 1559 (1962).
58. N. G. Ainslie, M. Pilkuhn, and H. Rupprecht, *J. Appl. Phys.*, <u>35</u>, 105 (1964).
59. H. Nelson and H. Kressel, *Appl. Phys. Lett.*, <u>15</u>, 7 (1969).
60. Zh. I. Alferov, V. M. Andreev, V. I. Korol'kov, E. L. Portnoi, and D. N. Tret'yakov, *Sov. Phys. Semicond.*, <u>2</u>, 1289 (1969).

61. M. B. Panish, I. Hayashi, and S. Sumski, *IEEE J. Quant. Electronics*, QE-5, 210 (1969).

62. I. Hayashi, M. B. Panish, and P. W. Foy, *IEEE J. Quant. Electronics*, QE-5, 211 (1969).

63. H. Kressel and H. Nelson, *RCA Review*, 30, 106 (1969).

64. W. N. Carr and G. E. Pittman, *Appl. Phys. Lett.*, 3, 173 (1963).

65. H. Rupprecht, J. M. Woodall, K. Konnerth, and G. D. Pettit, *Appl. Phys. Lett.*, 9, 221 (1966).

66. I. Hayashi, M. B. Panish, and F. K. Reinhart, *J. Appl. Phys.*, 42, 1929 (1971).

67. E. M. Philipp-Rutz and H. D. Edmonds, *Appl. Optics*, 8, 1859 (1969).

68. H. D. Edmonds and A. W. Smith, *IEEE J. Quant. Electronics*, QE-6, 356 (1970).

69. N. G. Basov, *IEEE J. Quant. Electronics*, QE-4, 855 (1968).

70. T. P. Lee and R. Roldan, *IEEE J. Quant. Electronics*, QE-6, 339 (1970).

71. R. F. Broom, E. Mohn, C. Risch, and R. Salathe, *IEEE J. Quant. Electronics*, QE-6, 328 (1970).

72. J. E. Ripper and T. L. Paoli, *IEEE J. Quant. Electronics*, QE-6, 326 (1970).

73. T. L. Paoli and J. E. Ripper, *IEEE J. Quant. Electronics*, QE-6, 335 (1970).

74. M. I. Nathan, *Appl. Optics*, 5, 1514 (1966).

75. W. E. Ahearn and J. Crowe, *IEEE J. Quant. Electronics*, QE-6, 377 (1970).

76. J. F. Butler and T. C. Harman, *Appl. Phys. Lett.*, 12, 347 (1968).

77. J. M. Besson, J. F. Butler, A. R. Calawa, W. Paul, and R. H. Rediker, *Appl. Phys. Lett.*, 7, 206 (1965).

78. J. M. Besson, W. Paul, and A. R. Calawa, *Phys. Rev.*, 173, 699 (1968).

79. K. Weiser and J. F. Woods, *Appl. Phys. Lett.*, 7, 225 (1965).

80. C. E. Hurwitz, *Appl. Phys. Lett.*, 9, 420 (1966).

81. O. V. Bogdankevich, N. A. Borisov, I. V. Krynkova, and B. M. Lavrushin, *Soviet Physics-Semiconductors*, 2, 845 (1969).

82. C. E. Hurwitz, *Appl. Phys. Lett.*, 9, 116 (1966).

83. N. G. Basov, O. V. Bogdankevich, A. N. Pechenov, A. S. Nasibov, and K. P. Fedoseev, *Soviet Physics JETP*, 28, 900 (1969).

84. N. G. Basov, O. V. Bogdankevich, and A. Z. Grasyuk, *IEEE J. Quant. Electronics*, QE-2, 594 (1966).

85. C. A. Klein, *IEEE J. Quant. Electronics*, QE-4, 186 (1968).

86. W. P. Dumke, *Advances in Lasers*, Vol. 2, A. K. Levine, Ed., Marcel Dekker, New York, 1968.

87. J. G. Skinner and J. E. Geusic, in *Quantum Electronics* 3, P. Grivet and N. Bloembergen, Eds., Columbia University Press, New York, 1964, p. 1437.

Chapter 7

OPTICAL SECOND HARMONIC GENERATION AND PARAMETRIC OSCILLATION

Amnon Yariv

Electrical Engineering Department
California Institute of Technology
Pasadena, California

7.1 INTRODUCTION

The usual propagation of electromagnetic radiation in linear media involves a polarization proportional to the electric field that induces it. In this treatment we consider some of the consequences of the nonlinear dielectric properties of certain classes of crystals in which, in addition to the linear response, a field produces a polarization proportional to the square of the field.

The nonlinear response can give rise, as will be shown below, to the exchange of energy between a number of electromagnetic fields of different frequencies. Two of the most important applications of this phenomenon are (a) second harmonic generation in which part of the energy of an optical wave of frequency ω propagating through a crystal is converted to that of a wave at 2ω; and (b) parametric oscillation in which a strong pump wave at ω_3 causes the simultaneous generation in a nonlinear crystal of radiation at ω_1 and ω_2, where $\omega_3 = \omega_1 + \omega_2$. These will be treated in the following.

7.2 ON THE PHYSICAL ORIGIN OF NONLINEAR POLARIZATION

The optical polarization of dielectric crystals is caused mostly by the displacement of the outer loosely bound valence electrons under the influence of the optical field. This is a result of the fact that the heavy nuclei cannot respond to an electric field varying at optical frequencies and remain nearly stationary. Denoting the electron deviation from the equilibrium position by x, and the density of electrons by N, we find that the polarization p is given by

$$p(t) = -Ne\,x(t)$$

In centrosymmetric crystals the potential energy of an electron must reflect the crystal symmetry so that, using a one-dimensional analogue, it has the form

$$V(x) = \frac{m}{2}\omega_0^2 x^2 + \frac{m}{4}Bx^4 + \cdots \tag{1}$$

where ω_0^2 and B are constants* and m is the electron mass. Because of the symmetry, $V(x)$ contains only even powers of x so that $V(-x) = V(x)$. The restoring force on an electron is

$$F = -\frac{\partial V}{\partial x} = -m\omega_0^2 x - mBx^3 \tag{2}$$

and is zero at the equilibrium position $x = 0$.

The linear polarization of crystals in which the polarization is proportional to the electric field is accounted for by the first term in Eq. 1. To see this, consider a "slowly varying" electric field, $E(t)$ (i.e., a field whose Fourier components are at frequencies small compared to ω_0). The excursion $x(t)$ caused by this field is found by equating the total force on the electron to zero†

$$-eE(t) - n\omega_0^2 x(t) = 0$$

so that

$$x(t) = -\frac{e}{m\omega_0^2}E(t) \tag{3}$$

thus resulting in a polarization that is proportional to the field.

Now, in an asymmetric crystal in which the condition $V(x) = V(-x)$ is no longer fulfilled, the potential function contains odd powers, so that

$$V(x) = \frac{m\omega_0^2}{2}x^2 + \frac{m}{3}Dx^3 \cdots \tag{4}$$

---

*The constant ω_0 corresponds to the resonance frequency of the electronic oscillator.

†The assumption of slow variation makes it possible to neglect the acceleration term $m\,d^2x/dt^2$ in the force equation.

which corresponds to a restoring force on the electron

$$F = \frac{\partial V(x)}{\partial x} = -(m\omega_0^2 x + mDx^2 + \cdots) \tag{5}$$

An examination of Eq. 5 reveals that a positive excursion $(x > 0)$ results in a larger restoring force (assuming $D > 0$), since the two terms have the same sign, than the same excursion in the opposite direction. It follows immediately that, if the electric force on the electron is positive (i.e., $E < 0$) the induced polarization is smaller than when the field direction is reversed. This situation is depicted in Fig. 1.

Next consider an alternating electric field at an (optical) frequency ω applied to the crystal. In a linear crystal the induced polarization will be proportional, at any moment, to the field resulting in a polarization oscillating at ω, as shown in Fig. 2a. In a nonlinear crystal we can use Fig. 1b to obtain the induced polarization corresponding to a given field and then plot it (vertically) as in Fig. 2b. The result is a distorted polarization wave in which the stiffer restoring force at $x > 0$ results in positive peaks (b) that are smaller than the negative ones (b′). A Fourier analysis of the nonlinear polarization wave 2b shows that it contains the second harmonic of ω, as well as an average (d-c) term. The average, fundamental, and second harmonic components are plotted in Fig. 3.

To relate the nonlinear polarization formally to the inducing field, we use Eq. 5 for the restoring force and take the driving electric field as $E^{(\omega)}\cos\omega t$. The equation of motion of the electron $F = m\ddot{x}$ is then

$$\frac{d^2 x(t)}{dt^2} + \sigma\frac{dx(t)}{dt} + \omega_0^2 x(t) + Dx^2(t) = -\frac{eE^{(\omega)}}{2m}(e^{i\omega t} + e^{-i\omega t}) \tag{6}$$

where we account for the losses by a frictional force $-m\sigma\dot{x}$. An inspection of Eq. 6 shows that the term Dx^2 gives rise to a component oscillating at 2ω, so that we assume the solution for $x(t)$ in the form to be*

$$x(t) = \tfrac{1}{2}(q_1 e^{i\omega t} + q_2 e^{2i\omega t} + \text{c.c.}) \tag{7}$$

where c.c. stands for "complex conjugate."

Substituting the last expression into Eq. 6, we obtain

$$-\frac{\omega^2}{2}(q_1 e^{i\omega t} + \text{c.c.}) - 2\omega^2(q_2 e^{2i\omega t} + \text{c.c.}) + \frac{i\omega\sigma}{2}(q_1 e^{i\omega t} - \text{c.c.})$$

---

*Here we must use the real form of $x(t)$ instead of the complex one, since, the differential equation involves x^2.

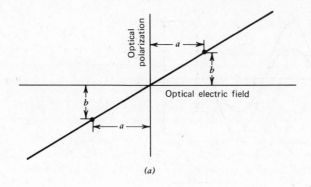

(a)

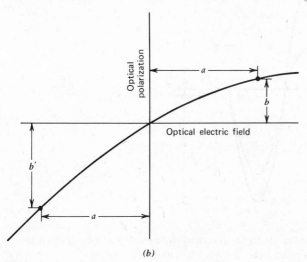

(b)

Fig. 1. The relation between the induced polarization and the electric field causing it: (a) in a linear dielectric; (b) in a crystal lacking inversion symmetry.

$$+ i\omega\sigma(q_2 e^{2i\omega t} - \text{c.c.}) + \frac{\omega_0^2}{2}(q_1 e^{i\omega t} + q_2 e^{2i\omega t} + \text{c.c.})$$

$$+ \frac{D}{4}(q_1^2 e^{2i\omega t} + q_2^2 e^{4i\omega t} + q_1 q_1^* + 2q_1 q_2 e^{3i\omega t}$$

$$+ 2q_1 q_2^* e^{-i\omega t} + q_2 q_2^* + \text{c.c.}) = \frac{-eE^{(\omega)}}{2m}(e^{i\omega t} + \text{c.c.}) \tag{8}$$

If Eq. 8 is to be valid for all times t, the coefficients of $e^{\pm i\omega t}$ and $e^{\pm 2i\omega t}$ on both sides of the equation must be equal. Equating first the coeffi-

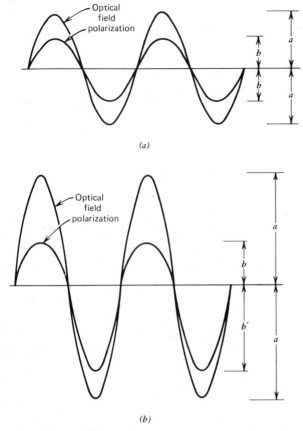

(a)

(b)

Fig. 2. An applied sinusoidal electric field and the resulting polarization in
(a) a linear crystal; (b) a crystal lacking an inversion symmetry.

cients of $e^{i\omega t}$ and assuming that $|Dq_2| \ll [(\omega_0^2 - \omega^2)^2 + \omega^2\sigma^2]^{1/2}$, we find
that

$$q_1 = -\frac{eE^{(\omega)}}{m}\frac{1}{(\omega_0^2 - \omega^2) + i\omega\sigma} \tag{9}$$

The polarization at ω is related to the electronic deviation at ω by

$$p^{(\omega)}(t) = -\frac{Ne}{2}(q_1 e^{i\omega t} + \text{c. c.})$$

$$= \frac{\epsilon_0}{2}[\chi(\omega)E^{(\omega)}e^{i\omega t} + \text{c. c.}] \tag{10}$$

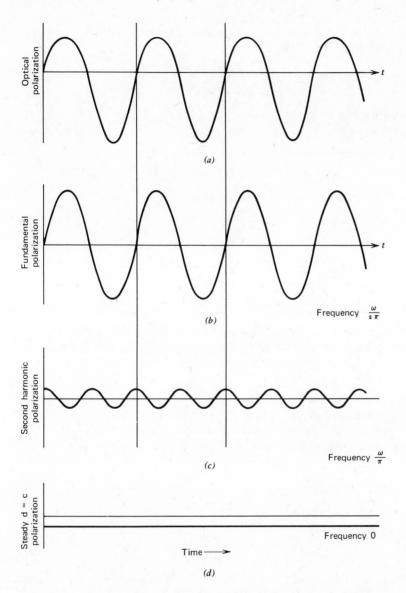

Fig. 3. Analysis of the nonlinear polarization wave (a) of Fig. 2b shows that it contains components oscillating at: (b) the same frequency (ω) as the wave inducing it; (c) twice that frequency (2ω); and (d) an average (d-c) negative component.

where $\chi(\omega)$ is thus the linear susceptibility. By using Eq. 9 in Eq. 10 and solving for $\chi(\omega)$, we obtain

$$\chi(\omega) = \frac{Ne^2}{m \epsilon_0 [(\omega_0^2 - \omega^2) + i\omega\sigma]} \tag{11}$$

We now proceed to solve for the amplitude q_2 of the electronic motion at 2ω. The equating of the coefficients of $e^{2i\omega t}$ on both sides of Eq. 8 leads to

$$q_2(-4\omega^2 + 2i\omega\sigma + \omega_0^2) = -\tfrac{1}{2} D q_1^2$$

and after substituting the solution (Eq. 9) for q_1,

$$q_2 = \frac{-De^2 [E^{(\omega)}]^2}{2m^2 [(\omega_0^2 - \omega^2) + i\omega\sigma]^2 [\omega_0^2 - 4\omega^2 + 2i\omega\sigma]} \tag{12}$$

In a manner similar to Eq. 10 the nonlinear polarization at 2ω is

$$p^{(2\omega)}(t) = -\frac{Ne}{2}(q_2 e^{2i\omega t} + \text{c. c. }) = \tfrac{1}{2}\{d^{(2\omega)}[E^{(\omega)}]^2 e^{2i\omega t} + \text{c. c.}\} \tag{13}$$

The second of Eqs. 13 defines the *nonlinear optical coefficient* $d^{(2\omega)}$. If we denote the complex amplitude of the polarization as $p^{(2\omega)}$, we have from Eq. 13,

$$p^{(2\omega)}(t) = \tfrac{1}{2}[P^{(2\omega)} e^{2i\omega t} + \text{c. c. }] \text{ and } P^{(2\omega)} = d^{(2\omega)} E^{(\omega)} E^{(\omega)} \tag{14}$$

i.e., $d^{(2\omega)}$ is the ratio of the (complex) amplitude of the polarization at 2ω to the square of the fundamental amplitude. Substituting Eq. 12 for q_2 in Eq. 13, and then solving for $d^{(2\omega)}$, we find

$$d^{(2\omega)} = \frac{-DNe^3}{2m^2 [(\omega_0^2 - \omega^2) + i\omega\sigma]^2 [\omega_0^2 - 4\omega^2 + 2i\omega\sigma)]} \tag{15}$$

Using Eq. 11, we can rewrite Eq. 15 as

$$d^{(2\omega)} = \frac{mD\chi^{(\omega)}\chi^{(2\omega)}\epsilon_0^3}{2N^2 e^3} \tag{16}$$

Equation 16 is of importance, since it relates the nonlinear optical coefficient d to the linear optical susceptibilities $\chi(\omega)$ and to the anharmonic coefficient D. Estimates based on this relation are quite successful in predicting the size of the coefficient d in a large variety of crystals.[1,2]

Equation 14 is scalar. In reality the second harmonic polarization along, say, the x direction, is related to the electric field at ω by

$$P_x^{(2\omega)} = d_{xxx}^{(2\omega)} E_x^{(\omega)} E_x^{(\omega)} + d_{xyy}^{(2\omega)} E_y^{(\omega)} E_y^{(\omega)} + d_{xzz}^{(2\omega)} E_z^{(\omega)} E_z^{(\omega)}$$

$$+ 2d_{xzy}^{(2\omega)} E_z^{(\omega)} E_y^{(\omega)} + 2d_{xzx}^{(2\omega)} E_z^{(\omega)} E_x^{(\omega)} + 2d_{xxy}^{(2\omega)} E_x^{(\omega)} E_y^{(\omega)} \tag{17}$$

Similar relations give $P_y^{(2\omega)}$ and $P_z^{(2\omega)}$. Considerations of crystal symmetry reduce the number of nonvanishing $d_{ijk}^{(2\omega)}$ coefficients and, in certain cases, to be discussed in the following, cause them to vanish altogether. Table 1 lists the nonlinear coefficients of a number of crystals.

Crystals are usually divided into two main groups, depending on whether the crystal structure remains unchanged upon inversion (i.e., replacing the coordinate $\vec{r}$ by $-\vec{r}$) or not. Crystals belonging to the first group are called centrosymmetric, while crystals of the second group are called noncentrosymmetric (see, e.g., Ref. 3). In Fig. 4 we show the crystal structure of NaCl, a centrosymmetric crystal, while an example of a crystal lacking inversion symmetry (noncentrosymmetric) is provided by crystals of the ZnS (zinc blende) class such as GaAs, CdTe, and others. The crystal structure of ZnS is shown in Fig. 5. The lack of inversion symmetry is evident in the projection of the atomic positions given by Fig. 6.

In crystals possessing an inversion symmetry, all the nonlinear optical coefficients $d_{ijk}^{(2\omega)}$ must be zero. This follows directly from the relation,

$$P_i^{(2\omega)} = \sum_{j,k=x,y,z} d_{ijk}^{(2\omega)} E_j^{(\omega)} E_k^{(\omega)} \tag{18}$$

which is a compact notation for Eq. 17. Let us reverse the direction of the electric field, so that in Eq. 18 $E_j^{(\omega)}$ becomes $-E_j^{(\omega)}$, and $E_k^{(\omega)}$

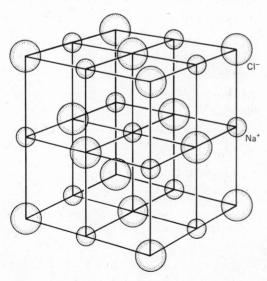

Fig. 4. The crystal structure of NaCl. The crystal is centrosymmetric, since an inversion of any ion about the central Na$^+$ ion, as an example, leaves the crystal structure unchanged.

TABLE 1. THE NONLINEAR OPTICAL COEFFICIENTS OF A NUMBER OF CRYSTALS. (THESE COEFFICIENTS ARE DEFINED BY EQ. 17 WITH $1 = x$, $2 = y$, $3 = z$.)

Crystal	$d_{ijk}^{(2\omega)}$ in Units of 9×10^{-22} (mks)
L_iIO_3	$d_{31} = d_{33} = 6 \pm 0.5$
$NH_4H_2PO_4$ (ADP)	$d_{36} = d_{312} = 0.45$ $d_{14} = d_{123} = 0.45 \pm 0.02$
KH_2PO_4 (KDP)	$d_{36} = d_{312} = 0.45 \pm 0.03$ $d_{14} = d_{123} = 0.45 \pm 0.03$
KD_2PO_4	$d_{36} = d_{312} = 0.42 \pm 0.02$ $d_{14} = d_{123} = 0.42 \pm 0.02$
KH_2ASO_4	$d_{36} = d_{312} = 0.48 \pm 0.03$ $d_{14} = d_{123} = 0.51 \pm 0.03$
Quartz	$d_{11} = d_{111} = 0.37 \pm 0.02$
$AlPO_4$	$d_{11} = d_{111} = 0.38 \pm 0.03$
ZnO	$d_{33} = d_{333} = 6.5 \pm 0.2$ $d_{31} = d_{311} = 1.95 \pm 0.2$ $d_{15} = d_{113} = 2.1 \pm 0.2$
CdS	$d_{33} = d_{333} = 28.6 \pm 2$ $d_{31} = d_{311} = 14.5 \pm 1$ $d_{15} = d_{113} = 16 \pm 3$
GaP	$d_{14} = d_{123} = 80 \pm 14$
GaAs	$d_{14} = d_{123} = 107 \pm 30$
$BaTiO_3$	$d_{33} = d_{333} = 6.4 \pm 0.5$ $d_{31} = d_{311} = 18 \pm 2$ $d_{15} = d_{113} = 17 \pm 2$
$LiNbO_3$	$d_{31} = d_{311} = 4.76 \pm 0.5$ $d_{22} = d_{222} = 2.3 \pm 1.0$
Te	$d_{11} = d_{111} = 730 \pm 230$
Se	$d_{11} = d_{111} = 130 \pm 30$
$Ba_2NaNb_5O_{15}$	$d_{33} = d_{333} = 10.4 \pm 0.7$ $d_{32} = d_{322} = 7.4 \pm 0.7$

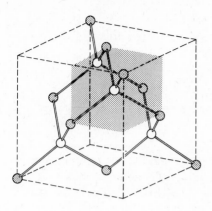

Fig. 5. The crystal structure of noncentro-symmetric cubic zinc sulfide.

becomes $-E_k^{(\omega)}$. Since the crystal is centrosymmetric, the reversed field "sees" a crystal identical to the original one, so that the polarization produced by it must bear the same relationship to the field as originally, i.e., the new polarization is $-P_i^{(2\omega)}$. Since the new polarization and the electric field causing it are still related by Eq. 18, we have

$$-P_i^{(2\omega)} = \sum_{j,k} d_{ijk}^{(2\omega)}(-E_j^{(\omega)})(-E_k^{(\omega)}) \tag{19}$$

Equations 18 and 19 can only hold simultaneously if the coefficients $d_{ijk}^{(2\omega)}$ are all zero. We may thus summarize: *In crystals possessing*

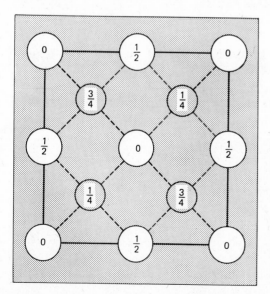

Fig. 6. The atomic positions in the unit cell of ZnS projected on a cube face. The fractions denote the height above base in units of a cube edge. The dark spheres correspond to zinc (or sulfur) atoms and are situated on a face centered cubic (fcc) lattice, while the white spheres correspond to sulfur (or zinc) atoms and are situated on another fcc lattice displaced by $(\frac{1}{4}, \frac{1}{4}, \frac{1}{4})$ from the first one. Note the lack of inversion symmetry.

an inversion symmetry, there is no second harmonic generation.

In the following work we shall ignore the vector nature of the optical nonlinearity and use the scalar form (Eq. 14).

7.3 THE FORMALISM OF WAVE PROPAGATION IN NONLINEAR MEDIA

In this section we derive the equations governing the propagation of electromagnetic waves in nonlinear media. These equations will then be used to describe second harmonic generation and parametric oscillation.

The starting point is Maxwell's equations,

$$\vec{\nabla} \times \vec{h} = \vec{i} + \frac{\partial \vec{d}}{\partial t} \tag{20a}$$

$$\vec{\nabla} \times \vec{e} = -\mu_0 \frac{\partial \vec{h}}{\partial t} \tag{20b}$$

using

$$\vec{d} = \epsilon_0 \vec{e} + \vec{p}$$

$$\vec{i} = \sigma \vec{e} \tag{21}$$

where σ is the conductivity, separating the total polarization $\vec{p}$ into its linear and nonlinear portions, according to

$$\vec{p} = \epsilon_0 \chi_e \vec{e} + \vec{p}_{NL} \tag{22}$$

Equation 20a becomes

$$\vec{\nabla} \times \vec{h} = \sigma \vec{e} + \epsilon \frac{\partial \vec{e}}{\partial t} + \frac{\partial}{\partial t} \vec{p}_{NL} \tag{23}$$

with $\epsilon = \epsilon_0 (1 + \chi_e)$. Taking the curl of both sides of Eq. 20b, and using the vector identity $\vec{\nabla} \times \vec{\nabla} \times \vec{e} = \vec{\nabla}\vec{\nabla} \cdot \vec{e} - \vec{\nabla}^2 \vec{e}$, replacing $\vec{\nabla} \times \vec{h}$ by Eq. 23, and taking $\vec{\nabla} \cdot \vec{e} = 0$, we get

$$\vec{\nabla}^2 \vec{e} = \mu_0 \sigma \frac{\partial \vec{e}}{\partial t} + \mu_0 \epsilon \frac{\partial^2 \vec{e}}{\partial t^2} + \mu_0 \frac{\partial^2}{\partial t^2} \vec{p}_{NL} \tag{24}$$

Next we go over to a scalar notation and, using Eq. 14, replace $\vec{p}_{NL}$ by de^2, so that Eq. 24 becomes

$$\nabla^2 e = \mu_0 \sigma \frac{\partial e}{\partial t} + \mu_0 \epsilon \frac{\partial^2 e}{\partial t^2} + \mu_0 d \frac{\partial^2 e^2}{\partial t^2} \tag{25}$$

where we assumed, for simplicity, that $\vec{p}_{NL}$ is parallel to $\vec{e}$. Let us limit our consideration to a field made up of three plane waves propagating in the z direction with frequencies ω_1, ω_2, and ω_3, according to

$$e^{(\omega_1)}(z, t) = \tfrac{1}{2}\left\{ E_1(z)e^{[i(\omega_1 t - k_1 z)]} + \text{c.c.} \right\}$$

$$e^{(\omega_2)}(z, t) = \tfrac{1}{2}\left\{ E_2(z)e^{[i(\omega_2 t - k_2 z)]} + \text{c.c.} \right\} \qquad (26)$$

$$e^{(\omega_3)}(z, t) = \tfrac{1}{2}\left\{ E_3(z)e^{[i(\omega_3 t - k_3 z)]} + \text{c.c.} \right\}$$

so that the total instantaneous field is

$$e = e^{(\omega_1)}(z, t) + e^{(\omega_2)}(z, t) + e^{(\omega_3)}(z, t) \qquad (27)$$

Then we substitute Eq. 27, using Eq. 26, into the wave equation Eq. 25, and separate the resulting equation into three equations, each containing only terms oscillating at one of the three frequencies. The last term in Eq. 25 will give rise to terms such as

$$\mu_0 d \frac{\partial^2}{\partial t^2} E_1 E_2 e^{\{i[(\omega_1 + \omega_2)t - (k_1 + k_2)z]\}}$$

or

$$\mu_0 d \frac{\partial^2}{\partial t^2} E_3 E_2^* e^{\{i[(\omega_3 - \omega_2)t - (k_3 - k_2)z]\}^{1/2}}$$

These oscillate at the new frequencies, $(\omega_1 + \omega_2)$ and $(\omega_3 - \omega_2)$, and will not, in general, be able to drive the oscillation at ω_1, ω_2, or ω_3. An exception to the last statement is the case when

$$\omega_3 = \omega_1 + \omega_2 \qquad (28)$$

In this case the term

$$\mu_0 d \left(\frac{\partial^2}{\partial t^2} \right) E_1 E_2 \, e^{\{i[(\omega_1 + \omega_2)t - (k_1 + k_2)z]\}}$$

oscillates at $\omega_1 + \omega_2 = \omega_3$ and can thus act as a source for the wave at ω_3. In physical terms we have power flow from the fields at ω_1 and ω_2 into that at ω_3, or vice versa. Assuming that Eq. 28 holds, we return to Eq. 25 and, writing it for the oscillation at ω_1, obtain

$$\nabla^2 e^{(\omega_1)} = \mu_0 \sigma_1 \frac{\partial e^{(\omega_1)}}{\partial t} + \mu_0 \epsilon \frac{\partial^2 e^{(\omega_1)}}{\partial t^2} + \mu_0 d \frac{\partial^2}{\partial t^2}$$

$$\times \left[\frac{E_3(z)E_2^*(z)}{4} \, e^{i[(\omega_3 - \omega_2)t - (k_3 - k_2)z]} + \text{c.c.} \right] \qquad (29)$$

Next we observe that, in view of Eq. 26,

$$\nabla^2 e^{(\omega_1)} = \frac{1}{2} \frac{\partial^2}{\partial z^2} \left\{ E_1(z)e^{[i(\omega_1 t - k_1 z)]} + \text{c.c.} \right\}$$

$$= -\frac{1}{2}\left[k_1^2 E_1(z) + 2ik_1 \frac{dE_1(z)}{dz}\right] e^{[i(\omega_1 t - k_1 z)]} + c.c.$$

where we assumed that

$$k_1 \frac{dE_1(z)}{dz} \gg \frac{d^2 E_1(z)}{dz^2} \tag{30}$$

Using Eq. 28 and the last result in Eq. 29, and taking $\partial/\partial t = i\omega_i$, we obtain

$$-\frac{1}{2}\left[k_1^2 E_1(z) + 2ik_1 \frac{dE_1(z)}{dz}\right] e^{[i(\omega_1 t - k_1 z)]} + c.c.$$

$$= [i\omega_1 \mu_0 \sigma_1 - \omega_1^2 \mu_0 \epsilon] \left\{\frac{E_1(z)}{2} e^{[i(\omega_1 t - k_1 z)]} + c.c.\right\}$$

$$- \left[\frac{\omega_1^2 \mu_0 d}{4} E_3(z)E_2^*(z)e^{\{i[\omega_1 t - (k_3 - k_2)z]\}} + c.c.\right] \tag{31}$$

Recognizing that $k_1^2 = \omega_1^2 \mu_0 \epsilon$, we can rewrite Eq. 31, after multiplying all the terms by

$$\frac{i}{k_1} \exp(-i\omega_1 t + ik_1 z)$$

as

$$\frac{dE_1}{dz} = -\frac{\sigma}{2}\sqrt{\frac{\mu_0}{\epsilon_1}} E_1 - \frac{i\omega_1}{2}\sqrt{\frac{\mu_0}{\epsilon_1}} dE_3 E_2^* \, e^{[-i(k_3 - k_2 - k_1)z]}$$

and similarly,

$$\frac{dE_2^*}{dz} = -\frac{\sigma_2}{2}\sqrt{\frac{\mu_0}{\epsilon_2}} E_2^* + \frac{i\omega_2}{2}\sqrt{\frac{\mu_0}{\epsilon_2}} dE_1 E_3^* \, e^{[-i(k_1 - k_3 + k_2)z]}$$

$$\frac{dE_3}{dz} = -\frac{\sigma_3}{2}\sqrt{\frac{\mu_0}{\epsilon_3}} E_3 - \frac{i\omega_3}{2}\sqrt{\frac{\mu_0}{\epsilon_3}} dE_1 E_2 \, e^{[-i(k_1 + k_2 - k_3)z]} \tag{32}$$

for the fields at ω_2 and ω_3. These are the basic equations describing nonlinear parametric interactions.[4] We notice that they are coupled to each other via the nonlinear constant d.

7.4 OPTICAL SECOND HARMONIC GENERATION

The first experiment in nonlinear optics[5] consisted of generating the second harmonic ($\lambda = 0.3470\ \mu$) of a ruby laser beam ($\lambda = 0.694\ \mu$) which was focused on a quartz crystal. The experimental arrangement is depicted in Fig. 7. The conversion efficiency of this first experi-

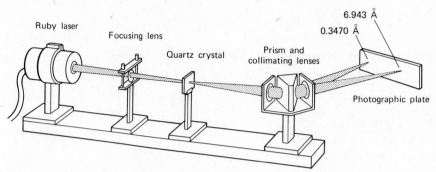

Fig. 7.   The experimental arrangement used in the first experimental demonstration of the second harmonic generation.[5]   A ruby laser beam at $\lambda_0 = 0.694\ \mu$
is focused on a quartz crystal, causing the generation of a (weak) beam at
$\lambda_0 = 0.347\ \mu$.   The two beams are then separated by a prism and detected on a
photographic plate.

ment ($\sim 10^{-8}$) was improved by methods to be described below to a point
at which approximately 30% conversion has been observed in a single
pass through a few centimeters length of a nonlinear crystal.   This
technique is finding important applications in generating short wave
radiation from longer wavelength lasers.

In the case of second harmonic generation, two of the three fields
that figure in Eq. 32 are of the same frequency.   We may thus put ω_1
$= \omega_2 = \omega$, for which case the first two equations are the complex conjugate of one another, and we need to consider only one of them.   We
take the input field at ω to correspond to E_1 in Eq. 32 and the second
harmonic field to E_3, we put $\omega_3 = \omega_1 + \omega_2 = 2\omega$ and neglect the absorption,
so that $\sigma = 0$.   The last equation becomes

$$\frac{dE^{(2\omega)}}{dz} = - i\omega \sqrt{\frac{\mu_0}{\epsilon}}\, d[E^{(\omega)}(z)]^2\, e^{i(\Delta k)z} \tag{33}$$

where

$$\Delta k = k_3 - 2k_1 = k^{(2\omega)} - 2k^{(\omega)} \tag{34}$$

To simplify the analysis further, we may assume that the depletion of
the input wave at ω, due to conversion of its power to 2ω, is negligible.
Under those conditions, which apply in the majority of the experimental
situations, we can take $E(z)^{(\omega)} = $ constant in Eq. 33 and neglect its dependence on z.   Assuming no input at 2ω, i.e., $E^{(2\omega)}(0) = 0$, we obtain
from Eq. 33 by integration the output field at the end of a crystal of
length l,

$$E^{(2\omega)}(l) = - i\omega \sqrt{\frac{\mu_0}{\epsilon}} \, d[E^{(\omega)}]^2 \frac{e^{i\Delta kl} - 1}{i\Delta k}$$

so that the output intensity is proportional to

$$E^{(2\omega)}(l)E^{(2\omega)*}(l) = \left(\frac{\mu_0}{\epsilon_0}\right) \frac{\omega^2 d^2}{n^2} |E^{(\omega)}|^4 l^2 \frac{\sin^2(\Delta kl/2)}{(\Delta kl/2)^2} \tag{35}$$

Here we used $\epsilon/\epsilon_0 = n^2$, where n is the index of refraction. If the input beam is confined to a cross section $A(m^2)$, then the power per unit area (intensity) is related to the field by

$$\frac{P_{2\omega}}{A} = \frac{1}{2} \sqrt{\frac{\epsilon}{\mu_0}} |E^{(2\omega)}|^2 \tag{36}$$

and Eq. 35 can be written as

$$\eta_{SHG} = \frac{P_{2\omega}}{P_\omega} = 2 \left[\frac{\mu_0}{\epsilon_0}\right]^{3/2} \frac{\omega^2 d^2 l^2}{n^3} \frac{\sin^2(\Delta kl/2)}{(\Delta kl/2)^2} \frac{P_\omega}{A} \tag{37}$$

for the conversion efficiency from ω to 2ω. We notice that the conversion efficiency is proportional to the intensity P_ω/A of the fundamental beam.

7.4.1 Phase Matching in Second Harmonic Generation

According to Eq. 37, a prerequisite for efficient second harmonic generation is that $\Delta k = 0$, or, using Eq. 34,

$$k^{(2\omega)} = 2k^{(\omega)} \tag{38}$$

where

$$k^{(\omega)} = \frac{\omega}{c_0} n^\omega$$

If $\Delta k \neq 0$, the second harmonic power generated at some plane, say z_1, having propagated to some other plane (z_2), is not in phase with the second harmonic wave which is generated at z_2. This results in the interference described by the factor $\sin^2(\Delta kl/2)/(\Delta kl/2)^2$ in Eq. 37. Two adjacent peaks of this spatial interference pattern are separated by the so-called "coherence length,"

$$l_c = \frac{2\pi}{\Delta k} = \frac{2\pi}{k^{(2\omega)} - 2k^{(\omega)}} \tag{39}$$

The coherence length l_c is thus a *measure* of the *maximum length of the crystal that is useful in producing the second harmonic power.* The length of l_c is determined by the dispersion of n. Taking

$$\Delta k = k^{(2\omega)} - 2k^{(\omega)} = \frac{2\omega}{c_0} [n^{2\omega} - n^\omega] \tag{40}$$

where we used the relation $k^{(\omega)} = \omega n^{\omega}/c_0$, we find that the coherence length is

$$l_c = \frac{\pi c_0}{\omega[n^{2\omega} - n^{\omega}]} = \frac{\lambda_0}{2[n^{2\omega} - n^{\omega}]} \tag{41}$$

where λ_0 is the free space wavelength of the fundamental beam. If we take a typical value of $\lambda_0 = 1$ μ and $n(2\omega) - n(\omega) \cong 10^{-2}$, we get $l_c \cong 100$ μ. If l_c were to increase from 100 μ to 2 cm, as an example, then according to Eq. 32 the second harmonic power would go up by a factor of 4×10^4.

The technique that is used widely[6,7] to satisfy the *phase matching* requirement, $\Delta k = 0$, takes advantage of the natural birefringence of anisotropic crystals. Recalling that $k^{(\omega)} = \omega\sqrt{\mu\epsilon_0}\,n^{\omega}$, Eq. 38 becomes

$$n^{2\omega} = n^{\omega} \tag{42}$$

so that the indices of refraction at the fundamental and second harmonic frequencies must be equal. In normally dispersive materials the index of the ordinary wave or the extraordinary wave along a given direction increases with ω, as can be seen from Table 2. This makes it

TABLE 2.   INDEX OF REFRACTION DISPERSION DATA OF KH_2PO_4.   (AFTER ZERNIKE.[8])

Wavelength, μ	Index	
	n_0, Ordinary Ray	n_e, Extraordinary Ray
0.2000	1.622630	1.563913
0.3000	1.545570	1.498153
0.4000	1.524481	1.480244
0.5000	1.514928	1.472486
0.6000	1.509274	1.468267
0.7000	1.505235	1.465601
0.8000	1.501924	1.463708
0.9000	1.498930	1.462234
1.0000	1.496044	1.460993
1.1000	1.493147	1.459884
1.2000	1.490169	1.458845
1.3000	1.487064	1.457838
1.4000	1.483803	1.456838
1.5000	1.480363	1.455829
1.6000	1.476729	1.454797
1.7000	1.472890	1.453735
1.8000	1.468834	1.452636
1.9000	1.464555	1.451495
2.0000	1.460044	1.450308

impossible to satisfy Eq. 42 when both the ω and 2ω beams are of the same type, i.e., when both are extraordinary or ordinary. We can, however, under certain circumstances, satisfy Eq. 42 by choosing one wave of each type. To illustrate the point, consider the dependence of the index of refraction of the extraordinary wave in a uniaxial crystal on the angle θ between the propagation direction and the crystal optic (z) axis. It is given by[11]

$$\frac{1}{n_e^2(\theta)} = \frac{\cos^2\theta}{n_0^2} + \frac{\sin^2\theta}{n_e^2} \tag{43}$$

If $n_e^{2\omega} < n_0^{\omega}$, there exists an angle θ_m at which $n_e^{2\omega}(\theta_m) = n_0^{\omega}$, so that if the fundamental beam (at ω) is launched along θ_m as an ordinary ray, the second harmonic beam will be generated along the *same direction* as an extraordinary ray. The situation is illustrated by Fig. 8. The angle θ_m is determined by the intersection between the sphere (shown as a circle in the figure), corresponding to the index surface of the ordinary beam at ω with the index surface of the extraordinary ray which gives $n_e^{2\omega}(\theta)$. The angle θ_m for negative uniaxial crystals (i.e., crystals in which $n_e^{\omega} < n_0^{\omega}$) is that satisfying $n_e^{2\omega}(\theta_m) = n_0^{\omega}$ or using Eq. 43,

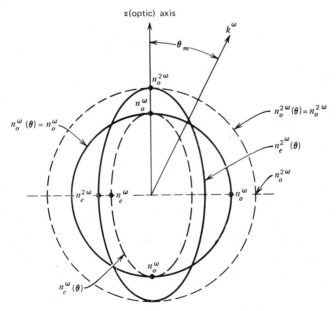

Fig. 8. The normal (index) surfaces for the ordinary and extraordinary rays in a negative ($n_e < n_0$) uniaxial crystal. If $n_e^{2\omega} < n_0^{\omega}$, the condition $n_e^{2\omega}(\theta) = n_0^{\omega}$ is satisfied at $\theta = \theta_m$. The eccentricities shown are greatly exaggerated.

$$\frac{\cos^2 \theta_m}{(n_0^{2\omega})^2} + \frac{\sin^2 \theta_m}{(n_e^{2\omega})^2} = \frac{1}{(n_0^{\omega})^2} \tag{44}$$

and solving for θ_m,

$$\sin^2 \theta_m = \frac{(n_0^{\omega})^{-2} - (n_0^{2\omega})^{-2}}{(n_e^{2\omega})^{-2} - (n_0^{2\omega})^{-2}} \tag{45}$$

7.4.2 Numerical Example—Second Harmonic Generation

Consider the problem of second harmonic generation, using the output of a pulsed ruby laser ($\lambda_0 = 0.6940 \ \mu$) in a KH_2PO_4 crystal (KDP) under the following conditions:

$$L = 1 \ cm$$

$$\frac{P_\omega}{A} = 10^8 \ W/cm^2$$

The appropriate d coefficient is, according to Table 1, $d = 5 \times 10^{-24}$ Mks units. Using these data in Eq. 37 and assuming $\Delta k = 0$, we obtain a conversion efficiency of

$$\frac{\overline{P}(\lambda_0 = 0.347 \ \mu)}{\overline{P}(\lambda_0 = 0.694 \ \mu)} \approx 15\%$$

The angle θ_m between the z axis and the direction of propagation for which $\Delta k = 0$ is given by Eq. 45. The appropriate indices are taken from Table 2, and are

$$n_e(\lambda_0 = 0.694 \ \mu) = 1.465 \quad n_e(\lambda_0 = 0.347 \ \mu) = 1.487$$

$$n_0(\lambda_0 = 0.694 \ \mu) = 1.505 \quad n_0(\lambda_0 = 0.347 \ \mu) = 1.534$$

The substitution of these data into Eq. 40 gives

$$\theta_m = 50.4$$

degrees. To obtain phase matching along this direction, the fundamental beam in the crystal must be polarized as appropriate to an ordinary ray in accordance with the discussion following Eq. 43.

We conclude from this example that very large intensities are needed to obtain high efficiency, second harmonic generation. This efficiency will, according to Eq. 37, increase as the square of the nonlinear optical coefficient d and will consequently improve as new materials are developed. Another approach is to take advantage of the dependence of η_{SHG} on P_ω/A and to place the nonlinear crystal inside the laser resonator, where the energy flux P_ω/A can be made very large.* This approach has been used successfully[9]; it will be discussed below in considerable detail.

7.4.3 Experimental Verification of Phase Matching

According to Eq. 37, if the phase matching condition $\Delta k = 0$ is violated, the output power is reduced by a factor

$$F = \frac{\sin^2\left[(\Delta k l/2)\right]}{(\Delta k l/2)^2} \tag{46}$$

from its (maximum) phase-matched value. The phase mismatch $\Delta k l/2$ is given according to Eq. 40 by

$$\frac{\Delta k l}{2} = \frac{\omega l}{c_0}\left[n_e^{2\omega}(\theta) - n_0^\omega\right] \tag{47}$$

and is thus a function of θ. If we use Eq. 43 to expand $n_e^{2\omega}(\theta)$ as a Taylor series near $\theta \cong \theta_m$, retain the first two terms only, and assume perfect phase matching at $\theta = \theta_m$ so that $n_e^{2\omega}(\theta_m) = n_0^\omega$, we obtain

$$\Delta k(\theta)l = -\frac{2\omega l}{c_0}\sin(2\theta_m)\frac{(n_e^{2\omega})^{-2} - (n_0^{2\omega})^{-2}}{2(n_0^\omega)^{-3}}(\theta - \theta_m) \tag{48}$$

$$\equiv 2\beta(\theta - \theta_m)$$

where β, as defined by Eq. 48, is a constant depending on $n_e^{2\omega}$, $n_0^{2\omega}$, n_0^ω, ω, and l. If we plot the output power at 2ω as a function of θ, we should expect, according to Eqs. 37 and 48, to find it varying as

$$P_{2\omega}(\theta) \propto \frac{\sin^2\left[\beta(\theta - \theta_m)\right]}{\left[\beta(\theta - \theta_m)\right]^2} \tag{49}$$

Figure 9 shows an experimental plot of $P_{2\omega}(\theta)$, as well as a plot of Eq. 49.

7.5 SECOND HARMONIC GENERATION INSIDE THE LASER RESONATOR

According to the numerical example of Section 3 we need to use large power densities at the fundamental frequency ω to obtain appreciable conversion from ω to 2ω in typical nonlinear optical crystals.

---

*The one-way power flow inside the optical resonator P_i is related to the power output P_e as $P_i = P_e/(1 - R)$, where R is the reflectivity.

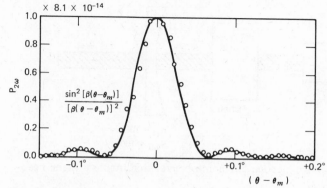

Fig. 9. Variation of the second harmonic power $P_{2\omega}$ with the angular departure $(\theta - \theta_m)$ from the phase matching angle. TEM_{00q}, $P_1 = 1.48 \times 10^{-3}$ W, $l = 1.23$ cm, KDP. (From Ashkin, Boyd, and Dziedzic.[10])

These power densities are not usually available from continuous lasers. The situation is altered, however, if the nonlinear crystal is placed within the laser resonator. The intensity (one-way power per unit area in W/m^2) inside the resonator exceeds its value outside a mirror by $(1 - R)^{-1}$, where R is the mirror reflectivity. If $R \cong 1$, the enhancement is very large; since the second harmonic conversion efficiency is, according to Eq. 37, proportional to the intensity, we may expect a far more efficient conversion inside the resonator. We shall show below that, under the proper conditions, we can extract the *total available power* of the laser at 2ω instead of at ω, and in that sense obtain 100% conversion efficiency. In order to appreciate the last statement, consider as an example the case of a continuous wave laser in which the maximum power output, at a given pumping rate, is available when the output mirror has a (optimal) transmission of 5%.

The output mirror is next replaced with one having 100% reflection at ω, and a nonlinear crystal is placed inside the laser resonator. If the crystal is chosen so that conversion efficiency from ω to 2ω in a *single pass* is 5%, the laser is loaded optimally as in the previous case, except that the coupling is caused by the loss of power by second harmonic generation instead of by the output mirror. It follows that the power generated at 2ω is the same as that coupled previously through the mirror and that the total available power of a laser can thus be converted to the second harmonic.

An experimental setup similar to the one used in the first internal second harmonic generation experiment[9] is shown in Fig. 10. The YAlG:Nd^{3+} laser emits a (fundamental) wave at $\lambda_0 = 1.06$ μ. The mir-

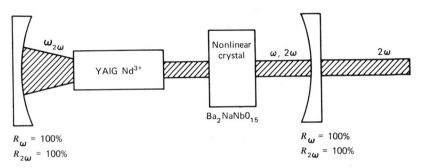

$R_\omega = 100\%$
$R_{2\omega} = 100\%$

$R_\omega = 100\%$
$R_{2\omega} = 100\%$

Fig. 10. A typical setup for second harmonic conversion inside a laser reso-nator. (After Geusic et al.[9])

rors are, as nearly as possible, totally reflecting at $\lambda_0 = 1.06\ \mu$. A Ba_2NaNbO_{15} crystal is used to generate the second harmonic at $\lambda_0 = 0.53\ \mu$. The latter is coupled through the mirror which, ideally, transmits all the radiation at this wavelength.

In the mathematical treatment of internal second harmonic genera-tion that follows, we use the results of the analysis of optimum power coupling in laser oscillators.[13]

The mirror transmission T_{opt}, which results in the maximum pow-er output from a laser oscillator, is given by

$$T_{opt} = \sqrt{g_0 L_i} - L_i \tag{50}$$

where L_i is the residual (i.e., unavoidable) fractional intensity loss per pass and g_0 is the fractional unsaturated gain per pass.* The use-ful power output under optimum coupling is

$$P_0 = S(\sqrt{g_0} - \sqrt{L_i})^2 \tag{51}$$

where the saturation power† of the laser transition S is given by

$$S = \frac{8\pi hc\Delta\nu A}{\lambda^3} \tag{52}$$

In the present problem the conversion from ω to 2ω can be consid-ered, as far as the ω oscillation is concerned, as just another loss

---

*We may recall here that the residual losses include all loss mechanisms, except those representing useful power coupling. The unsaturated gain g_0 is that exercised by a very weak wave and represents the maximum available gain at a given pumping strength.

†S/A is the optical intensity (W/m²) that reduces the inversion, and hence the gain, to one-half its zero intensity (unsaturated) value.

mechanism. We may think of it as caused by a mirror with a transmission T' taken as equal to the conversion efficiency (from ω to 2ω) per pass which, according to Eq. 37, is

$$T' = \frac{P_{2\omega}}{P_\omega} = 2\left(\frac{\mu_0}{\epsilon_0}\right)^{3/2} \frac{\omega^2 d^2 l^2}{n^3} \left[\frac{\sin^2(\Delta k l/2)}{(\Delta k l/2)^2}\right] \frac{P_\omega}{A} \tag{53}$$

where d is the crystal nonlinear coefficient, l the length, A the cross-sectional area, Δk the wave vector mismatch, and P_ω the one-way traveling power *inside* the laser. We can rewrite T' in the form

$$T' \equiv \eta P_\omega \tag{54}$$

where η is defined by the last two equations. The equivalent mirror transmission T' is thus proportional to the power.

Using the last result in Eq. 50, we find immediately that at optimum conversion the product ηP_ω must have the value,

$$(\eta P_\omega)_{\text{opt}} = \sqrt{g_0 L_i} - L_i \tag{55}$$

The total loss per pass seen by the fundamental beam is the sum of the conversion loss (ηP_ω) and the residual losses which, under optimum coupling, become

$$L_{\text{opt}} = L_i + (\eta P_\omega)_{\text{opt}} = \sqrt{g_0 L_i} \tag{56}$$

Our next problem is to find the internal power P_ω at optimum coupling, so that using Eq. 53 we may calculate the second harmonic power. We start with the expression for the total power P_e, extracted from the laser atoms under optimum coupling conditions[13]

$$(P_e)_{\text{opt}} = \frac{8\pi h \Delta \nu V}{\lambda^3 (t_c)_{\text{opt}}} \left[\sqrt{\frac{g_0}{L_i}} - 1\right] = L_{\text{opt}} S\left[\sqrt{\frac{g_0}{L_i}} - 1\right] \tag{57}$$

The fraction of the total power P_e emitted by the atoms that is available as useful output is T/L. This power is also given by the product $P_\omega T$ of the one-way internal power P_ω and the fraction T of this power which is coupled per pass. The equating of these two forms gives

$$P_\omega = \frac{P_e}{L}$$

and using Eq. 57, we get

$$(P_\omega)_{\text{opt}} = S\left[\sqrt{\frac{g_0}{L_i}} - 1\right] \tag{58}$$

for the one-way fundamental power inside the laser under optimum coupling conditions. The amount of second harmonic power generated under optimum coupling is

$$(P_{2\omega})_{opt} = (\eta P_\omega)_{opt}(P_\omega)_{opt}$$

which, using Eqs. 55 and 58, results in

$$(P_{2\omega})_{opt} = S(\sqrt{g_0} - \sqrt{L_i})^2 \tag{59}$$

which is equal to the maximum available power output from a laser oscillator.[13]

The nonlinear coupling constant η was defined by Eqs. 53 and 54 as

$$\eta = 2\left(\frac{\mu_0}{\epsilon_0}\right)^{3/2} \frac{\omega^2 d^2 l^2}{n^3 A} \left[\frac{\sin^2(\Delta k l/2)}{(\Delta k l/2)^2}\right] \tag{60}$$

Its value under optimum coupling can be derived from Eqs. 55 and 58 and is

$$\eta_{opt} = \frac{(\eta P_\omega)_{opt}}{(P_\omega)_{opt}} = \frac{L_i}{S} \tag{61}$$

and is thus *independent of the pumping strength*.* It follows that, once η is adjusted to its optimum value L_i/S, it remains optimal at any pumping level. This is quite different from the case of optimum coupling in ordinary lasers, in which optimum mirror transmission was found to depend on the pumping strength.[13]

In closing we may note that, apart from its dependence on the crystal length l, the nonlinear coefficient d and the beam cross section A, η depends also on the phase mismatch $\Delta k l$. Since Δk was shown in Eq. 47 to depend on the direction of propagation in the crystal, we can use the crystal orientation as a means of varying η.

7.5.1 Numerical Example of Internal Second Harmonic Generation

Consider the problem of designing an internal second harmonic generator of the type illustrated in Fig. 10. The YAlG:Nd^{3+} laser is assumed to have the following characteristics:

1. $\lambda_0 = 1.06 \ \mu = 1.06 \times 10^{-6}$ m.
2. $\Delta\nu = 1.35 \times 10^{11}$ Hz.
3. Beam diameter (averaged over entire resonator length) = 2000 μ.
4. L_i = internal loss per pass = 2×10^{-2}.
5. $n = 1.5$.

---

*We recall here that the pumping strength in our analysis is represented by the unsaturated gain g_0.

The crystal used for second harmonic generation is $BaNaNb_5O_{15}$, whose second harmonic coefficient (see Table 1) is $d \cong 1.1 \times 10^{-22}$ mks units.

Our problem is to calculate the length l of the nonlinear crystal which results in a full conversion of the optimally available fundamental power into the second harmonic at $\lambda_0 = 0.53\ \mu$. The crystal is assumed to be oriented at the phase matching condition, so that $\Delta k \equiv k^{2\omega} - 2k^{\omega} = 0$.

The optimum coupling parameter is given by Eq. 61 as $\eta_{opt} = L_i/S$, where S is the saturation power parameter defined by Eq. 52. The use of the above data in Eq. 52 gives

$$S = 8\ \text{W}$$

which, taking $L_i = 2 \times 10^{-2}$, yields

$$\eta_{opt} = 2.5 \times 10^{-3}$$

Next we use the definition (Eq. 60)

$$\eta = 2 \left(\frac{\mu_0}{\epsilon_0} \right)^{3/2} \frac{\omega^2 d^2 l^2}{n^3 A}$$

where we put $\Delta k = 0$ and take the beam diameter at the crystal as $50\ \mu$. (The crystal can be placed near a beam waist, where the beam diameter is a minimum.) By equating the last expression to $\eta_{opt} = 2.5 \times 10^{-3}$, using the numerical data given above, and solving for the crystal length, we get

$$l_{opt} = 0.2\ \text{cm}$$

7.6 PHOTON MODEL OF SECOND HARMONIC GENERATION

A very useful point of view and one that follows directly from the quantum mechanical analysis of nonlinear optical processes,[11] is based on the photon model illustrated in Fig. 11. According to this picture the basic process of second harmonic generation can be viewed as an annihilation of two photons at ω and a simultaneous creation of a photon at 2ω. Recalling that a photon has an energy $\hbar\omega$ and a momentum $\hbar\vec{k}$, we find that, if the fundamental conversion process is to conserve momentum as well as energy,

$$\vec{k}^{(2\omega)} = 2\vec{k}^{(\omega)} \tag{62}$$

which is a generalization to three dimensions of the condition $\Delta k = 0$ shown in Section 7.3 to lead to maximum second harmonic generation.

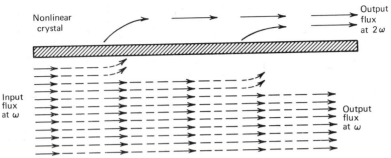

Fig. 11. A schematic representation of the process of second harmonic generation. Input photons (each arrow represents one photon) at ω are "annihilated"
by the nonlinear crystal in pairs with a new photon at 2ω being "created" for
each annihilated pair. (Note that in reality both ω and 2ω occupy the same
space inside the crystal.)

7.7 PARAMETRIC AMPLIFICATION AND OSCILLATION

Optical parametric amplification in its simplest form involves the
transfer of power from a "pump" wave at ω_3 to waves at frequencies
ω_1 and ω_2, where $\omega_3 = \omega_1 + \omega_2$. It is fundamentally similar to the case
of second harmonic generation treated in Section 7.4. The only dif-
ference is in the direction of power flow. In second harmonic generation, power is fed from the low-frequency optical field at ω to the field
at 2ω. In parametric amplification power flow, as indicated above, is
from the high-frequency field (ω_3) to the low-frequency fields at ω_1 and
ω_2. In the special case in which $\omega_1 = \omega_2$, we have the exact reverse of
second harmonic generation. This is the case of the so-called "degen-
erate parametric amplification."

Before embarking on a detailed analysis of the optical case, it may
be worthwhile to review some of the low-frequency beginnings of parametric oscillation.

Consider a classical nondriven oscillator, whose equation of motion
is given by

$$\frac{d^2v}{dt^2} + \kappa \frac{dv}{dt} + \omega_0^2 v = 0 \tag{63}$$

The variable v may correspond to the excursion of a mass M which is
connected to a spring with a constant $\omega_0^2 M$ or to the voltage across a
parallel RLC circuit, in which case $\omega_0^2 = (LC)^{-1}$ and $\eta = (RC)^{-1}$. The
solution of Eq. 63 is

$$v(t) = v(0) \exp\left[-\frac{\kappa t}{2}\right] \exp\left[\pm i \sqrt{\omega_0^2 - \frac{\kappa^2}{4}}\, t\right] \tag{64}$$

i.e., a damped sinusoid.

Lord Rayleigh[12] in 1883 investigating parasitic resonances in pipe organs, considered the consequences of the following equation:

$$\frac{d^2v}{dt^2} + \kappa \frac{dv}{dt} + (\omega_0^2 + 2\alpha \sin\omega_p t)v = 0 \tag{65}$$

This equation may describe an oscillator in which an energy storage parameter (mass or spring constant in the mechanical oscillator, L or C in the RLC oscillator) is modulated at a frequency ω_p. As an example consider the case of the RLC circuit shown in Fig. 12, in which the capacitance is modulated according to

$$C = C_0\left(1 - \frac{\Delta C}{C_0}\sin\omega_p t\right) \tag{66}$$

The equation of the voltage across the RLC circuit is given by Eq. 63 with $\omega_0^2 = (LC)^{-1}$.

Using Eq. 66 and assuming $\Delta C \ll C_0$, we find that Eq. 63 becomes

$$\frac{d^2v}{dt^2} + \kappa \frac{dv}{dt} + \frac{1}{LC_0}\left(1 + \frac{\Delta C}{C_0}\sin\omega_p t\right)v = 0 \tag{67}$$

which, if we make the identification

$$\omega_0^2 = \frac{1}{LC_0}, \qquad \alpha = \frac{\omega_0^2 \Delta C}{2C_0} \tag{68}$$

is identical to Eq. 65.

The most important feature of the parametrically driven oscillator described by Eq. 65 is that it is capable of *sustained oscillation* at ω_0. To show this, let us assume a solution

$$v = a\cos[\omega t + \phi] \tag{69}$$

Expanding $\sin\omega_p t$ in Eq. 65 in terms of exponentials, substituting Eq. 69, and neglecting nonsynchronous terms oscillating at $(\omega_p + \omega)$, we obtain

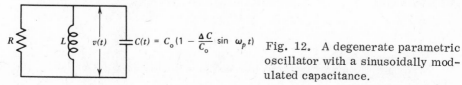

Fig. 12. A degenerate parametric oscillator with a sinusoidally modulated capacitance.

$$(\omega_0^2 - \omega^2) e^{[i(\omega t + \phi)]} + i\omega\kappa \, e^{[i(\omega t + \phi)]} - i\alpha e^{\{i[(\omega_p - \omega)t - \phi]\}} = 0 \qquad (70)$$

From Eq. 70 it follows that steady-state oscillation is possible if

$$\omega_p = 2\omega \quad \text{(so that } \omega_p - \omega = \omega\text{)}$$

$$\omega = \omega_0, \quad \phi = 0 \text{ or } \pi, \quad \alpha = \omega\kappa \qquad (71)$$

or, in words: The pump frequency ω_p is twice the oscillation frequency ω_0. The oscillation phase* is $\phi = 0$ or π, and the strength of the pumping α must satisfy $\alpha = \omega\kappa$. The last condition is referrred to as the "start-oscillation" condition or "threshold condition," since it gives the pumping strength (α) needed to overcome the losses (κ) at the oscillation threshold. In the case of the RLC circuit, whose capacitance is pumped according to Eq. 66, the threshold oscillation condition $\alpha = \omega\kappa$ can be written with the aid of Eq. 68 as

$$\frac{\Delta C}{2C_0} = \frac{\kappa}{\omega_0} = \frac{1}{Q} \qquad (72)$$

where the quality factor $Q = \omega_0 RC$ is related to the decay rate η by $\eta = \omega_0/Q$.

In practice, if the capacitance of the circuit shown in Fig. 12 is modulated so that Eq. 72 is satisfied, the circuit will break into spontaneous oscillation at a frequency $\omega_0 = \omega_p/2$. This constitutes a transfer of energy from ω_p to $\omega_0 = \omega_p/2$.

The physical nature of this transfer may become clearer if we consider the time behavior of the voltage $v(t)$, the charge $q(t)$, and the capacitance $C(t)$, as illustrated in Fig. 13.

The capacitance $C(t)$ is a parallel-plate capacitor whose capacitance is periodically varied. Assume first that $C(t)$ is varied as in Fig. 13a by pulling the capacitor plates apart and pushing them together again, $C \propto \text{(plate separation)}^{-1}$. At the same time the circuit is caused to oscillate, so that the charge $q(t)$ on the capacitor plates varies as in Fig. 13b. Now according to Fig. 13a, when the charge on the plates is a maximum, the plate separation is increased suddenly and they are pulled apart slightly. The charge cannot change instantaneously, but since work must be done (against the Coulomb attraction of the opposite charges on the capacitor plates) to separate the plates, energy is fed into the capacitor and appears as a sudden increase in the stored energy ϵ, and because of the relation $\epsilon = \frac{1}{2} CV^2$, in the voltage. One quarter of a period later, the charge and thus the field between the

---

*The phase ϕ is relative to that of the pump oscillation as given by Eq. 66.

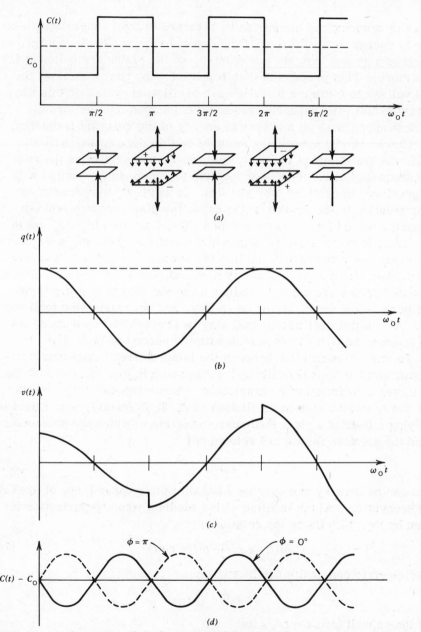

Fig. 13. A physical model of a capacitively pumped parametric oscillator. (*a*) Square wave capacitance variation at twice the circuit oscillation frequency. Also shown is the motion of the capacitor plates, the charge and the forces on the plates. (*b*) The charge on one of the capacitor plates. (*c*) The voltage ac-across the circuit. (*d*) Variation of the capacitance $C(t)$ at two phases, relative to that of the charge.

307

plates is zero and the plates can be returned to their original position with no energy expenditure. At the end of half a cycle, the charge has reversed sign and is again a maximum, so the plates are pulled apart once more. This process is then repeated many times, causing the total voltage to increase twice in each oscillation cycle. In this way, energy at *twice* the resonant frequency is pumped into the circuit, where it appears as an increase in energy of the resonant frequency.

There are two noteworthy features to this degenerate oscillator. First, the frequency of the pump *must* be very nearly twice the resonant frequency of the oscillator for gain to occur, in agreement with the previous conclusions (see Eq. 71). In addition, the phase of the pump relative to the charge on the capacitor plates must be chosen properly. Consider the case in which $C(t) = C_0 \pm (\Delta C) \sin(2\omega_0 t)$, as in Fig. 13$d$. If we take the $(-)$ sign which corresponds to the solid curve, then energy is continuously fed *into* the system, as described above. If, however, the pumping phase is inverted, i.e., the $(+)$ sign, then the capacitor plates are pushed together when the charge is a maximum, thus performing work, giving up energy, and decreasing the total voltage. Any initial oscillations that may be present will be damped out. The phase condition $(\phi = 0)$ agrees with the second of Eqs. 71.

To make a connection between the lumped circuit parametric oscillator and the optical nonlinearity discussed in Eq. 14, we show that the (time) modulation of a capacitance at some frequency ω_p, which was shown to give rise to oscillation at $\omega_p/2$, is formally equivalent to applying a field at ω_p to a nonlinear dielectric in which the polarization p and the electric field e are related by

$$p = \epsilon_0 \chi e + d e^2 \qquad (73)$$

This can be done by considering a parallel plate capacitance of area A and separation s which is filled with a medium whose polarization is given by Eq. 73. Using the relations,*

$$d(t) = \epsilon_0 e(t) + p(t) = \epsilon e(t) \qquad (74)$$

the dielectric constant ϵ can be written as

$$\epsilon = \epsilon_0 (1 + \chi) + d e$$

and the capacitance $C = \epsilon A/s$ as

---

*The electric displacement $d(t)$ should not be confused with the nonlinear constant d in Eq. 73.

$$C = \frac{\epsilon_0(1+\chi)A}{s} + \frac{Ad}{s}e \tag{75}$$

If the electric field is given by

$$e = - E_0 \sin\omega_p t$$

the capacitance becomes

$$C = \frac{\epsilon_0(1+\chi)A}{s} - \frac{AdE_0}{s}\sin\omega_p t \tag{76}$$

which is of a form identical to Eq. 66. It follows that the two points of view used to describe the parametric processes—the one represented by Eq. 66 in which an energy storage parameter is modulated, and that in which the electric (or magnetic) response is nonlinear as in Eq. 73, are equivalent.

We return now to the basic nonlinear parametric equations (Eqs. 32) to analyze the case of optical parametric amplification. We find it convenient to introduce a new field variable defined by

$$A_l = \sqrt{\frac{n_l}{\omega_l}}\, E_l, \quad l = 1,\ 2,\ 3 \tag{77}$$

so that power flow per unit area at ω_l is given by

$$\frac{P_l}{A} = \frac{1}{2}\sqrt{\frac{\epsilon_0}{\mu_0}}\, n_l\, |E_l|^2 = \frac{1}{2}\sqrt{\frac{\epsilon_0}{\mu_0}}\, \omega_l\, |A_l|^2 \tag{78}$$

If we describe the power flow P_l/A per unit area in terms of the photon flux N_{ω_l} (photons/m^2-sec), we have

$$\frac{P_l}{A} = N_{\omega_l}\hbar\omega_l = \frac{1}{2}\sqrt{\frac{\epsilon_0}{\mu_0}}\, |A_l|^2 \omega_l \tag{79}$$

so that $|A_l|^2$ is proportional to the photon flux at ω_l. The equations of motion (Eqs. 32) for the A_l variables become

$$\frac{dA_1}{dz} = -\frac{1}{2}\,\alpha_1 A_1 - \frac{i}{2}\lambda A_2^* A_3 e^{-i(\Delta k)z}$$

$$\frac{dA_2^*}{dz} = -\frac{1}{2}\,\alpha_2 A_2^* + \frac{i}{2}\lambda A_1 A_3^* e^{i(\Delta k)z} \tag{80}$$

$$\frac{dA_3}{dz} = -\frac{1}{2}\,\alpha_3 A_3 - \frac{i}{2}\lambda A_1 A_2 e^{i(\Delta k)z}$$

where

$$\Delta k = k_3 - (k_1 + k_2)$$

$$\lambda = d \sqrt{\left(\frac{\mu_0}{\epsilon_0}\right) \frac{\omega_1 \omega_2 \omega_3}{n_1 n_2 n_3}}$$

$$\alpha_l = \sigma_l \sqrt{\frac{\mu_0}{\epsilon_l}}, \quad l = 1, 2, 3$$

(81)

The advantage of using the A_l instead of E_l is now apparent since, unlike Eq. 32, Eq. 80 involves a single coupling parameter λ.

We shall now use Eq. 80 to solve for the field variables $A_1(z)$, $A_2(z)$, and $A_3(z)$ for the case in which three waves with amplitudes $A_1(0)$, $A_2(0)$, and $A_3(0)$ at frequencies ω_1, ω_2, and ω_3, respectively, are incident on a nonlinear crystal at $z = 0$. We take $\omega_3 = \omega_1 + \omega_2$, $\alpha_1 = \alpha_2 = \alpha_3 = 0$ (no losses), and $\Delta k = k_3 - k_1 - k_2 = 0$. In addition, we assume that $\omega_1 |A_1(z)|^2$ and $\omega_2 |A_2(z)|^2$ remain small compared to $\omega_3 A_3(0)^2$ throughout the interaction region. This last condition, in view of Eq. 79, is equivalent to assuming that the power is drained off the "pump" (at ω_3) by the "signal" (ω_1) and that the idler (ω_2) is negligible compared to the input power at ω_3. This enables us to view $A_3(z)$ as a constant. With the assumptions stated above, Eq. 80 becomes

$$\frac{dA_1}{dz} = -\frac{ig}{2} A_2^*, \quad \frac{dA_2^*}{dz} = \frac{ig}{2} A_1$$

(82)

where

$$g = \lambda A_3(0) = \sqrt{\left(\frac{\mu_0}{\epsilon_0}\right) \frac{\omega_1 \omega_2}{n_1 n_2}} \, d E_3(0)$$

(83)

The solution of the coupled equations (Eq. 82) subject to the initial conditions $A_1(z = 0) = A_1(0)$, $A_2(z = 0) = A_2(0)$ is

$$A_1(z) = A_1(0) \cosh \frac{g}{2} z - i A_2^*(0) \sinh \frac{g}{2} z$$

$$A_2^*(z) = A_2^*(0) \cosh \frac{g}{2} z + i A_1(0) \sinh \frac{g}{2} z$$

(84)

Equation 84 describes the growth of the signal and idler waves under phase matching conditions. In the case of parametric amplification the input will consist of the pump (ω_3) wave and one of the other two fields, say ω_1. In this case $A_2(0) = 0$, and using the relation $N_{\omega_i} \propto A_i A_i^*$ for the photon flux, we obtain from Eq. 84

$$N_{\omega_1}(z) \propto A_1^*(z) A_1(z) = |A_1(0)|^2 \cosh^2 \frac{gz}{2} \xrightarrow[gz \gg 1]{} \frac{|A_1(0)|^2}{4} e^{gz}$$

$$N_{\omega_2}(z) \propto A_2^*(z) A_2(z) = |A_1(0)|^2 \sinh^2 \frac{gz}{2} \xrightarrow[gz \gg 1]{} \frac{|A_1(0)|^2}{4} e^{gz}$$

(85)

so that, for $gz \gg 1$, the photon fluxes at ω_1 and ω_2 grow exponentially. If we limit our attention to the wave at ω_1, it undergoes an amplification by a factor

$$\frac{A_1^*(z)A_1(z)}{A_1^*(0)A_1(0)}\bigg|_{gz \gg 1} \frac{1}{4}e^{gz} \tag{86}$$

7.7.1 Numerical Example—Parametric Amplification

The magnitude of the gain coefficient g, available in a traveling wave parametric interaction, is estimated for the following case involving the use of a LiNbO$_3$ crystal.

$$d_{311} = 5 \times 10^{-23}$$

$$\nu_1 \cong \nu_2 = 3 \times 10^{14}$$

$$P_{\omega_3} = 5 \times 10^6 \text{ W/cm}^2$$

$$n_1 \cong n_2 = 2.2$$

Converting P_{ω_3} to $|E_3|^2$ with the use of Eq. 78, and then substituting in Eq. 83, we find that

$$g = 0.7$$

cm^{-1}. This shows that traveling wave parametric amplification is not expected to lead to large values of gain, except for extremely large pump power densities. The main attraction of the parametric amplification described above is probably in giving rise to parametric oscillation which will be described in Section 7.9.

7.8 PHASE MATCHING IN PARAMETRIC AMPLIFICATION

The analysis in the last section of parametric amplification assumed that the phase matching condition

$$k_3 = k_1 + k_2 \tag{87}$$

is satisfied. It is important to determine the consequences of violating this condition. We start with Eq. 18, taking the loss coefficients $\alpha_1 = \alpha_2 = 0$,

$$\frac{dA_1}{dz} = -i\frac{g}{2}A_2^* e^{-i(\Delta k)z}$$

$$\frac{dA_2^*}{dz} = +i\frac{g}{2}A_1 e^{i(\Delta k)z} \tag{88}$$

The solution of Eq. 88 is facilitated by the substitution,

$$A_1(z) = m_1 e^{\{[s-i/2(\Delta k)]z\}}$$

$$A_2^*(z) = m_2 e^{\{s+i/2(\Delta k)]z\}} \tag{89}$$

where m_1 and m_2 are coefficients independent of z. The exponential growth constant s is to be determined. The substitution of Eq. 89 in Eq. 88 leads to

$$s - i\frac{\Delta k}{2} \ m_1 + i\frac{g}{2} m_2 = 0$$

$$-i\frac{g}{2}m_1 + \ s + i\frac{\Delta k}{2} \ m_2 = 0 \tag{90}$$

By equating the determinant of the coefficients of m_1 and m_2 in Eq. 90 to zero, we obtain the two solutions

$$s_\pm = \pm \tfrac{1}{2} \sqrt{g^2 - (\Delta k)^2} \equiv \pm b \tag{91}$$

The general solution of Eq. 88 is the sum of the two independent solutions:

$$A_1(z) = m_1^+ e^{\{[s_+ - i/2(\Delta k)]z\}} + m_1^- e^{\{[s_- - i/2(\Delta k)]z\}}$$

$$A_2^*(z) = m_2^+ e^{\{s_+ + i/2(\Delta k)]z\}} + m_2^- e^{\{[s_- - i/2(\Delta k)]z\}} \tag{92}$$

The coefficients m_1^+, m_1^-, m_2^+, m_2^- are next determined by requiring that, at $z = 0$, the solution (Eq. 92) agree with the complex input amplitudes $A_1(0)$ and $A_2^*(0)$. This leads straightforwardly to the result,

$$A_1(z)e^{[i(\Delta k/2)z]} = A_1(0)\left[\cosh(bz) + \frac{i(\Delta k)}{2b} \sinh(bz)\right]$$

$$- i\frac{g}{2b}A_2^*(0)\sinh(bz)$$

$$A_2^*(z)e^{[-i(\Delta k)/2z]} = A_2^*(0)\left[\cosh(bz) - \frac{i(\Delta k)}{2b}\sinh(bz)\right]$$

$$+ i\frac{g}{2b} A_1(0)\sinh(bz) \tag{93}$$

The last result reduces, as it should, to Eq. 84 if we put $\Delta k = 0$.

The most noteworthy feature of Eqs. 91 and 93 is that the exponential gain coefficient b is a function of Δk and that, unless

$$g \geq \Delta k \tag{94}$$

no sustained growth of the signal (A_1) and idler (A_2) waves is possible, since in this case the cosh and sinh functions in Eq. 93 become

$$\sin\left\{\tfrac{1}{2}[(\Delta k)^2 - g^2]^{1/2} z\right\}$$
$$\cos\left\{\tfrac{1}{2}[(\Delta k)^2 - g^2]^{1/2} z\right\}$$

respectively, and the energies at ω_1 and ω_2 oscillate as a function of the distance z.

The problem of phase matching in parametric amplification is fundamentally the same as that in second harmonic generation. Instead of satisfying the condition (Eq. 33), $k^{2\omega} = 2k^{\omega}$, we have to satisfy the condition, according to Eq. 87:

$$k^{\omega_3} = k^{\omega_1} + k^{\omega_2} \tag{95}$$

This is done, as in second harmonic generation, by using the dependence of the phase velocity of the extraordinary wave in axial crystals on the direction of propagation. In a negative uniaxial crystal $(n_e < n_0)$, we can, as an example, choose the signal and idler waves as ordinary, while the pump at ω_3 is applied as an extraordinary wave. Using Eq. 43 and the relation $k^{\omega} = (\omega/c_0)n^{\omega}$, the phase matching condition (Eq. 95) is satisfied when all three waves propagate at an angle θ_m to the z (optic) axis, where

$$n_e^{\omega_3}(\theta_m) \equiv \left[\left(\frac{\cos\theta_m}{n_0^{\omega_3}}\right)^2 + \left(\frac{\sin\theta_m}{n_e^{\omega_3}}\right)^2\right]^{-1/2} = \frac{\omega_1}{\omega_3} n_0^{\omega_1} + \frac{\omega_2}{\omega_3} n_0^{\omega_2} \tag{96}$$

7.9 PARAMETRIC OSCILLATION*

In the discussions of the last two sections we have demonstrated that a pump wave at ω_3 can cause a simultaneous amplification in a nonlinear medium of "signal" and "idler" waves at frequencies ω_1 and ω_2, respectively, where $\omega_3 = \omega_1 + \omega_2$. If the nonlinear crystal is placed within an optical resonator, as shown in Fig. 11, which provides high Q resonances for the signal or idler waves (or both), the parametric gain will, at some threshold pumping intensity, cause a simultaneous oscillation at the signal and idler frequencies. The threshold pumping corresponds to the point at which the parametric gain just balances the losses of the signal and idler waves. This is the physical basis of the optical parametric oscillator. Its practical importance derives from

---

*The first demonstration of optical parametric oscillation is that of J. A. Giordmaine and R. C. Miller (See Ref. 14). Some of the early theoretical analysis of optical parametric oscillation may be found in Refs. 15 and 16.

its ability to convert the power output of the pump laser to power at the signal and idler frequencies which, as will be shown below, can be tuned continuously over large ranges.

To analyze this situation, we return to Eq. 80. We take $\Delta k = 0$ and neglect the depletion of the pump waves, so that $A_3(z) = A_3(0)$. The result is

$$\frac{dA_1}{dz} = -\frac{1}{2}\alpha_1 A_1 - i\frac{g}{2}A_2^*$$

$$\frac{dA_2^*}{dz} = -\frac{1}{2}\alpha_2 A_2^* + i\frac{g}{2}A_1 \tag{97}$$

where, as in Eq. 83,

$$g = \sqrt{(\mu_0/\epsilon_0)(\omega_1\omega_2/n_1 n_2)}\, dE_3(0)$$

$$\alpha_{1,2} = \sigma_{1,2}\sqrt{\mu_0/\epsilon_{1,2}} \tag{98}$$

Equation 97 describes traveling wave parametric interaction. We shall use them below to describe the interaction inside a resonator such as the one shown in Fig. 14. This procedure seems plausible if we think of propagation inside an optical resonator as a folded optical path. The magnitude of the spatial distributed loss constants α_1 and α_2 must then be chosen, so that they account for the actual losses in the resonator. The latter will include losses due to the less than perfect reflection at the mirrors, as well as distributed loss in the nonlinear crystal and that from diffraction.*

If the parametric gain is sufficiently high to overcome the losses, steady-state oscillation results. When this is the case,

$$\frac{dA_1}{dz} = \frac{dA_2^*}{dz} = 0 \tag{99}$$

so that the power gained via the parametric interaction just balances the losses.

The addition of $d/dz = 0$ to Eq. 97 gives

$$-\frac{\alpha_1}{2}A_1 - i\frac{g}{2}A_2^* = 0$$

$$i\frac{g}{2}A_1 - \frac{\alpha_2}{2}A_2^* = 0 \tag{100}$$

---

*The effective loss constant α_i is chosen, so that $\exp(-\alpha_i l)$ is the total attenuation in intensity per resonator pass at ω_i, where l is the crystal length.

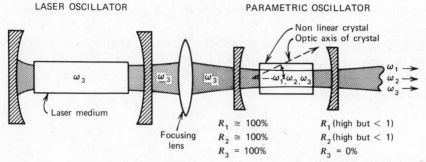

Fig. 14. A schematic diagram of an optical parametric oscillator in which the laser output at ω_3 is used as the pump, giving rise to oscillations at ω_1 and ω_2 (where $\omega_3 = \omega_1 + \omega_2$) in an optical cavity that contains the nonlinear crystal and resonates at ω_1 and ω_2.

The condition for nontrivial solutions for A_1 and A_2* is that the determinant at Eq. 100 vanish

$$\det \begin{vmatrix} -\dfrac{\alpha_1}{2} & -i\dfrac{g}{2} \\[2mm] i\dfrac{g}{2} & -\dfrac{\alpha_2}{2} \end{vmatrix} = 0$$

so that

$$g^2 = \alpha_1 \alpha_2 \tag{101}$$

This is the *threshold condition* for *parametric oscillation*.

If we choose to express the mode losses at ω_1 and ω_2 by the quality factors Q_1 and Q_2, respectively, we have*

$$\alpha_i = \frac{\omega_i n_i}{Q_i c_0} \tag{102}$$

Using Eq. 98, Eq. 101 can be written as

$$\frac{d(E_3)_t}{\sqrt{\epsilon_1 \epsilon_2}} = \frac{1}{\sqrt{Q_1 Q_2}} \tag{103}$$

where $(E_3)_t$ is the value of E_3 at threshold. This relation can be shown to be formally analogous to that obtained in Eq. 72 for the lumped circuit parametric oscillator. According to Eq. 76 $\Delta C/C_0 = dE/\epsilon$ so that,

---

*This relation follows from recognizing that the temporal decay rate $\sigma = \omega/Q$ is related to α by $\sigma = \alpha c_0/n$.

apart from a factor of two, if we put $Q_1 = Q_2$ and $\epsilon_1 = \epsilon_2$ Eq. 103 is the same as Eq. 72.

Another useful form of the threshold relation results from representing the quality factor Q in terms of the (effective) mirror reflectivities. If, furthermore, we express E_3 in terms of the power flow per unit area, according to

$$E_3^2 = 2\frac{P_{\omega 3}}{A}\sqrt{\mu_0/\epsilon_0 n_3^2}$$

we can rewrite Eq. 103 as

$$\left(\frac{P_{\omega 3}}{A}\right)_t = \frac{1}{2}\left(\frac{\epsilon_0}{\mu_0}\right)^{3/2}\frac{n_1 n_2 n_3(1 - R_1)(1 - R_2)}{\omega_1 \omega_2 l_c^2 d^2} \tag{104}$$

where l_c is the length of the crystal.

7.9.1 Numerical Example of Parametric Oscillation Threshold

Let us estimate the threshold pump requirement $P_{\omega 3}/A$ (W/cm^2) of a parametric oscillation of the kind shown in Fig. 14 which utilizes a LiNbO$_3$ crystal. We use the following set of parameters: $(1 - R_1)$ $= (1 - R_2) = 2\times 10^{-2}$ (i.e., total loss per pass at ω_1 and $\omega_2 = 2\%$), $(\lambda_0)_1$ $= (\lambda_0)_2 = 1$ μ, $l_c = 0.5$ m, $n_1 = n_2 = n_3 = 1.5$, and $d_{311}(L_iNbO_3) = 5\times 10^{-23}$.

Substitution in Eq. 104 yields

$$\left(\frac{P_{\omega 3}}{A}\right)_t = 5000$$

W/cm^2. This involves a modest amount of total power, so that the example helps us appreciate the attractiveness of optical parametric oscillation as a means for generating coherent optical frequency at new optical frequencies.

7.10 PHASE MATCHING AND FREQUENCY TUNING IN PARAMETRIC OSCILLATION

We have shown in the analysis of the last section that a pump wave at ω_3, which is applied to a nonlinear crystal inside an optical resonator, can give rise to simultaneous oscillation at ω_1 and ω_2, where ω_3 $= \omega_1 + \omega_2$. The condition $\omega_3 = \omega_1 + \omega_2$ does not define ω_1 and ω_2 uniquely. The particular set of ω_1 and ω_2 (out of the infinite number of such combinations), which is the one to be generated, is that satisfying the momentum conservation condition.

$$\Delta k = k_3 - (k_1 + k_2) = 0 \qquad (105)$$

This can be shown by rederiving the expression (Eq. 101) for the threshold of parametric oscillation, this time allowing for $\Delta k \neq 0$. Instead of Eq. 97, we obtain

$$\frac{dA_1}{dz} = -\frac{1}{2}\alpha_1 A_1 - i\frac{g}{2}A_2^* e^{-i(\Delta k)z}$$

$$\frac{dA_2^*}{dz} = -\frac{1}{2}\alpha_2 A_2^* + i\frac{g}{2}A_1 e^{i(\Delta k)z} \qquad (106)$$

Assuming that both waves grow as e^{sz}, and proceeding in a manner identical to that used to derive Eq. 91, we obtain the following expression for the exponential gain constant s:

$$s_{\pm} = -\tfrac{1}{4}(\alpha_i + \alpha_s) \pm \sqrt{\tfrac{1}{4}(\alpha_i - \alpha_s)^2 + i\Delta k(\alpha_i - \alpha_s) + g^2 - (\Delta k)^2} \qquad (107)$$

The threshold for parametric oscillation occurs when $s_+ = 0$, at which point the gain just balances the losses. From Eq. 107 this condition is satisfied, e.g., when $\alpha_i = \alpha_s$ and

$$(g^2)_t = (\Delta k)^2 + \alpha_s^2 \qquad (108)$$

which for $\Delta k = 0$ agrees with Eq. 101. Equation 108 can be written as

$$(g^2)_t = [(g^2)_t]_{\text{min threshold}}\left[1 + \frac{(\Delta k)^2}{[(g^2)_t]_{\text{min threshold}}}\right] \qquad (109)$$

where the increase in the pump threshold intensity, relative to its minimum $(\Delta k = 0)$ value, is represented by the factor inside the square brackets.

It is clear from Eq. 109 that, as the pumping intensity is gradually increased from zero, the threshold condition will first be satisfied for the pair of frequencies ω_1 and ω_2 for which, in addition to the condition $\omega_1 + \omega_2 = \omega_3$, $\Delta k = k_3 - (k_1 + k_2) = 0$. This pair of frequencies will thus be the one to break into oscillation.

7.10.1 Frequency Tuning

We have shown above that the pair of signal (ω_1) and idler frequencies which are caused to oscillate by parametric pumping at ω_3 satisfy the condition $k_3 = k_1 + k_2$. Using $k_i = \omega_i n_i / c_0$, we can write it as

$$\omega_3 n_3 = \omega_1 n_1 + \omega_2 n_2 \qquad (110)$$

In a crystal the indices of refraction depend, in general, as shown in Section 7.4, on the frequency, the crystal orientation (if the wave is extraordinary), the electric field (in electrooptic crystals), and on the temperature. If, e.g., we change the crystal orientation in the oscillator shown in Fig. 14, the oscillation frequencies ω_1 and ω_2 will change, in order to compensate for the change in indices, so that condition (Eq. 110) will be satisfied at the new frequencies.

To be specific, we consider the case of a parametric oscillator pumped by an extraordinary beam at a fixed frequency ω_3. The signal (ω_1) and idler (ω_2) are ordinary waves. At some crystal orientation θ_0, the oscillation takes place at frequencies ω_{10} and ω_{20}. Let the indices of refraction at ω_{10}, ω_{20}, and ω_3 under those conditions be n_{10}, n_{20} n_{20}, and n_{30}, respectively. We want to find the change in ω_1 and ω_2 caused by a small change $\Delta\theta$ in the crystal orientation.

From Eq. 110 we have at $\theta = \theta_0$

$$\omega_3 n_{30} = \omega_{10} n_{10} + \omega_2 n_{20} \tag{111}$$

After changing the crystal orientation from θ_0 to $\theta_0 + \Delta\theta$, the following changes occur:

$$n_{30} \rightarrow n_{30} + \Delta n_3$$

$$n_{10} \rightarrow n_{10} + \Delta n_1$$

$$n_{20} \rightarrow n_{20} + \Delta n_2$$

$$\omega_{10} \rightarrow \omega_{10} + \Delta\omega_1$$

Since $\omega_1 + \omega_2 = \omega_3 = \text{constant}$,

$$\omega_{20} \rightarrow \omega_{20} + \Delta\omega_2 = \omega_{20} - \Delta\omega_1 \; (\text{i.e.,} \quad \Delta\omega_2 = -\Delta\omega_1)$$

Since Eq. 110 must be satisfied at $\theta = \theta_0 + \Delta\theta$, we have

$$\omega_3(n_{30} + \Delta n_3) = (\omega_{10} + \Delta\omega_1)(n_{10} + \Delta n_1) + (\omega_{20} - \Delta\omega_1)(n_{20} + \Delta n_2)$$

Neglecting the second-order terms $\Delta n_1 \Delta\omega_1$ and $\Delta n_2 \Delta\omega_1$ and using Eq. 111, we get

$$\Delta\omega_1 \bigg|_{\substack{\omega_1 \cong \omega_{10} \\ \omega_2 \cong \omega_{20}}} = \frac{\omega_3 \Delta n_3 - \omega_{10} \Delta n_1 - \omega_{20} \Delta n_2}{n_{10} - n_{20}} \tag{112}$$

According to our starting hypotheses the pump is an extraordinary ray, so that is depends on the orientation θ, giving

$$\Delta n_3 = \frac{\partial n_3}{\partial\theta}\bigg|_{\theta_0} \Delta\theta \tag{113}$$

The signal and idler are ordinary rays, so that their indices depend on the frequencies but not on the direction. It follows that

$$\Delta n_1 = \frac{\partial n_1}{\partial \omega_1}\bigg|_{\omega_{10}} \Delta \omega_1$$

$$\Delta n_2 = \frac{\partial n_2}{\partial \omega_1}\bigg|_{\omega_{20}} \Delta \omega_2 \tag{114}$$

Using Eq. 112, we obtain

$$\frac{\partial \omega_1}{\partial \theta} = \frac{\omega_3 (\partial n_3 / \partial \theta)}{(n_{10} - n_{20}) + [\omega_{10}(\partial n_1 / \partial \omega_1) - \omega_{20}(\partial n_2 / \partial \omega_2]} \tag{115}$$

for the rate of change of the oscillation frequency with respect to the crystal orientation. Using the relation $d(1/x^2) = -(2/x^3) dx$, we find

$$\frac{\partial n_3}{\partial \theta} = -\frac{n_3^{\,3}}{2} \sin(2\theta) \left[\left(\frac{1}{n_e^{\,\omega 3}} \right)^2 - \left(\frac{1}{n_0^{\,\omega 3}} \right)^2 \right]$$

which, when substituted in Eq. 115, gives

$$\frac{\partial \omega_1}{\partial \theta} = \frac{-\frac{1}{2}\omega_3 n_{30}^{\,3}[(1/n_e^{\,\omega 3})^2 - (1/n_0^{\,\omega 3})^2] \sin(2\theta)}{(n_{10} - n_{20}) + [\omega_{10}(\partial n_1 / \partial \omega_1) - \omega_{20}(\partial n_2 / \partial \omega_2)]} \tag{116}$$

An experimental curve, showing the dependence of the signal and idler frequencies on θ in $NH_4H_2PO_4$ (ADP), is shown in Fig. 15. Also shown is a theoretical curve based on a quadratic approximation of Eq. 110 which was plotted using the dispersion (i.e., n versus ω) data of ADP.[17]

Reasoning similar to that used to derive the angle tuning expression Eq. 116 can be applied to determine the dependence of the oscillation frequency on temperature. Here we need to know the dependence of the various indices on temperature. An experimental temperature tuning curve is shown in Fig. 16.

7.11 POWER OUTPUT AND PUMP SATURATION IN OPTICAL PARAMETRIC OSCILLATORS

From the theory of the laser oscillator (Eq. 116) we know that, in steady state, the gain cannot exceed the threshold value regardless of the intensity of the pump. A closely related phenomenon exists in the case of parametric oscillation. The pump field E_3 gives rise to amplification of the signal and idler waves. When E_3 reaches its critical (threshold) value, given Eq. 103, the gain just equals the losses and the device is on the threshold of oscillation. If the pump field E_3 is increased beyond its threshold value, the gain can no longer follow it and must be "clamped" at its threshold value. This follows from the fact that, if the gain constant g exceeds its threshold value Eq. 101, a

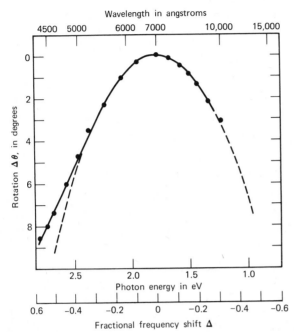

Fig. 15. The dependence of the signal (ω_1) frequency on the angle between the pump propagation direction and the optic axis of the ADP crystal. The angle θ is measured with respect to the angle for which $\omega_1 = \omega_3/2 = (\omega_1 - \omega_3/2)/(\omega_3/2)$.

steady-state solution is no longer possible, and the signal and idler intensities will increase with time. Since the gain g is proportional to the pump field E_3, it follows that, above threshold, the *pump field inside* the optical resonator must saturate at its level just prior to oscillation. As power is conserved, it follows that any additional pump power input must be diverted into power at the signal and idler fields. Since $\omega_3 = \omega_1 + \omega_2$, it follows that, for each input pump photon above threshold, we generate one photon at the signal (ω_1) and one at the idler (ω_2) frequencies, so that[18]

$$\frac{P_1}{\omega_1} = \frac{P_2}{\omega_2} = \frac{(P_3)_t}{\omega_3}\left[\frac{P_3}{(P_3)_t} - 1\right] \tag{117}$$

where P_1 and P_2 are the power generated at ω_1 and ω_2, respectively; and P_3 and $(P_3)_t$ are the actual value of the pump power input and its threshold value, respectively.

The last argument shows that, in principle, the parametric oscillator can attain high efficiencies. This requires operation well above

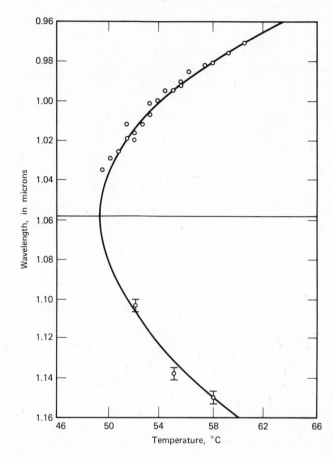

Fig. 16. Signal and idler wavelength as a function of the temperature of the oscillator crystal. (After Giordmaine and Miller.[14])

threshold so that $P_3/(P_3)_t \gg 1$. These considerations are borne out by actual experiments.[19]

Figure 17 shows experimental confirmation of the phenomenon of pump saturation.[18,21] After a transient buildup the pump intensity inside the resonator settles down to its threshold value.

Figure 17b shows that the signal power is proportional to the excess (above threshold) pump input power. This is in agreement with Eq. 117.

7.12 FREQUENCY UP-CONVERSION

Parametric interactions in a crystal can be used to convert a signal from a low frequency ω_1 to a high frequency ω_3 by mixing it with a strong laser beam at ω_2, where

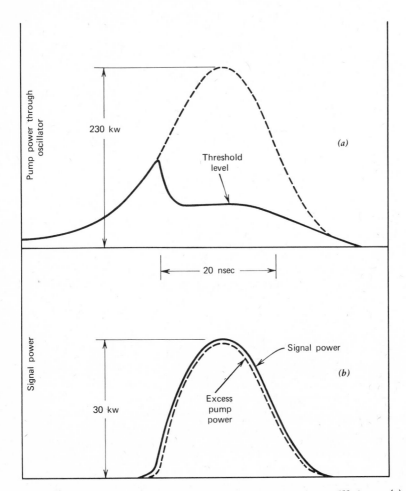

Fig. 17. Power levels and pumping in a parametric oscillator. (a) shows wave forms of P_3, the pump power passing through the oscillator. The dashed wave form was obtained when the crystal was rotated, so that oscillation did not occur; the solid wave form was obtained when oscillation took place. In (b) the dashed wave form is the normalized difference between the wave forms in (a); the solid wave form is the measured signal power. (After Bjorkholm.[19])

$$\omega_1 + \omega_2 = \omega_3 \qquad (118)$$

Using the quantum mechanical photon picture described in Section 7.6, we can consider the basic process taking place in frequency up-conversion as one in which a signal (ω_1) photon and a pump (ω_2) photon are annihilated while, simultaneously, one photon at ω_3 is gener-

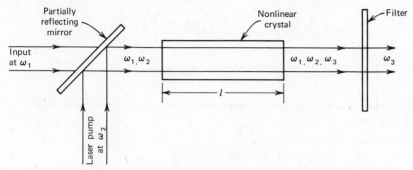

Fig. 18. Parametric up-conversion in which a signal at ω_1 and a strong laser beam at ω_2 combine in a nonlinear crystal to generate a beam at the sum frequency $\omega_3 = \omega_1 + \omega_2$.

ated.[22-25] Since a photon energy is $\hbar\omega$, conservation of energy dictates that $\omega_3 = \omega_1 + \omega_2$, while, in a manner similar to Eq. 62 the conservation of momentum leads to the relationship

$$\vec{k}_3 = \vec{k}_1 + \vec{k}_2 \tag{119}$$

between the wave vectors at the three frequencies. This point of view also suggests that the number of output photons at ω_3 cannot exceed the input number of photons at ω_1.

The experimental situation is demonstrated by Fig. 18. The ω_1 and ω_2 beams are combined in a partially transmissive mirror (or prism), so that they traverse together (in near parallelism) the length l of a crystal possessing nonlinear optical characteristics.

The analysis of frequency up-conversion starts with Eq. 80. As-suming negligible depletion of the pump wave A_2 and no losses ($d = 0$) at ω_1 and ω_3, we can write the first and third of these equations as

$$\frac{dA_1}{dz} = -i\frac{g}{2}A_3$$
$$\frac{dA_3}{dz} = -i\frac{g}{2}A_1 \tag{120}$$

where, using Eqs. 77 and 81, and choosing without loss of generality the pump phase as zero, so that $A_2(0) = A_2^*(0)$,

$$g = \sqrt{(\omega_1\omega_3/n_1 n_3)(\mu_0/\epsilon_0)}\, dE_2 \tag{121}$$

E_2 being the amplitude of the electric field of the pump laser. If the input waves have (complex) amplitudes $A_1(0)$ and $A_3(0)$, the general so-lution of Eq. 120 is

$$A_1(z) = A_1(0)\cos\frac{g}{2}z - iA_3(0)\sin\frac{g}{2}z$$

$$A_3(z) = A_3(0)\cos\frac{g}{2}z - iA_1(0)\sin\frac{g}{2}z$$

(122)

In the case of a single (low) frequency input at ω_1, we have $A_3(0) = 0$. In this case,

$$|A_1(z)|^2 = |A_1(0)|^2\cos^2\frac{g}{2}z$$

$$|A_3(z)|^2 = |A_1(0)|^2\sin^2\frac{g}{2}z$$

(123)

so that

$$|A_1(z)|^2 + |A_3(z)|^2 = |A_1(0)|^2$$

In the discussion following Eq. 79, we pointed out that $|A_l(z)|^2$ is proportional to the photon flux (photons/m²-sec) at ω_l. Using this fact, we may interpret Eq. 123 as stating that the photon flux at ω_1 plus the photon flux at ω_3 at any plane z is a constant equal to the input ($z = 0$) flux at ω_1. If we rewrite Eq. 123 in terms of powers, we obtain

$$P_1(z) = P_1(0)\cos^2\frac{g}{2}z$$

$$P_3(z) = \frac{\omega_3}{\omega_1}P_1(0)\sin^2\frac{g}{2}z$$

(124)

In a crystal of length l the conversion efficiency is thus

$$\frac{P_3(l)}{P_1(0)} = \frac{\omega_3}{\omega_1}\sin^2\left(\frac{g}{2}l\right)$$

(125)

and can have a maximum value of ω_3/ω_1, corresponding to the case in which all the input (ω_1) photons are converted to ω_3 photons.

In most practical situations the conversion efficiency is small (see the numerical example below), so that using $\sin x \cong x$, $x \ll 1$, we get

$$\frac{P_3(l)}{P_1(0)} \cong \frac{\omega_3}{\omega_1}\left(\frac{g^2 l^2}{4}\right)$$

which, using Eqs. 121 and 78, can be written as

$$\frac{P_3(l)}{P_1(0)} \cong \frac{\omega_3^2 l^2 d^2}{2n_1 n_2 n_3}\left(\frac{\mu_0}{\epsilon_0}\right)^{3/2}\left(\frac{P_2}{A}\right)$$

(126)

where A is the cross-sectional area of the interaction region.

7.12.1 Numerical Example of Frequency Up-Conversion

The main practical interest in parametric frequency up-conversion is because it offers a means of detecting ir radiation (a region in which detectors are either inefficient, very slow, or require cooling to cryogenic temperatures) by converting the frequency into the visible or near-visible part of the spectrum. The radiation can then be detected by means of efficient and fast detectors such as photomultipliers or photodiodes.[23-26]

As an example of this application, consider the problem of up-converting a 10.6 μ signal, originating in a CO_2 laser to 0.96 μ by mixing it with the 1.06 μ output of a Nd^{3+}: YAG laser. The nonlinear crystal chosen for this application has to have low losses at 1.06, 10.6, as well as 0.96 μ. In addition, its birefringence has to be such as to make phase matching possible. The crystal proustite (Ag_3AsS_3) listed in Table 8.1 of Ref. 26 meets these requirements.

We use the following data:

1. $\dfrac{P_{1.06\mu}}{A} = 10^4 \text{ W/cm}^2 = 10^8 \text{ W/m}^2$.

2. $l = 10^{-2}$ m.

3. $n_1 \cong n_2 \cong n_3 = 2.6$ (an average number based on the data of Ref. 26).

4. $d_{eff} = 1.1 \times 10^{-21}$. (This is taken conservatively as a little less than half the value given in Table 1 for d_{22}.)

Then we obtain from Eq. 126

$$\frac{P_{\lambda = 0.96\,\mu}(l = 1 \text{ cm})}{P_{\lambda = 10.6\,\mu}(l = 0)}$$

indicating a useful amount of conversion efficiency.

REFERENCES

1. R. C. Miller, "Optical Second Harmonic Generation in Piezoelectric Crystals," *Appl. Phys. Lett.*, 5, 17 (1964).
2. C. G. B. Garret and F. N. H. Robinson, "Miller's Phenomenological Rule for Computing Nonlinear Susceptibilities," *J. Quant. Electronics*, 2, 328 (1966).
3. J. F. Nye, *Physical Properties of Crystals*, Oxford University Press, New York, 1957.
4. J. A. Armstrong, N. Bloembergen, J. Ducuing, and P. S. Pershan, "Interactions between Light Waves in a Nonlinear Dielectric," *Phys. Rev.*, 127, 1918 (1962).

5. P. A. Franken, A. E. Hill, C. W. Peters and G. Weinreich, "Generation of Optical Harmonics," *Phys. Rev. Lett.*, 7, 118 (1961).

6. P. D. Maker, R. W. Terhune, M. Nisenoff, and C. M. Savage, "Effects of Dispersion and Focusing on the Production of Optical Harmonics," *Phys. Rev. Lett.*, 8, 21 (1962).

7. J. A. Giordmaine, "Mixing of Light Beams in Crystals," *Phys. Rev. Lett.* 8, 19 (1962).

8. F. Zernike, Jr., "Refractive Indices of Ammonium Dihydrogen Phosphate and Potassium Dihydrogen Phosphate between 2000 Å and 1.5 μ," *J. Opt. Soc. Amer.*, 54, 1215 (1964).

9. J. E. Geusic, H. J. Levinstein, S. Singh, R. G. Smith, and L. G. Van Uitert, "Continuous 0.53 μ Solid-State Source Using $Ba_2NaNb_5O_{15}$," *IEEE J. Quant. Electronics*, 4, 352 (1968).

10. A. Ashkin, G. D. Boyd, and J. M. Dziedzic, "Observation of Continuous Second Harmonic Generation with Gas Lasers," *Phys. Rev. Lett.*, 11, 14 (1963).

11. A. Yariv, *Quantum Electronics*, John Wiley & Sons, New York, 1967.

12. Lord Rayleigh, "On Maintained Vibrations," *Phil. Mag.*, 15, Ser. 5, Pt. I, 229 (1883).

13. A. Yariv, *Introduction to Optical Electronics*, Holt, Rinehart and Winston, New York, Chap. 6, in press.

14. J. A. Giordmaine and R. C. Miller, "Tunable Optical Parametric Oscillation in $LiNbO_3$ at Optical Frequencies," *Phys. Rev. Lett.*, 14, 973 (1965).

15. R. H. Kingston, "Parametric Amplification and Oscillation at Optical Frequencies," *Proc. IRE*, 50, 472 (1962).

16. N. M. Kroll, "Parametric Amplification in Spatially Extended Media and Applications to the Design of Tunable Oscillators at Optical Frequencies," *Phys. Rev.*, 127, 1207 (1962).

17. D. Magde and H. Mahr, "Study in Ammonium Dihydrogen Phosphate of Spontaneous Parametric Interaction Tunable from 4400 to 16000 Å," *Phys. Rev. Lett.*, 18, 905 (1967).

18. A. Yariv and W. H. Louisell, 5A2 "Theory of the Optical Parametric Oscillator," *IEEE J. Quant. Electronics*, 2, 418 (1966).

19. J. E. Bjorkholm, "Efficient Optical Parametric Oscillation Using Doubly and Singly Resonant Cavities," *Appl. Phys. Lett.*, 13, 53 (1968).

20. L. B. Kreuzer, "High-Efficiency Optical Parametric Oscillation and Power Limiting in $LiNbO_3$," *Appl. Phys. Lett.*, 13, 57 (1968).

21. A. E. Siegman, "Nonlinear Optical Effects: An Optical Power Limiter," *Appl. Optics*, 1, 739 (1962).

22. W. H. Louisell, A. Yariv, and A. E. Siegman, "Quantum Fluctuations and Noise in Parametric Processes," *Phys. Rev.*, 124, 1646 (1961).

23. F. M. Johnson and J. A. Durado, "Frequency Up-Conversion," *Laser Focus*, 3, 31 (1967).

24. J. E. Midwinter and J. Warner, "Up-Conversion of Near Infrared to Visible Radiation in Lithium-Meta-Niobate," *J. Appl. Phys.*, 38, 519 (1967).

25. J. Warner, "Photomultiplier Detection of 10.6 μ Radiation Using Optical Up-Conversion in Proustite," *Appl. Phys. Lett.*, 12, 222 (1968).

26. K. F. Hulme, O. Jones, P. H. Davies, and M. V. Hobden, "Synthetic P Proustite (Ag_3AsS_3): A New Material for Optical Mixing," *Appl. Phys. Lett.*, 10, 133 (1967).

Chapter 8

LASER SYSTEMS

Monte Ross

McDonnell Douglas Astronautics
St. Louis, Missouri

8.1 INTRODUCTION

There are two major types of laser systems: one dealing with information transfer, the other with energy transfer. In the former category are communications, radar, tracking, velocity and distance measurements, information storage and display, and specialized instrumentation. In the second category are laser thermal applications:
welding, drilling, cutting, etc.

For either category, in order to be useful and perform tasks previously impossible or impractical, laser systems use one of the following significant properties of the laser:

1. Spatial coherence.
2. Spectral coherence.
3. Bandwidth potential.

Because of the laser's spatial coherence, extremely narrow beamwidths are generated which are much narrower than any other source
of electromagnetic radiation can achieve. Basically, this results from
the beamwidth's being directly proportional to the wavelength of radiation. The beamwidth relationship is given by the Rayleigh diffraction
limit criterion,

$$\theta = 1.22 \frac{\lambda}{D} \tag{1}$$

where θ = the beamwidth in radians
     λ = the wavelength in micrometers
     D = the diameter of the transmitter aperture.
The importance of the beamwidth can be shown if one puts it in terms
of the antenna gain. A 4-in.-diameter laser system, using a diffraction limited laser, has an antenna gain equal to a microwave antenna
diameter of 400,000 in. Since the higher the gain, the less power necessary to transmit, narrow beamwidth enables the low-power laser to
be used for applications such as communications.

    The narrow beamwidth also enables high resolution to be achieved
for radar and tracking purposes. Since the beamwidth of 1 milliradian
means that in 1000 ft the beam will spread but 1 ft in diameter, it becomes feasible to detect objects separated by more than 1 ft as separate objects. In tracking applications the object or target position can
be measured to accuracies previously unattainable. The achievable
beamwidth as a function of aperture and wavelength is given in Fig. 1.

    This spatial coherence property of lasers is also very important
in laser thermal applications. The focusing of the energy to a fine
spot, thus achieving enormous energy densities, is a result of the spatial coherence capability of the laser. Ordinary light, no matter how
intense, cannot be focused to the same extent.

    The property of spectral purity is of value to communications,
ranging, and other applications, in which background light and other
noise sources must be rejected. The extremely narrow spectrum of
frequencies emitted by the laser enables optical filtering techniques to
be employed to reduce other light sources. Very high spectral purity
in some lasers enables receiver heterodyning techniques to be used to

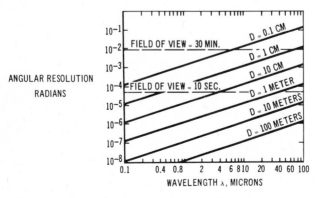

Fig. 1. Rayleigh diffraction resolution (aperture) versus wavelength.
(D = diameter of limiting aperture.)

enhance signal power and reduce other noise effects.

Another important property, bandwidth, enables short pulses to be employed for ranging, tracking, and communication purposes. The bandwidth capability of laser systems is mainly limited by technology.

In the following pages we provide basic considerations and recent results in information systems and thermal systems. Each laser application makes practical use of at least one of the laser systems properties discussed briefly above. In the application discussions below, it will be evident how the laser's properties are essential to the system. A very simple test to establish if a laser is really necessary for an application is to see if the system requires one of the attributes derived from these laser system properties: spatial and spectral coherence, and bandwidth capability.

8.2 LASER INFORMATION SYSTEMS

It is of value in discussing laser information systems to present first considerations on noise and detection in the optical spectrum. Noise is the ultimate limitation on system capability, since it prevents the signal from being recognized. In the optical spectrum, noise considerations are different than at radio frequencies; resultant system configurations may well reflect this difference. Although, at rf, thermal noise (kTB) is the dominant factor, quantum noise becomes most significant in the optical spectrum. Figure 2 illustrates the nature of radiation noise as a function of the wavelength. The quantum noise (hf) increases directly with frequency, whereas the thermal noise drops sharply at optical frequencies. A detailed treatment of this is not possible in this limited space but can be found in Refs. 1–3; however, the quantum noise is increasing simply because the energy per photon is increasing.

The nature of radiation is that of discrete particles; a random change in the number of particles represents a discontinuous change in energy. The minimum possible change is one particle or photon. At higher frequencies, this is a higher-energy value. In the visible, f is 10^5 higher than at S-band in the microwave region; therefore, the photon energy is 10^5 higher. To phrase it differently, for the same power, there will be 10^5 fewer photons in the visible band than at S-band. The exact number of photons being received in any measurement time varies because of the statistical nature of radiation. The photon fluctuation is approximately proportional to the square root of the mean; i.e., if the mean is 100 photons, a rms of 10 photons will

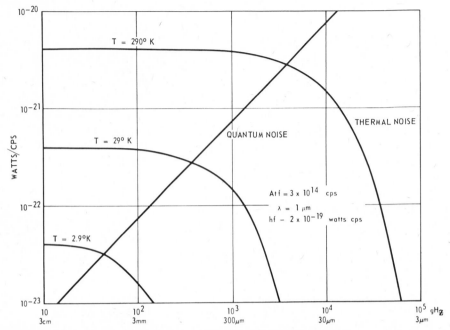

Fig. 2. Thermal and quantum noise versus frequency.

result. Therefore, for a single measurement, reasonable probabilities of measuring values other than 100 exist. This random photon fluctuation is the quantum noise which manifests itself in detectors as radiation-induced shot noise. If we consider the radiation as the signal, then we are measuring signal-induced shot noise. If we consider the radiation as unwanted background energy, we are measuring background-induced shot noise.

The well-known noise expression is given by[3]

$$i_n^2 = 2qiB \qquad (2)$$

where i_n = the shot noise current (rms)

i = the average detected current

B = the detector bandwidth

q = the electronic charge.

The current may represent the sum of any currents in the photodetector due to signal (i_s), background (i_b), and dark current (i_d), such that

$$i = i_s + i_b + i_d \qquad (3)$$

We note that the detected current is

$$i_s = \eta q \bar{n}_s \qquad (4)$$

where η is the quantum efficiency, and $\bar{n}_s$ is the average number of signal photons per second, since $\bar{n}_s = P_s/hf$, where P_s is the incident signal power. Then, using expressions (3) and (4), we have

$$\frac{i_s^2}{i_n^2} = \frac{\eta \bar{n}_s}{2B} \qquad (5)$$

However, $\bar{n}_s/2B$ can be interpreted as $\bar{n}$, the average number of photons in a measurement interval τ; there being $2B$ such measurement intervals every second such that the frequency time plane $2B\tau = 1$. By substituting $\bar{n} = \bar{n}_s/2B$ and taking the square root, we have

$$\frac{i_s}{i_n} = \eta\bar{n} = \frac{\eta\bar{n}}{\sqrt{\eta\bar{n}}} \qquad (6)$$

From Eq. 6 we note that the signal-to-noise ratio is proportional to the number of photoelectrons measured each interval, divided by the square root of that number, which is the quantum noise in the signal that manifests itself as signal-induced shot noise in the detector.

Thus the most significant difference in noise from rf is that the signal itself supplies noise and the background noise is also quantum in character. This significantly affects the system design, as will be evident in application discussions.

There are two basic types of laser detection systems: (a) direct detection (commonly referred to as noncoherent detection), and (b) coherent detection, also denoted as optical heterodyning. The heterodyne system is directly analogous to typical rf heterodyne systems. The components work on different physical principles, but the system concept and techniques are direct transfers from established rf and microwave concepts. The received signal, which has been modulated at the transmitter, is mixed with a local oscillator; the beat frequency passes through the if to a second detector, where the signal is recovered.

In direct detection, the system is more analogous to a crystal video detector at rf. The laser transmitter is pulse-modulated, and the received energy is collected and focused on a photodetector, usually a photomultiplier. The receiver responds only to the laser's intensity changes (not to the phase). The signal is amplified within the photomultiplier, so that at the output the signal is much larger than any thermal noise. The sensitivity limitations are signal quantum noise and background noise.

The proven desirability of heterodyning over video detection for rf systems does not necessarily apply to laser frequencies. In fact, there are systems in which pulsed direct detection is more effective than heterodyning, due to the different noise considerations at optical frequencies.

It is of value to discuss direct photodetection first, since until recently this was the only available technique. The technique consists simply of detecting the incident energy within the spectral response of the detector, with the resultant detected signal being able to follow the amplitude variations induced by the incoming signal modulation. In direct photodetection, all optical frequency and phase information are lost. The detector cannot respond to frequency or phase modulation of the optical carrier. It will reproduce amplitude variations of the incident power, as long as the rate of the variations is less than the frequency response of the detector.

The direct photodetector can make no distinctions between signal photons or nonsignal (background) photons that are within the relatively broad spectral response characteristics. It has no special arrival angle requirement, except that the photon be intercepted by the photosensitive area. Thus, in order to achieve spectral discrimination, we must insert an optical filter; similarly, if we are to achieve spatial filtering, we must reduce the field of view by optical means.

The advantage of direct photodetection by itself is its simplicity. Hence, in general, one would not utilize the other techniques, unless the system requirements were such that some technical advantage would be attained through their implementation. The advantage that is usually looked for as a justification for using photomixing or addition of a quantum amplifier is sensitivity, i.e., the ability to detect a weaker signal than can be detected with direct photodetection. In order to make clear the technical differences between photomixing and photodetection capability, we must first establish both (a) the ultimate, and (b) the presently attainable sensitivities of direct photodetection techniques.

An ideal direct photodetection device can be defined as one that can detect each photon within the signal frequency spectrum and no photons outside the signal frequency spectrum, and that contributes no internal noise current carriers or photoelectrons. Note that, in terms of any currently known photodetector, none of the three requirements is met.

The photodetectors that come closest to possessing one characteristic usually are worse on the others. However, the first characteristic can be partially solved with an optical filter and is unnecessary in the cases in which the background is not significant.

The sensitivity of the photodetector can be limited in three ways: it may be (a) photon-limited, (b) internal-noise-limited, and (c) background-limited.

Ideally, the photodetector could respond, so that one photoelectron or its equivalent is activated for each photon, such that $\eta = 1$. When the quantum efficiency is less than one, a portion of the signal is irrevocably lost, so that $n_d = \eta n_r$, where n_r is the received photons and n_d the detected photoelectrons. When a photodetector has little internal noise, so that its sensitivity depends only on the quantum efficiency, the detector can be considered photon-limited. For this condition the sensitivity of the photodetector falls short of the theoretically possible sensitivity by the quantum efficiency factor.

Internal noise consists primarily of shot noise and the thermal noise of the output resistance. The internal shot noise such as that in photomultipliers is given by $2qi_dBR$. It can be seen that the frequency response B and the output resistance R enter directly into the amount of the internal shot noise—to the same extent, in fact, as the dark current, i_d. However, the frequency response B and the output resistance R are not independent. Given the minimum output capacity of the detector C, the output resistance R can be only as large as the time constant RC which is limited by the frequency response B. (For the case of narrowband subcarrier systems, this is not strictly true, since conjugate matching, which increases the output power substantially, can be accomplished.) The ratio of signal power to signal, and background and internal shot noise power, is given by

$$\frac{S}{N} = \frac{i_s^2}{2q(i_d + i_b + i_s)B} \tag{7}$$

A major consideration is the detector thermal output noise power. The total signal-to-noise ratio is then

$$\frac{S}{N} = \frac{i_s^2 R}{FKTB + 2qBR(i_b + i_d + i_s)} \tag{8}$$

Any reduction in R will reduce the output signal power but will not affect the thermal noise power in the output resistance which is independent of the value of R.

The advantage of detectors with inherent postdetection gain can be seen from the analysis of this equation. If i_s is the output current that has been multiplied by postdetection gain, it improves the S/N ratio over no postdetection gain and improves it dramatically in the cases in which the thermal noise is greater than the total shot noise.

The shot noise is not discriminated against by postdetection gain, since it also is amplified. The thermal noise at the output is not amplified by the detector, however; in this fact lies the advantage of postdetection gain such as occurs in the photomultiplier and, to some extent, in avalanche photodiodes.

In many low-level applications, the thermal noise will be much greater than the output signal power, unless detectors with postdetection gain are employed or a quantum amplifier is employed before the photodetection. The amount of gain necessary to attain signal-photon-limited performance has been analyzed and is dependent on output capacities, signal bandwidth, and noise temperature of the following amplifier. In general a current gain of greater than 10^3 or 10^4 is necessary.

8.2.1 Photomixing

Because of the laser's narrow spectral line, it has become possible to obtain mixing action at optical frequencies between two laser sources, one of which can be considered the signal and the other a local oscillator. Thus it is possible to build an optical heterodyne or homodyne receiver. (The operation is called homodyne when the local oscillator is the same frequency as the optical carrier.) Figure 3 illustrates the photomixing technique. Complex experiments prior to laser development indicated that photomixing action can occur; however, the lack of a sufficiently narrow spectral source with adequate

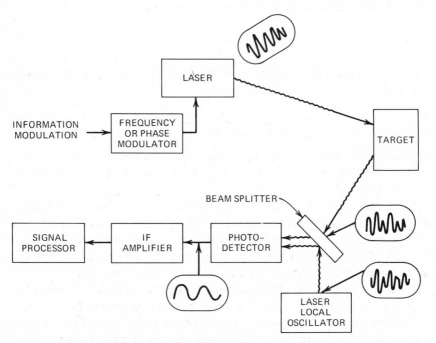

Fig. 3. Example of optical heterodyning (photomixing).

power made measurements extremely difficult.

Photomixing offers certain advantages as a receiving technique compared to direct photodetection. There are also a number of additional complexities to utilization of a photomixing system. Photomixing is best understood by considering the wave nature of light. This is consistent with the dual nature of radiation, represented by waves and particles. Photomixing, as in the rf heterodyning processes, is a form of coherent detection, whereas direct photodetection can be regarded as noncoherent detection. In noncoherent detection, there is no negative output, the detector functioning as a rectifying element. The noncoherent detector response can be expressed as an even infinite series, as follows:

$$e_0 = ae_1^2 + be_1^4 + ce_1^6 + \cdots \tag{9}$$

Generally, we can ignore the higher-order terms and get a good approximation of the action by assuming that $e_0 = ae_i^2$, i.e., that the rectifier is a square-law detector. Most photodetectors are ideal square-law devices, in that higher-order terms are not present.

8.3 SOME COMPARISONS BETWEEN PHOTOMIXING AND DIRECT PHOTODETECTION

It is clear that many factors must be considered in deciding which receiving technique is more useful for a particular application. It would be fruitful, however, to present some fundamental comparisons between the two, with the realization that any judgment must depend on how the comparison applies to a particular use. Table 1 is a summary of the comparisons.

First, the fundamental difference is that photomixing requires a laser local oscillator. In some radarlike systems, in which the transmitter and receiver are at the same physical location, it may be possible to utilize the same device as both the transmitter source and the local oscillator. In communication systems, however, it is clear that a laser must be used in a photomixer receiver, although one is not necessary in direct photodetection. The use of a laser local oscillator requires a stable, narrow linewidth source to avoid additional noise and interference.

The spatial requirements for photomixing have been shown to be much more severe than for microwave mixing.[3] The basic reason is that the light wavelength is small, compared to the photomixing area. Consider Fig. 4. The local oscillator and signal should be in phase

TABLE 1. COMPARISON BETWEEN PHOTOMIXING AND DIRECT
PHOTODETECTION

Photomixing	Photodetection
Needs laser local oscillator.	No local oscillator required.
Spatial phasing requirements (superposition).	No spatial phase problem.
Expected Doppler shifts require wide optical bandwidth and cause IF problems.	Expected Doppler shifts only require wider optical bandwidth.
Phase distortion limits effective receiver area.	Restriction several orders of magnitude less severe (effective receiver areas can be built much larger).
3-dB less noise (hfB) theoretically possible than in photodetection.	Minimum noise is 2 hfB.
Large conversion gain possible with no decrease in input S-N ratio.	Postdetection secondary emission multiplication with no decrease in input S-N ratio, current gains of 10^5 to 10^6.
Background directional and frequency discrimination without use of optical filter	No frequency discrimination in photodetection without use of optical filter. Additional discrimination with optical filter.
Medium can affect system through loss of coherence.	Medium not as critical, loss of coherence not catastrophic.

across the whole photosurface. If they are not, beat currents at one part of the surface will be out of phase with beat currents at another part of the surface, resulting in signal loss.

Since λ is of the order of 10^{-4} cm and the photosurface might be about 1 cm, it turns out that the angle between oscillator and signal must be less than 10^{-4}, in order not to reduce the term involving the difference frequency.

The stringent spatial requirements on photomixing given above lead to serious practical problems in a receiver system that differs considerably from the problems connected with the use of photomixing as an experimental laboratory tool. The atmosphere, for example, creates corruption of the phase front, making it most difficult to perform heterodyning, especially at visible wavelengths.

This requirement, in turn, puts constraints on how large a photon-collecting area one can use, i.e., lens or parabolic mirror size. Op-

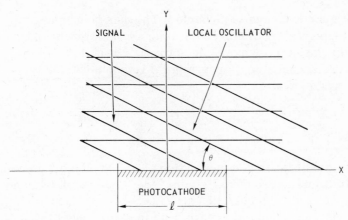

Fig. 4. Spatial relationships in photomixing.

tical techniques will restrict the receiver lens or reflecting parabola
to the size at which phase distortion begins to become appreciable.
When the path lengths from points on the collection area to the photo-
cathode become significantly different by the order of the wavelength
of the optical frequency, spatial phase coincidence with the local oscillator will be lost and rapid deterioration in photomixing performance
will occur. Hence, for systems in which large collecting areas are
possible, photodetection offers the ability to gather more useful signal
photons.

The possible large Doppler shift at the optical frequencies can present serious problems in a photomixing system subject to large velocity changes. As the radial component of velocity changes, the Doppler
shift frequency changes, causing a different frequency. In some cases,
either the if must be extremely broad or it must be tunable over a broad
range, or the laser local oscillator must be tunable over the necessary
range.

The heterodyne SNR is given for continuous wave signals, by

$$\frac{s}{n} = \frac{P_s}{hfB} \qquad (10)$$

In direct detection, in the most sensitive condition of being signal-
photon-limited, we obtain for continuous wave operation,

$$\frac{s}{n} = \frac{P_s}{2hfB} \qquad (11)$$

However, it is generally not good modulation design to use continuous wave signals with direct detection. By using short pulses, substan-

tial improvement in performance is possible. This has been under
analytical and experimental study for some time; in the following pages
we treat some of the possible formats and error rate results. It is
appropriate and more accurate to discuss digital communications in
error rate terms, rather than SNR.[3,4]

8.4 COMMUNICATIONS

Recently, the field of laser communications has traveled two main
routes: (a) CO_2 $10:6$ μm heterodyne systems, and (b) direct detection,
visible, or near ir systems, useful with Nd:YAG and semiconductor
lasers. The reasons for this division have been the efficiency of CO_2
lasers and the feasibility of coherent detection at 10.6 μ, and analytical
and practical advantages to direct detection in the visible and near ir
over coherent detection.

The CO_2 heterodyne communication system is directly analogous
to typical rf communications. The components work on different physical principles, but the system concept and techniques are direct transfers from the established rf concept. The laser signal is modulated,
either by orthogonal phase shift modulation, or by FM or pulse code
modulation. The received signal is mixed with a local oscillator, and
the beat frequency passes through the rf for a second detection, where
the signal is recovered.

The CO_2 laser must be exceptionally stable in frequency because
a heterodyne system depends on a known local oscillator frequency and
a known carrier frequency. Single-mode, diffraction-limited performance is required to achieve the narrow beamwidth necessary for long-
range communications. Useful aperture size for heterodyne detection
is limited on the earth because of the atmosphere. This limitation, due
to phase corruption of the wave by the atmosphere, is a significant disadvantage because the received beam will be weak and the detector
needs all the intercepting area it can reasonably attain.

Direct detection systems do not suffer from phase corruption of the
atmosphere, because the phase of the optical signal is of no consequence. Because of this, one is not limited by a coherent aperture as
in heterodyne systems. The collector is essentially what has been referred to in the literature as a photon bucket. There are limitations
on this kind of collector, but receiving apertures much larger than coherent apertures can be achieved before these limitations take place.
The coherent aperture for the CO_2 wavelength is approximately 1 to
1.5 m; for visible wavelengths it is a few inches.[5,6] The improve-

ment in coherent aperture size at 10.6 μm is approximately 20, the ratio of the wavelengths (10/0.5). At present, it appears that noncoherent detection apertures can be 15 m or greater.[7]

In near space-earth-orbiting and synchronous satellites, both CO_2 lasers and direct detection visible lasers appear feasible. Decisions as to which is more desirable depend on the particular mission, technology development, and experimental verification; for example, earth-orbiting satellites create a Doppler shift problem for heterodyne systems. However, if one examines the deep-space problem, one finds that two factors weigh heavily against the use of CO_2 lasers in deep space: one is the limited coherent aperture on earth; the second is the longer wavelength which requires much larger, heavier optics to attain a beam the same width as with visible lasers. The limited collector size and weight force the CO_2 laser power for deep space use to such large values as to make it impractical.

Two major problems exist for direct detection laser communications. One is the need for enough signal photoelectrons to permit a high probability of signal detection. The second is the need to minimize the background noise level to make false detection highly unlikely. Ideally, the background noise should be reduced, so that the number of signal photoelectrons required to distinguish between "signal" and "no signal" is not much larger than the number of signal photoelectrons established for the "no background" case.

The background noise level can be appreciably reduced by using a narrow filter and reducing the field of view at the receiver. There are limitations to this approach, however. Further background discrimination, denoted as temporal discrimination, can be achieved by optical receiver, pulse-gating techniques. This discrimination method is uniquely applicable only to short-pulse, low-duty cycle modulation formats.

The effect of temporal noise discrimination techniques on communication efficiency is dependent on the system's modulation format. In general, the shorter the sampling interval, the greater will be the noise discrimination. This is true because fewer background photons are received in the shorter interval, reducing the probability that statistical fluctuation of the background level would exceed a preset threshold in the receiver and introduce false detection errors.

Poisson statistics for low-level detection show that the probability of no detection decreases substantially with an increase in signal level, whereas a decrease in background level leads to an extreme decrease in the probability of false detection. Therefore, it is advantageous to have the sampling interval as short as possible and the peak power as high as possible.

Another potential advantage to short pulse modulation is that M-ary techniques can be readily employed when each signal pulse contains more than a single bit of information.

In general, for visible and near infrared laser systems it is desirable to utilize short pulse laser modulation techniques. Several system concepts have evolved which are in this category. They are pulse interval modulation (PIM), pulse-gated binary modulation (PGBM), pulse polarization binary modulation (PPBM), and pulse position modulation (PPM).

Pulse-gated binary modulation is similar to normal, continuous wave binary PCM, in that one bit per pulse is transmitted by a 1 or 0 output. The difference is that, in PGBM, the laser operates at a low-duty cycle, in which the pulse width is only a small fraction of the interpulse spacing. A significant feature of PGBM is the fixed pulse spacing that is compatible with the inherently regular-spaced pulse output of a mode-locked laser. Pulses are gated on or off with a high-speed electrooptic modulator to provide coding. In addition, a fixed pulse spacing permits the use of a pulse-gated optical receiver for noise discrimination. These two features, coupled with the low-duty-cycle operation, make PGBM a powerful and efficient modulation technique for high data rates. A block diagram is shown in Fig. 5.

Pulse polarization binary modulation is identical to PGBM, except for the method of modulation of the 1 and 0 pulse output. Instead of gating the mode-locked laser pulses on and off as in PGBM, the polarization of the laser pulse radiation is rotated in PPBM for 1 and 0 differentiation. The advantage of PPBM is that a signal is expected in every time interval at the receiver. The disadvantage is that the receiver must have a dual channel, one for each polarization.

Pulse interval modulation is an M-ary process in which one pulse conveys information representing many bits in an ordinary binary system.[8] The normal interpulse time interval is divided into M discrete time slots. One and only one pulse is sent in the time interval of M slots (see Fig. 6). The specific time slot in which a pulse occurs is representative of a code symbol. The number of bits transmitted per pulse is therefore $\log_2 M$.

By using short pulses, many time slots (M large) are possible in a given time interval, thereby achieving many bits per pulse. Noise discrimination results from discrete sampling of each narrow time slot in the optical receiver. Very low duty cycle ($1/M$) is achieved because only one pulse occurs in the M time slots.

Pulse interval modulation, like PGBM, utilizes analog-to-digital conversion prior to encoding the laser pulsed output. However, in PIM

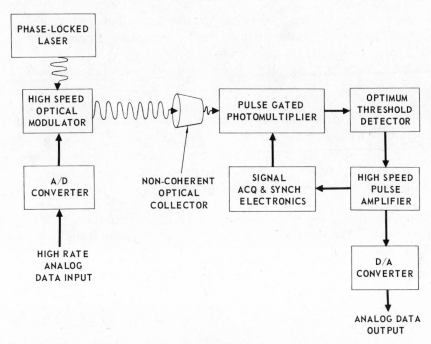

Fig. 5.   High-data-rate laser system block diagram.

a further conversion of the binary digital information into PIM format
is required.   The advantages of many bits per pulse and temporal noise
discrimination make PIM an extremely efficient pulse modulation format.   However, state-of-the-art, high-speed encoding and decoding
electronics technology limits the achievable bit rate.

Pulse position modulation is useful for direct analog information
inputs.   The technique is similar in form to the PIM technique, except that pulses are not assigned digital time slots and rates must be
at least twice the information bandwidth to meet sampling theory requirements.   Pulses are sent with interval spacing which are the dis-

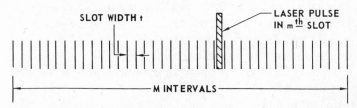

Fig. 6.   Pulse interval modulation (PIM).

crete value of the sampled analog information to be conveyed. The analog information is handled without format processing; hence high-speed, analog-to-digital conversion is avoided.

Because PPM is a sampled analog technique, each transmitted pulse conveys many bits of information. Therefore, the duty cycle is generally lower than PGBM but higher than PIM. The advantage of PPM for pulsed optical communications is the potential achievement of very high data rates.

8.5 DECISION MAKING AND ERROR RATES

In the receiver, the decision is made as to whether the signal is present or not. The probability of receiving m photoelectrons in a time interval t is

$$P_t(m) = \frac{a^m e^{-a}}{m!} \tag{12}$$

where (m) is the average number of photoelectrons received in the time interval t.

There exists in the decision-making receiver a threshold level, n_t, above which it is decided that a signal has been received. Obviously, by appropriate raising or lowering of the threshold, one can change the probabilities of exceeding the threshold. An analysis shows that an optimum threshold exists for each combination of signal, nonsignal, and duty cycle as given by[9]

$$n_{opt} = \frac{n_s + \log\{[P(0)]/[P(1)]\}}{\log[1 + (\bar{n}_s/\bar{n}_b)]} \tag{13}$$

where n_{opt} = the optimum threshold value (where optimum is defined as reducing the likelihood of an error to a minimum)

$P(0)$ = the probability of transmitting a zero "0" (i.e., no signal sent)

$P(1)$ = the probability of transmitting a one "1" (i.e., signal sent)

$\bar{n}_b$ = the average number of nonsignal photoelectrons/sec, where the nonsignal includes both internally and externally generated photoelectrons that are not from the signal source. Thus $a = \bar{n}_b t$, where there is no signal

$\bar{n}_s$ = the average number of signal photoelectrons in a single pulse. Thus, if one signal pulse is present within a measurement period t, $a = \bar{n}_b t + \bar{n}_s$.

An equal weighting of errors, i.e., an error of false detection costs as much as an error of no detection, is assumed in Eq. 13. However, we may arrange our system such that, in time T, we have M equal periods of duration t. If there is only one pulse to be transmitted in M periods, each of time t, the probability of transmitting a zero will be $(1 - 1/M)$; for $M \gg 1$, this will essentially be unity.

When we are making a decision every t, Eq. 13 can be written

$$n_{opt} = \frac{\bar{n}_s + \log M}{\log(1 + \bar{n}M/n_b T)} \tag{14}$$

Thus the threshold value will be a function of the number of short intervals in which a signal pulse might be present. Short-pulse transmission will enable easier signal detection, but since there are more of these periods in which to make a decision, is it not likely that more frequent false detection might result?

The Poisson statistics are such that quite the contrary result occurs. There are fewer false detections because of the enormous gain in signal detection capability. This can be accomplished by raising the threshold (as given in Eq. 14), thereby reducing the false detection probability each interval t to a low value.

The advantage of having lower false detections in time period T is further enhanced by the many bits/pulse that can be conveyed. This improves the error rate per bit in comparison with the long pulse (high-duty-cycle) case.

Consider Fig. 6, in which M intervals or slots are present. Each slot can represent a unique number. If we restrict our system to only one pulse occurring in the M intervals in time T (a so-called M-ary system), then each pulse represents $\log_2 M$ bits. If we send on the average F pulses/sec, we find that the bit rate R is

$$R = F \log_2 M \tag{15}$$

bits/sec. The number of intervals, M, is not independent of the pulse rate once a pulse width is established. With a pulse time interval, t, we have

$$R = \frac{1}{Mt} \log_2 M \tag{16}$$

bits/sec.

We can establish for PIM the amount of information per unit of signal energy received (bits/J) for fixed error rate/bit as a function of duty cycle. We note that the signal energy per pulse is $E = hf\bar{n}_s$, where h is Planck's constant $(6.62 \times 10^{-34}$ J/sec$)$ and f is the optical frequency. The average signal power is given by $P_{av.} = E_s F$. Thus, using Eq. 16 and $\lambda = 1$ μm, we obtain

$$\frac{R}{P_{\text{av.}}} = 5 \times 10^{18} \frac{\log_2 M}{n_s} \tag{17}$$

bits/J.

We have determined Eq. 17 as a function of duty cycle $(1/M)$, keeping the error rate/bit fixed. We have fixed the bit rate and nonsignal power, so that the only independent variable is the duty cycle. The requirement of fixed error rate/bit will force $\bar{n}_s$ to change as a function of the duty cycle. Calculations have been made for a variety of conditions; one set of data is plotted in Fig. 7.

Because PCM is a common form of pulse-time modulation used in communications, it is of value to compare PIM with PCM. We shall consider the PCM system to be a simple on-off type of 50%-duty cycle. Energy sent represents a 1, energy not sent represents a 0, and the noncoherent detector as in PIM must discriminate between 1s and 0s. Each pulse interval in PCM represents one bit. In this comparison, for simplicity, we assume that all pulse intervals contain information. Thus, for a bit rate of R, the PCM system pulse interval is $1/R$.

The large improvements of the PIM systems over the PCM-type systems are illustrated in Table 2. If we design a PIM system of error rate/bit of 10^{-4}, we find that, for the same average power and average background levels, the PCM signal level is so deep in the noise that the error rate is near its maximum 0.5. If we wish to obtain the same error rate/bit for the PCM system as for PIM, then we must increase the PCM average power by the levels indicated in row B.

Experimental results[10] for heterodyne systems are shown in Fig. 8. Experimental results[11,12] for pulsed laser systems include Fig. 9

INFORMATION EFFICIENCY IN BITS/JOULE

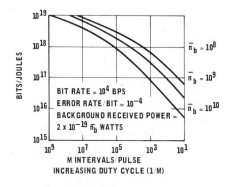

Fig. 7. Deep space, high-data-rate laser communications.

TABLE 2. COMPARISON OF PIM[a] and PCM[b] SYSTEMS INFORMATION RATE

| | System Bit Rate | | | | | |
| | I 260 BPS | | II 20,000 BPS | | III $1.3 \cdot 10^6$ BPS | |
	PIM	PCM	PIM	PCM	PIM	PCM
(a) Error rate/bit same average power and same average background	10^{-4}	0.5	10^{-4}	0.5	10^{-4}	0.49
(b) Relative increase in PCM power to equal PIM error rate/bit	1	10^4	1	10^3	1	85

[a]PIM System—pulse width = 1 NSEC   [b]PCM—50% duty cycle
I.   10 PPS, duty cycle 10^{-8}
II. 1000 PPS, duty cycle 10^{-6}
III. 10^5 PPS, duty cycle 10^{-4}

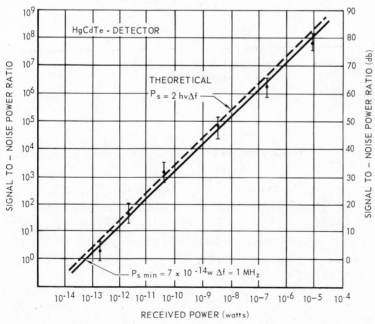

Fig. 8. Signal–to–noise ratio for Hg Cd Te–detector.

345

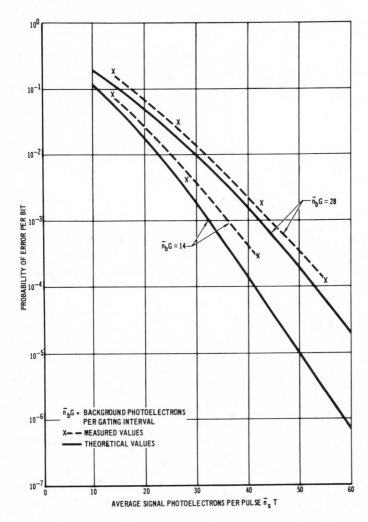

Fig. 9. PGBM error rate data.

for PGBM and Fig. 10 for PIM. A 10.6-μm communications experiment is planned by NASA in 1975 on ATS-G.[13,14] Direct detection links for specialized applications on earth should become common in the next few years. A Nd: YAG communication space experiment is planned in 1975 by the United States Air Force.

8.6 RANGING

There are two basic types of laser ranging systems: (a) the time of flight pulse systems,[15] and (b) the frequency and phase systems. The

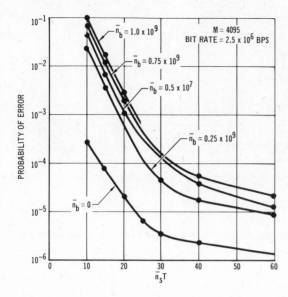

Fig. 10. PIM error measurements. 12 bits/pulse system.

latter may take two forms, in that frequency and phase techniques may be applied to (*a*) the carrier frequency, or (*b*) the subcarrier frequency.

Since light travels approximately 1 ft in 1 nsec, each nsec time difference between start and pulse return represents $\frac{1}{2}$ ft. Range resolution for a single pulse, dependent on rise time accuracy, is improved proportional to $\sqrt{N}$, where N is the number of pulses occurring (during which time nothing has changed). The shortest range measurable is restricted by the pulse width, i.e., no pulse overlap should occur between start and stop pulses. The maximum range is determined by the radar range equation applied to the optical case. Since the laser beam is quite narrow, it is possible for short ranges and large targets that all the energy will hit the target. For this special case, the space attenuation is a function of R^2, and not the usual R^4. The fourth power relationship of radar range is valid, in general, since the energy spreads out from the transmitter proportional to R^2 and the incident energy on the target returns toward the receiver, also spreading out proportional to R^2.

In frequency and phase techniques, the range is measured by the change in frequency or phase between a reference and the return signal. For example, 5-MHz subcarrier has a 200 nsec cycle time which represents the 100-ft range. Thus a 90-degree phase difference of the 5-MHz subcarrier between the reference and return signal indicates a 25-ft range. This method is used for short ranges. The velocity is measured by measuring the phase difference change with time. Range

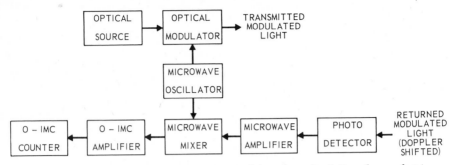

Fig. 11.   Block diagram of microwave-modulated, optical Doppler radar.

ambiguities that occur (every 100 ft in the example) can be resolved by other techniques (such as pulsed systems). Velocity can be measured by the Doppler shift

$$f_{ds} \cong \frac{2Vf}{c} \tag{18}$$

where f_{ds} = the Doppler shift
       V = the radial velocity
      c = the velocity of light
      f = either the carrier or subcarrier frequency.

If f is the carrier, the Doppler shift frequency can be quite large, resulting in the capability to measure very small velocities and accelerations. For $f = 3 \times 10^{14}$ Hz,

$$V = \frac{f_{ds}}{2 \times 10^6} \tag{19}$$

If f_{ds} could be measured to 1 cycle, a velocity less than 1 μm/sec could be measured. Since lasers are not that stable, the accuracy is not that good. However, the use of the carrier enables accuracies many times more than usually required. A velocity of 1 mm/sec will give a Doppler shift of 2 kHz which is in measurable range.

    The laser stability problems make carrier Doppler shift measurements, which involve coherent detection, quite difficult. Helium-neon and CO_2 lasers have been used, with He-Ne used in laboratory applications and CO_2 in operational systems.

    If we use a microwave subcarrier f_{sc} and measure its Doppler shift, then for $f_{sc} = 3 \times 10^9$, in m/sec,

$$V = \frac{f_{ds \text{ of subcarrier}}}{2 \times 10} \tag{20}$$

Now, a one-cycle shift, which is detectable in the microwave sub-carrier, measures a velocity of 0.05 m/sec which is more than adequate for most ranging applications. A block diagram is shown in Fig. 11.

In the radar system the received power will be a function of the target cross section, as well as the transmitter gain, power, receiver photon collecting area, range, and optical efficiency. Specifically, we can write

$$P_R = \eta_T \frac{P_T G_T \sigma}{4\pi R^2} \left(\frac{A_R}{4\pi R^2} \right) \tag{21}$$

where η_T = the overall optical transmission factor or efficiency
G_T = the target optical cross section
A_R = the effective receiver antenna area.
Noting that $G_T = 4\pi/\Omega$, we can write

$$P_R = \eta_T \frac{P_T \sigma A_R}{R^4 (4\pi) \Omega} \tag{22}$$

When the beamwidth is given by the diffraction-limited capability of the optical system, and the plane angle $\Theta = \lambda/D_T$ the solid angle for the diffraction limited case is

$$\Omega = \left(\frac{\lambda}{D_T} \right)^2 \approx \frac{\lambda^2}{A_T} \tag{23}$$

$$P_R = \frac{\eta_T P_T A_T A_R \sigma}{R^4 (4\pi) \lambda^2} \tag{24}$$

and

$$\eta_T = \eta_{0T} \eta_P \eta_{0R} \tag{25}$$

where η_p = the transmission optical efficiency of the two-way path
η_{0T} = the efficiency of the transmitter optics
η_{0R} = the efficiency of the receiving optics.
Within the earth's atmosphere, an η_T of only 10% can easily occur. The loss due to the medium is often given by $\exp(-2\alpha R)$, where α is the medium attenuation coefficient. In Table 3 we present the ranging expression with appropriate description of the terms and briefly describe the appropriate laser for pulse ranging.

In the earth's atmosphere a restriction on the beamwidth achievable is placed by the effects of the air on beam dispersion. It is likely that beams narrower than a few seconds will not be feasible in earth systems.[16]

In a radar system, the detection problem is compounded by two factors, (a) the fourth-power relationship with range, and (b) the area

TABLE 3. RANGING

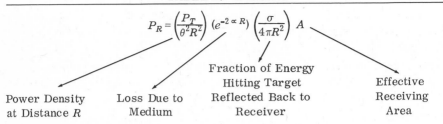

$$P_R = \left(\frac{P_T}{\theta^2 R^2}\right)(e^{-2\alpha R})\left(\frac{\sigma}{4\pi R^2}\right)A$$

Power Density at Distance R	Loss Due to Medium	Fraction of Energy Hitting Target Reflected Back to Receiver	Effective Receiving Area

P_R = received power
P_T = transmitted power
θ = laser beamwidth
σ = target scattering cross section
A = effective receiving area
α = medium attenuation coefficient.

Pulse Ranging

Short	GaAs	Very high rep rates	Small, efficient with pulsed outputs
Medium–ranging	Nd : YAG	High-rep rates	High-peak power
Long-range	Ruby Nd : Glass	Low-rep rate	Very high peak power

that must be searched. In some applications the latter problem does not exist. The fourth-power relationship of radar range is valid for all radar systems, since the energy spreads out from the transmitter proportional to R^2 and the incident energy on the target returns toward the receiver, spreading out proportional to R^2. In cases in which the target is larger than the incident beam, all of the transmitted energy is intercepted. For this special case the space attenuation is a function of R^2, and not R^4.

To calculate the target reflected power, it is necessary to know the optical cross section of the target. The optical cross section can be considered to be the target area multiplied by some reflection efficiency. If the target area is A_{ta}, then

$$\sigma = (A_{ta})\eta_{ta} \tag{26}$$

where η_{ta} is the optical cross-sectional efficiency factor. This factor η_{ta} will be a function of the wavelength and the target surface material and the angle at which the energy strikes the target.

The maximum range for a single pulse will be that in which the minimum detectable signal, P_{mds} is received. Thus, utilizing Eq. 24

$$R_{max} = \sqrt[4]{\frac{P_T \sigma A_R \eta_T}{P_{mds}(\Omega)4\pi}} \tag{27}$$

The SNR is improved by receiving multiple pulses, as in microwave radar. This improvement can be considered for a square law detector (energy detection) to be approximately the square root of the number of integrated pulses for low SNR. (At high SNR the improvement will be closer to linear.) Thus, if the radar system integrates n_r pulses returning from the target, the resultant SNR is given by

$$\left(\frac{S}{N}\right)_{n_r} = \sqrt{n_r}\left(\frac{S}{N}\right)_1 \tag{28}$$

where $(S/N)_1$ is the SNR for a single pulse.

The number of integrated pulses, as the radar scans through its beamwidth, is

$$n_r = \frac{\theta_b(\text{prf})}{(\text{scan rate})} \tag{29}$$

where Θ_b is the antenna beamwidth in degrees. The repetition frequency (prf) is in pulses per second, and the scan rate in degrees per second.

The presentation of performance curves valid for an energy-detection radar is complicated by the fact that performance depends not only on the signal-to-noise ratio but also on the absolute levels of signal and noise. Figures 12 and 13 illustrate one way of presenting the data.[17, 18] Detection probability (in percent) is plotted versus $10 \log (\eta W/hf)$ for various noise levels and for a fixed false alarm probability of 10^{-6}. In

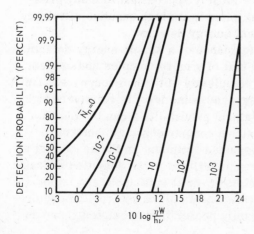

Fig. 12. Detection statistics—energy detection of specular or well-resolved rough targets.

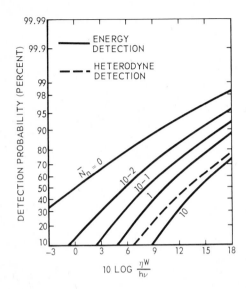

Fig. 13. Comparative performances of heterodyne and energy detection. Nonresolved rough target.

Fig. 12, a Poisson-distributed signal is assumed; therefore, the curves apply for a specular target or a well-resolved rough target.

In general the performance curves depend on the particular false-alarm probability that can be tolerated, with the curves shifting to the left as higher false-alarm probabilities are allowed. Notable exceptions are the curves valid in the absence of noise (i.e., when $\overline{N}_n = 0$) which we shall refer to as the "photon-limited" curves. The photon-limited curves represent absolute limits to the performance of an energy detection radar and are entirely independent of the false-alarm probability that can be tolerated. Furthermore, these limits are meaningful ones, in the sense that they can be closely approached in many present-day applications. A photomultiplier in its high quantum efficiency spectrum serves as an almost ideal energy detector.

The range measurement performance of a simple energy detection receiver can significantly exceed that of a more complex and critical heterodyne receiver. The extra complexity of the heterodyne system is justified only when (a) the background noise level is relatively high, and/or (b) relatively large false-alarm probabilities can be tolerated. We again emphasize that, in the above comparisons, the heterodyne-receiver performance has been assumed optimum, in the sense that all of the incident target-return energy is utilized in the detection process. If a balanced mixing system is not used, or if sufficient care is not taken to detect both polarization components of the return, then heterodyne performance can be significantly poorer than predicted above and

energy detection will be a superior technique over an even wider range
of conditions than implied above.

8.7 TRACKING

Laser systems are useful for high-resolution tracking because,
(a) the narrow beam enables target selection from multiple targets, and
(b) high-accuracy measurement of the range and range rate is feasible.
The beamspread of beamwidth, Θ, at a distance R will be approximately θR, so that a 1-milliradian beam will be spread only 10 ft at
10,000 ft. Therefore, targets 10 ft apart can be resolved, any detail
smaller than 10 ft cannot be resolved. At closer ranges the resolution
in actual feet improves. Profiles of targets such as automobiles at
hundreds of feet have been obtained. A 1-milliradian beam at 100 ft
can resolve close to 1 in., sufficient to readily provide a profile of
any large object.
The atmospheric limitation restricts the useful beamwidth on earth
to perhaps 20 to 100 microradians. In space, microradian beamwidths
are possible, limited only by the laser's coherence and external optics.
However, in space one is usually dealing with rapidly moving objects,
and the narrower the beam, the more difficult acquisition becomes, so
that system limitations may limit useful beamwidths to much larger
than a microradian.

8.8 OTHER LASER INFORMATION SYSTEMS

There are many other laser information system applications beside those discussed above. Laser systems include such variety as
plasma diagnostic techniques using laser interferometry, LIDAR techniques for measuring atmospheric constituents, and the well-publi-
cized holographic systems. All of these and other applications make
use of some laser property already discussed. To consider each general class of application in detail is beyond the scope of the article.
The references have been made extensive to include a number of specific applications of laser information systems. We discuss, briefly,
holography, since it is an area of large potential applications.

8.9 HOLOGRAPHY

The laser has generated enormous interest in holography, a method of reproducing not only the intensity but the phase of the signal re-

ceived from an object. The hologram work thus far indicates that three-dimensional pictures can be achieved in which one can literally see around an object by moving its position. There are severe problems in accomplishing this with moving objects, but, for stationary objects, remarkable pictures have been achieved using lasers. The hologram concept can be explained briefly as follows.

In ordinary photography, the intensity of the received light is recorded and the random phase of sunlight or artificial light is discarded. In holography, by the use of a coherent light source, such as a stable laser to illuminate the object, both the intensity and the relative phase can be recorded.[19]

The hologram appears as a collection of swirls and random lines when viewed with ordinary light. However, if properly illuminated by a monochromatic source or by a laser, a true three-dimensional picture results.

Moving objects present a much greater difficulty in the making of holograms, since the phase of the coherent light is recorded. Any movement on the order of the wavelength of the light (approximately 10^{-4} cm), during the time the hologram is being made, effectively destroys the phase relationship. The allowable movement of the system becomes a function of how fast one can take a hologram, which in turn is a function of the laser power and sensitivity and response of the film used. The helium-neon laser was utilized early because of its high coherence and continuous wave capability. Its power output (milliwatts), however, is insufficient to take rapid holograms. In many cases it required minutes to take a good hologram. Holograms have been constructed with pulsed ruby lasers, the argon-ion laser, and others. Three-dimensional holographic motion pictures are difficult, since velocities of merely 1 m/sec cause traversals of 1 wavelength in 1 μsec. Therefore, very narrow pulses of high coherence and high power are required.

The previous discussion has been simplified and elemental in approach. A much broader range of applications are possible than simply three-dimensional pictures. In general, holographic applications can be grouped into three sections relating to the laser use of the process.[20] In the first group are those applications involving image formation when, for a variety of reasons, normal image formation is not satisfactory. These applications include particle size analysis, holographic microscopy; data storage and retrieval, including displays; and image formation through a random medium.

The second group are those applications that are not image forming. Holographic interferometry for nondestructive testing on which one records a hologram of the object and then causes interferometry with this image with the laser illuminated object itself (or another image from a second hologram). Differences between the two can indicate structural and other changes in the object.

The third group are those applications that use the hologram in optical system because of its characteristics as an optical element. Examples are accurate specialized gratings and holographic filters in coherent optical data processing.

8.10 LASER THERMAL SYSTEMS

There are a wide variety of applications using lasers as a "directed energy" source.[21,22] A short, rather general list, includes:

Welding	Controlled fracturing
Drilling	Material surface heating
Component trimming	Solid state transformations
Cutting	and diffusion
Scribing	Zone melting

Lasers produce different effects, dependent on power densities, when applied to materials as sources of noncontact energy. Traditional methods of processing materials have depended primarily upon bringing various forms of mechanical, electrical, or chemical energy into contact with the workpiece. However, the "directed energy source" concept enables operations without intimate contact between the energy source and the workpiece. Such sources include high-power lasers capable of supplying power densities in excess of 10^6 W/cm^2 to surfaces remote from the source. This raises the attractive possibility of eliminating such items as knives, drills, abrasive wheels, flames, chemicals, and electrodes from some material processing operations, thereby reducing maintenance, replacement, and direct labor costs. Those effects of interest to machining and welding will be considered here, although many other useful thermally induced phenomena (i.e., optical absorption edge shift, refractive index changes, electrical conductivity changes, etc.) are possible by laser action.

1. *Melting*. If sufficient heat is applied to a solid, its temperature is increased to the melting point of the material, thus for the material enabling solid material to be transformed into the liquid state. Cutting and drilling are accomplished if the liquid is removed by gravity or convective flow, or by a jet or gas. Welding is accomplished if the liquid formed at the juncture of two separate pieces is permitted to solidify.

2. *Evaporation.* By the application of sufficient heat, the liquid created upon the melting of a solid can be increased in temperature to its evaporation point. The liquid can be transformed into a gas or vapor at the evaporation temperature by the addition of the heat of evaporation.

3. *Sputtering.* During the process of evaporation, bathing vapors can impart kinetic energy to small volumes of liquid within the melt such that these globules of liquid can acquire sufficient energy to overcome surface tension forces and be ejected from the heated zone.

4. *Heating of Entrapped Gases.* Porous materials contain gaseous voids that, upon heating, can produce large internal pressures due to the expansion of the heated gases. These forces can irreversibly weaken the sample by causing microfractures and the ejecting particulate matter.

5. *Chemical Reactions.* Laser heating can induce irreversible chemical reactions that result in the weakening or removal of material. Oxidation (i.e., burning) converts solids and liquids into volatile gases which readily escape from the heated zone. Steel can be cut by the exothermic chemical reaction between the steel and an oxygen jet, with the laser merely initiating and sustaining the reaction.

6. *Phase Changes.* Concentrated heat can alter the molecular arrangement of a material and so change the metallurgical phase. These products are often weaker or of increased volume, causing fracture from thermal or expansive stresses.

7. *Thermal Stresses.* If a material is heated unevenly, gradients in local temperature are produced. Most solid materials expand upon heating. Large temperature gradients in a material therefore produce large expansion differentials. Thermally induced stresses within a material can cause it to fail through fracture.

8. *Shock Wave Propagation.* It is possible that high-peak, power-pulsed laser systems cause beam energy to propagate into the irradiated material by means of a traveling shock wave. This impulse of acoustical energy, if excessive, can cause structural weakening and microfractures contributing to material failure.

Pulsed and continuous lasers have found a wide variety of applications in the processing of materials commonly used in industry. It appears that by now almost every type of material has been thrust into the focused beam of a high-power laser, and the gross characteristics of various laser effects are well established.

Many materials would appear to be unsuitable for laser processing because of their high reflectivity at certain laser wavelengths. Table 4 summarizes the nominal reflectivity of several metals at laser wave-

TABLE 4. REFLECTANCE OF SEVERAL METALS FOR NORMAL INCIDENCE
OF LIGHT AT COMMON LASER WAVELENGTHS

Wavelength, μ	Au	Cu	Mo	Ag	Al	Cr	Fe	Ni
0.4880 (Ar II)	0.415	0.437	0.455	0.952	—	—	—	0.597
0.6943 (Cr^{3+})	0.930	0.831	0.498	0.961	—	0.555	0.575	0.676
1.06 (Nd^{3+})	0.981	0.901	0.582	0.964	0.733	0.570	0.650	0.741
10.6 (CO_2)	0.975	0.984	0.945	0.989	0.970	0.930	—	0.941

lengths of interest. In particular, most metals are so highly reflecting
at the 10. 6-μ carbon dioxide wavelength that only a few percent of the
incident beam energy is transmitted into the material.

However, surface finish and the dynamics of material removal
greatly affect the absorption of energy within a material. The reflectivity of polished copper at 694.3 mμ could be reduced from 95% to
less than 20% by oxidizing the surface, and that surface roughness on
the order of the laser wavelength greatly increased the depth of energy
penetration. Once some absorption of energy has occurred, handbook
reflectivity data become almost meaningless. At high-power densities
a plasma plume of ejected vapors is formed, and the absorption of laser energy in the vapor can significantly reduce the effective reflectivity.
The removal of surface material by the initial pulse of laser energy
forms a small crater that "traps" the incident light within the hole; this
entrapment of radiation also increases the effective absorption rate at
the surface. Thus published reflectivity data can be employed to estimate absorption in smooth, uncontaminated surfaces at low-power
levels but should not be extrapolated to high-power densities in which
craters and plasma plumes are formed.

Incident energy that is not reflected at the surface is absorbed as
the light propagates into the medium. This absorption is described by
Lambert's law which states that

$$I(x) = I(0) \exp(-\alpha x)$$
$I(x) =$ light intensity in watts
after propagation of xm
$I(0) =$ light intensity at $x = 0$, in watts (30)
$\alpha =$ absorption coefficient in m^{-1}
$x =$ propagation distance in meters

Thus most of the energy propagating into the material is absorbed
in a few "skin depths" δ, where

$$\delta = \alpha^{-1}$$ (31)

For normal incidence, materials that strongly reflect light are highly absorbent as well. For most metals the skin depths δ are less than 0.1 μm for visible and ir wavelengths, and for most organic compounds far infrared 10.6-μ CO_2 radiation is absorbed in less than 1 μm. Therefore, laser absorption may be considered to be surface effect in most materials of interest, with energy propagating into the bulk primarily through heat conduction.

The power density within a laser spot is given by

$$P = \frac{4E}{\pi f^2 \theta^2 t} \tag{32}$$

where P = the power density at the focal plane of the lens
     E = the energy output from the laser
     f = the local length
     θ = the beam divergence (full angle)
     t = the laser pulse length.

The focused spot size

$$S = f\theta \tag{33}$$

The laser requirements in thermal systems for high power and high coherence are thus apparent, in that the power density is greatly depen-

TABLE 5.   CUTTING POWER OF JET-ASSISTED CO_2 LASER

Material	Thickness, in.	Cutting Rate, ipm	Max. Kerf Width, in.	Laser Power on Material, W	Gas Jet Assistance
Steel, C1010	0.125	22	0.040	190	Oxygen
Stainless steel, 321	0.020	92	0.012	165	Oxygen
Stainless steel, 321	0.050	30	0.020	165	Oxygen
Zircaloy	0.018	600	0.020	230	Oxygen
Boron fiber-resin composite	0.050	9	0.030	165	Argon
Alumina, 99.5%	0.025	50	0.012	250	Nitrogen
Carpet, polyester	3/8	120	0.020	200	Argon
ABS plastic	0.100	150	0.030	240	Nitrogen

dent on small angles and high energy per pulse (or high continuous wave power).

In Table 5 we list the cutting rates for various materials, attained with jet-assisted CO_2 lasers. It is apparent the present commercially available powers can do significant material-processing work. Dependent on the application, Nd: YAG, pulsed or continuous wave, pulsed ruby, argon-ion, Nd: glass, or CO_2 may be the best choice. Each case must be examined separately. The high continuous wave powers now available for Nd: YAG and CO_2 tend to make those two lasers the most likely candidates for use.

REFERENCES

1. J. P. Gordon, *Proc. IRE*, 50, 1898 (1962).
2. B. M. Oliver, *Proc. IEEE*, 53, 436 (1965).
3. M. Ross, *Laser Receivers*, John Wiley & Sons, New York, 1966.
4. W. Pratt, *Laser Communication Systems*, John Wiley & Sons, New York, 1969.
5. H. Hodara, "Effects of Turbulent Atmosphere on Phase and Frequency of Optical Waves," *Proc. IEEE*, 56, 2130–2136 (1968).
6. D. L. Fried, "Optical Heterodyne Detection of an Atmospherically Distorted Signal Wavefront," *Proc. IEEE*, 55, 57–67 (1967).
7. M. Ross, "Deep Space Laser Communications," *Laser Focus*, 51–59 (October 1969).
8. M. Ross, *IEEE EASCON Proc.*, October 1967. Also special supplement to *IEEE AES Trans.*, AES-3, 324 (November 1967).
9. T. Curran and M. Ross, *Proc. IEEE*, 53, 1770 (1965).
10. F. Goodwin and T. Nassmeier, "Optical Heterodyne Experiments at 10.6 Microns," Paper 10J-5 presented at IEEE Quantum Electronics Conference, Miami, Fla., May 1968.
11. M. Ross, S. I. Green, and J. Brand, *Proc. IEEE*, 58, (October 1970).
12. R. Denton and T. S. Kinsel, *Proc. IEEE*, 56, 140 (1968).
13. N. McAvoy, H. Richard, J. H. McElroy, and W. E. Richards, "10.6 Micron Laser Communications System Experiment for ATS-F and ATS-G," NASA Goddard Space Flight Center Preprint X-524-68-206, May 1968.
14. J. H. McElroy, N. McAvoy, and H. L. Richard, "Carbon Dioxide Laser Intersatellite Communication System," *Laser Journal*, 2, No. 1, 20–21. (January/February 1970).
15. T. S. Johnson, H. H. Plotkin, and P. L. Spedin, *IEEE J. Quant. Electronics*, QE-3, 440 (November 1967).
16. "Optical Space Communications," NASA SP-217, Proc. NASA-MIT Workshop, Williamstown, Mass., August 1968.
17. J. W. Goodman, *IEEE Trans. Aerospace Electronic Systems*, AES-2, No. 5, 526 (September 1966).
18. J. W. Goodman, *Proc. IEEE*, 53, 1688 (1965).

19. H. M. Smith, "Principles of Holography," John Wiley & Sons, New York, 1969.
20. B. Thompson, "Holographic Applications," in "Laser Applications," Vol. 1, M. Ross, Ed., Academic Press, New York, 1971.
21. F. P. Gagliano, K. M. Lumley, and L. S. Watkens, *Proc. IEEE*, 57, 114 (1969).
22. L. Weaver, "Thermal Applications," in "Laser Applications," Vol. 1, M. Ross, Ed., Academic Press, New York, 1971.

Chapter 9

IMAGE PICKUP AND DISPLAY DEVICES

B. Kazan

International Business Machines Corporation
Thomas J. Watson Research Center
Yorktown Heights, New York

9.1 INTRODUCTION

Although the fields of image pickup and display devices are still dominated by electron beam tubes, developments in solid state technology are resulting in significant improvements in these areas. Changes in image devices are evident in two ways. The availability of new materials and solid state structures has made possible the design of electron beam devices, particularly camera tubes, with improved characteristics. At the same time, solid-state circuit technology has progressed sufficiently to allow the fabrication of all-solid-state, circuit-scanned image devices whose performance (although still limited, compared to tube devices) already suggests the usefulness of such devices for special purposes.

Although solid-state image pickup and display devices both involve the common problem of scanning and addressing a two-dimensional array containing as many as 10^5 to 10^6 elements, in other respects their requirements and problems differ considerably. In camera pickup devices, for example, the aim is to sense small variations in the potential or conductivity of individual elements. Power dissipation in the sensor array itself is thus generally small. To enable the design of cameras that are compact and light in weight, photosensor arrays with a small area (of the order of 1 cm^2) are desired. Such devices can thus profit from the availability of small monolithic structures. In the case of display devices, however, it is necessary to distribute significant amounts of electrical power to the picture elements to produce an image of adequate brightness. Also, a relatively large area (of the order of 1000 cm^2) is usually required for viewing purposes. Here the use of small monolithic structures interconnected in large numbers does not necessarily provide a satisfactory solution. For such devices the use of large-area microcrystalline layers, for example, may be more desirable.

9.2 IMAGE PICKUP (CAMERA) DEVICES

9.2.1 Electron-Beam-Scanned Tubes

Introduction. Electron-beam pickup tubes fall into two broad classes, depending on whether their operation is based on the photoemission of electrons into vacuum or on the internal processes of photoconductivity. The first practical television camera tube, the iconoscope,[1] developed in about 1934, was based on the use of a mosaic of isolated photoemitting elements on the surface of a mica sheet. Since then, camera tubes based on photoemission have undergone many stages of development and refinement, resulting in today's image orthicon[1] and secondary emission conductivity[1] (SEC) pickup tubes. Despite their very good performance, however, such tubes are relatively expensive and bulky, and must be carefully adjusted to optimize their operation. Because of this, their use is frequently restricted to special applications, involving, e.g., the pickup of images at very low light levels.

By comparison photoconductive pickup tubes are in widespread use in many applications. They are not only capable of high-quality performance but are also relatively simple, compact, easy to use, and low in cost. From a technical point of view such tubes are of particular interest, having not only profited most directly from conventional semiconductor technology but also providing the basis for the development of the newer circuit-scanned devices. In view of these factors, only tubes based on photoconductivity will be considered in the following discussion.

Homogeneous Photoconductor Camera Tubes (Vidicon). The arrangement of a conventional photoconductive camera tube,[1,2] commonly referred to as the vidicon, is shown in cross section in Fig. 1. For simplicity, some of the auxiliary electrodes and components used for deflection and focusing have been omitted. The photoconductor is in the form of a continuous film supported on a transparent backplate, B, which is maintained at a potential, V. In operation the photoconductor surface is scanned in raster fashion by the electron beam, i_b, acting as a switch that sequentially charges the surface of each element to

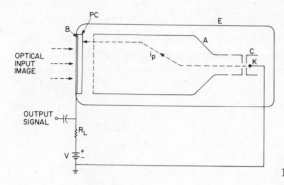

Fig. 1.   Vidicon camera tube.

ground potential. (Although the photoconductor is not divided into discrete elements, the effective size of the elements is determined by the
diameter of the electron beam.) Since, in the dark, the resistivity of
the photoconductor is high (e.g., $>10^{12}$ Ω-cm), the RC time constant
of the elements is substantially longer than the frame repetition rate
of the beam and the layer remains uniformly charged. If an optical image is projected onto the photoconductor, however, increasing the local conductivity, the illuminated elements will discharge to some degree between scans. When contacted by the electron beam, the discharged elements are then recharged, thus producing a capacitive current flow through them and the series load resistor, R_L, generating a
time-varying output signal across the load resistor. It is important to
note that, in this type of operation, a signal integration occurs, since
the output signal corresponds to the integrated loss of charge of the elements between scans.

The first commercial vidicon tubes employed Sb_2S_3 as the photoconductive material. Despite the fact that many photoconductive materials are known, very few exhibit the combined properties essential for
vidicon operation, namely, high dark resistivity, adequate light sensitivity over the visible spectrum, short response time (not exceeding,
e.g., $\frac{1}{30}$ sec), and physical characteristics to allow operation in a vacuum environment. Although important new materials (as described in
the following section) have emerged in the last few years, Sb_2S_3 is still
extensively used, despite some of its limitations. It is, therefore, of
value to discuss briefly the performance of tubes with Sb_2S_3 targets.

The Sb_2S_3 layer is produced by evaporation and deposited as a porous
polycrystalline layer about 5 μ thick and usually about 2.5 cm in diameter. Of importance is the fact that the contacts made to the photoconductive layer by the backplate and the electron beam are essentially
ohmic in nature, i.e., no barrier exists at the surfaces to prevent charges

from freely entering or leaving the material. Characteristic of photoconductors with ohmic contacts, the sensitivity or quantum yield (defined as the number of carriers traversing the layer per absorbed photon) is given by the relation[3]:

$$q = \frac{\mu \tau V}{L^2} \tag{1}$$

where V = the applied voltage
L = the electrode spacing
μ = the carrier mobility (only one carrier assumed mobile)
τ = the carrier lifetime.
Although this relation indicates that the sensitivity can be increased by raising the voltage, in practice (assuming that breakdown does not occur), the voltage is limited by problems of dark current, as mentioned below.

At low applied voltages, charge neutrality is maintained in the material, i.e., as each carrier is withdrawn, it is replaced by a new carrier at the opposite electrode. The dark current in this case is determined by the thermal generation rate. At high-applied fields, a significant number of carriers may be drawn into the material in excess of charge neutrality (limited only by space-charge considerations). Corresponding to these carriers a space-charge-limited current flow, I, results, given by the relation[4]:

$$I = \frac{10^{-13} V^2}{L^3} \mu k \tag{2}$$

in A/cm^2, where k is the dielectric constant. (This relationship is valid for the case in which one carrier is mobile and any trapped electrons present are in thermal equilibrium with the conduction electrons, μ being the effective mobility. For other trap distributions the space-charge-limited current rises at a much higher power of the voltage.)

In practice, because of the rapid rise in dark current with voltage, potentials not in excess of 20 to 30 V are applied across the photoconductive layer. At these potentials, quantum yields of the order of 0.1 are obtained. Despite this limited yield, however, good operation can be obtained with vidicon tubes in ordinary room light or somewhat lower.

It is of interest that, on theoretical grounds, the maximum quantum yield, q_m, obtainable for a photoconductor with ohmic contacts (except for special trapping conditions) is given by the relation[5]:

$$q_m = \frac{\tau_0}{R_d C} \tag{3}$$

where τ_0 is the response time of the photoconductor to changes in illumination (associated with carrier lifetime and trapping) and $R_d C$ is the capacitive time constant of the photoconductor in the dark. In practice, if the photoconductor is to be useful in the charge integration mode, $R_d C$ must be equal to a frame time ($\frac{1}{30}$ sec) or longer. At the same time the photoconductor must respond to changes in light within a frame time or less if moving images are to be viewed. If both these conditions are satisfied, q_m must be less than unity.

Aside from the limitations in gain, a number of problems exist with vidicons. Because of variations in thickness of the photoconductor or surface potential variations produced by differences in the landing angle of the beam, variations in field occur over the layer surface. This results in nonuniformities in dark current and sensitivity across the target which may appear as objectionable shading in the output picture.

Because of the characteristics of the material itself and also the reduction in photosensitivity as the potential across the layer falls during discharge, the output signal is less than linear with respect to the input light. Typically, the output voltage is proportional to the 0.65 power of the input light ($\gamma = 0.65$). Where signals from two or more tubes are to be added (as in color TV camera) such a nonlinear transfer characteristic may be objectionable.

Because of the high density of traps, the Sb_2S_3 photoconductor exhibits a noticeable lag in its response to changes in light, particularly at low-light levels. This lag tends to become progressively worse as the light level is reduced, causing an objectionable smear in the case of moving images.

Another problem associated with vidicon tubes is the susceptibility of the Sb_2S_3 target to damage. After scanning a specific area for an extended time, this area tends gradually to change its characteristics. If a new area, overlapping the old area, is scanned, a background pattern corresponding to the old area may be seen. Also, exposure of the target to very strong light may produce permanent local changes or damage to the photoconductor.

Extended-Junction Camera Tubes (*Plumbicon*). Despite the considerable amount of effort devoted to improving the Sb_2S_3 target and investigating alternative photoconductors for the vidicon, relatively few improvements were made until about 1963, when the Plumbicon tube was announced.[6] In the Plumbicon, instead of a "homogeneous" photoconductor with ohmic contacts, a target structure having rectifying or blocking contacts is used. This approach has made possible a number of significant improvements, as discussed below.

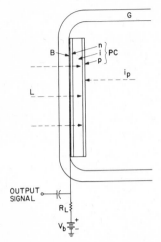

Fig. 2. Photoconductive (PbO) target of Plumbicon tube.

The target structure of the Plumbicon (exaggerated in thickness) is shown in cross section in Fig. 2. Instead of a uniform single layer, an evaporated triple-layer structure of PbO about 10 μ thick is employed. Adjacent to the n-type SnO_2 transparent backing electrode, B, is a thin layer of n-type PbO. This is followed by a thicker layer, i, of intrinsic PbO, and in turn by a thin layer of p-type PbO which is scanned by the electron beam, i_p. Although the internal operation of this layer structure is different from that of the Sb_2S_3 layer used in the vidicon, the circuit aspects of signal generation are the same. As in the vidicon, a positive bias, V, is maintained on the backplate, and the opposite surface of the layer is scanned by the electron beam which sequentially charges the elements to zero potential.

The operation of the target itself can be understood with the aid of Fig. 3 showing a simplified energy band diagram for a p-i-n structure. (Since the intrinsic material may vary from slightly p-type to n-type

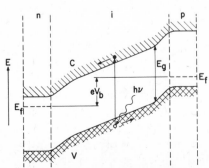

Fig. 3. Energy-band diagram of reverse-biased p-i-n diode.

across its thickness, the band edges are shown somewhat curved, rather than straight.) As a result of the charging of the p-layer to zero potential by the electron beam, the Fermi levels of the p- and n-type layers are vertically displaced by an amount, eV_b. The structure is thus maintained in the reverse-biased condition, blocking the flow of majority carriers from both the n-type and p-type layers into the intrinsic layer. Since the band gap of the PbO is about 2.0 eV (the crystal form being about 90% tetragonal), the dark current due to thermally generated carriers in the intrinsic material is also low.

As the bias voltage across the layers is increased, the dark current rises until about 10 V is applied. As the voltage is raised beyond this, however, there is practically no further increase in current. In practical tubes, whose scanned area is about 12×16 mm, the total dark or reverse current is of the order of 10^{-9} A. This should be compared to the vidicon, whose dark current in typical operation is at least an order of magnitude higher. Equally important, since the dark current in the Plumbicon is very low and not field-dependent, background disturbances due to this are minimized.

Since a relatively high bias voltage (for example, 50 V) can be applied across the target layer of the Plumbicon, a large fraction of the holes and electrons optically generated in the intrinsic layer are swept out to the n- and p-layers, respectively, before they can recombine. The photocurrent, as well as the dark current, is thus saturated, resulting in a quantum yield at the peak of the spectral sensitivity curve of about 50%. Because of the efficient extraction of carriers, the transfer characteristic is also relatively linear (corresponding to a γ of 0.95), making such tubes useful for color TV cameras.

An additional advantage of the PbO layer is its relatively rapid response to changes in input light. As already indicated, the vidicon has an objectionable photoconductive lag at low-light levels. In the Plumbicon, trapping effects are much smaller, reducing the lag to about 20% or less of that of the vidicon. Also, the lag is relatively independent of the light level.

Because of the 2.0 eV band gap of the PbO, light of wavelengths greater than about 6200 Å is not absorbed. Since the eye response extends to about 6700 Å, such tubes are not adequate for normal TV use. To provide sensitivity at wavelengths longer than this, some of the oxygen of the intrinsic PbO layer (on the electron beam side) is replaced with sulfur.[7] This reduces the band gap and allows light of wavelengths up to about 7200 Å to be absorbed in this region. Extension of the spectral response by this technique results in only a small increase in the dark current.

The PbO layers of the Plumbicon are prepared by the evaporation of the PbO in a suitable gas atmosphere. The resultant layer consists of crystallites having the form of platelets about 1.0×1.0 μ in area and 0.1 μ thick. These platelets are oriented perpendicular to the layer surface and occupy about 30 to 50% of the layer volume. To make the PbO material n-type, it can be prepared either with an excess of Pb or doped with Bi. To make it p-type, it can be prepared with an excess of oxygen or doped with thallium, copper, or silver. In practice it is essential to prevent the p-type layer scanned by the electron beam from being too conductive; otherwise, the resolution of the output image will be degraded by surface leakage and the signal amplitude reduced. To avoid this, the p-type layer is made as thin as possible and its doping level limited. The latter approach, however, cannot be carried too far, since sufficient doping of this layer is necessary to obtain good blocking action and low dark current.

Silicon-Diode-Array Camera Tubes. Although the PbO target of the Plumbicon is somewhat less delicate than the target of the vidicon, it can also be damaged in operation. For applications such as the Picturephone, a rugged camera tube is required that can operate unattended under varying illumination conditions, particularly high-light levels, without degradation of the target. Also, electronic zooming is desired, in which the magnification of the output image is increased by reducing the scanned area of the camera tube. As already mentioned, with tubes employing evaporated targets, scanned areas are gradually aged, making zooming operation unsatisfactory. Problems of the above type have led to the development of the silicon diode array tube[8] described below.

The cross section of the target portion of such a tube is shown in Fig. 4. The target here consists of a self-supported wafer of 10 Ω-cm, n-type silicon about 2.2 cm in diameter and 10 to 20 μ thick. For support purposes the thickness of the edges is increased to about 100 μ. On the side facing the electron beam, the surface is coated with a 0.5 μ-thick layer, D, of SiO_2, having an array of small holes 8 μ in diameter and 20 μ on centers. Through these holes, boron is diffused to a depth of about 2 μ, forming an array of p-type islands. The resultant structure consists of 660×660 diodes within 1.32 cm^2 on the wafer surface. As shown in Fig. 4, the surface of the SiO_2 and the p-type islands is coated with a thin layer, S, of resistive material. For contacting the substrate, the edge of the wafer is made n^+ and coated with gold, G.

In operation, a positive bias, V_b, of 5 to 10 V is maintained on the silicon wafer while it is scanned with the low-energy electron beam. If we assume for the present that the resistive film is absent, the SiO_2

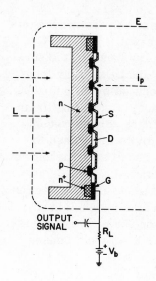

Fig. 4. Silicon diode-array camera-tube target.

layer will be charged to zero potential (i.e., cathode potential of the electron gun), remaining at this potential because of its high resistivity. Since the p-type islands are also charged to zero potential, the p-n junctions are reverse-biased. The reverse current, however, of the junctions is sufficiently low in the dark, allowing the junction capacitance of the diodes to remain almost fully charged between scans.

If a pattern of light, L, is projected onto the opposite side of the wafer, the absorbed photons will generate electron-hole pairs close to the input surface, the absorption coefficient of the silicon being > 3000 cm^{-1} for visible radiation. Since the depletion width of the p-n junctions at the opposite side of the wafer is about 5 μ (considerably less than the total thickness of the wafer), most of the carriers will originate in the field-free region outside the junction. Photogenerated holes that can diffuse across the wafer to the opposite side, however, will act to discharge the back-biased diodes, producing a potential pattern on the array of p-type islands. As in the vidicon, scanning of the surface by the electron beam shifts the p-type islands back to zero potential, producing a time-varying current corresponding to the input image through the load resistor.

Since minority carrier diffusion lengths of 30 to 100 μ (i.e., greater than the wafer thickness) can be obtained, most of the photogenerated holes can diffuse across to the junction regions before recombining. In practice, recombination at the input surface is the primary factor limiting the collection efficiency of the junctions. This recombination is especially significant in the case in which carriers are generated

very close to the input surface by shorter wavelengths. To reduce this
recombination, the surface is coated with a layer that is more n-type
than the wafer. This causes a bending of the energy bands at the surface, resulting in a local field that tends to draw the holes away from
the surface. In practice the surface layer is produced by diffusing
phosphorus into the wafer to a depth of about 0.4 μ. By coating the input surface with a thin layer of silicon monoxide, the surface recombination is further reduced and the optical reflectivity is reduced to
less than 0.1 over the visible spectrum. By the use of such measures,
collection efficiencies in excess of 0.5 can be obtained.

One of the problems arising with the above targets is the retention
of negative charges on the SiO_2 layer surrounding the p-type islands.
This charge tends to prevent the scanning electron beam (whose energy
in the neighborhood of the target is several eV or less) from landing
on the more positive p-type islands. One solution is to coat the entire
surface with a thin film of high-resistance material, S, as shown in
Fig. 4. This film, referred to as a resistive sea, prevents the buildup of excess negative charge on the SiO_2 surface by providing a controlled leakage path to the adjacent diode elements. In practice the
surface resistivity of the conductive sea should be $\gtrsim 50 \times 10^{13}$ ohms/
square, to prevent signal degradation. A possible material for the
resistive sea is evaporated Sb_2S_3 (similar to the photoconductive material used in the vidicon). However, as already mentioned, this material may change its characteristics in operation. Another possible
material, although involving problems of reproducibility, is evaporated
GaAs.

The total dark current of the diode array target does not exceed
about 5 to 50 nA (corresponding to a dark current of less than 10^{-13} A
per diode). Since the capacitance of the array is about 2000 pF/cm^2,
the time constant in the dark is longer than a TV frame time, as required for charge storage operation.

One of the advantages of the silicon diode array targets is its negligible photoconductive lag. Since the minority carriers (holes) diffuse
across the silicon substrate within a few microseconds, image smear
due to this effect is completely eliminated. As in the vidicon and Plumbicon, however, some image lag occurs because of the inability of the
scanning beam to shift fully the target elements to zero potential in a
single scan.

In practice, the beam diameter is made somewhat larger than the
diode spacing, avoiding the necessity of registering the beam with the
individual diodes in scanning. Aside from the resolution limit due to
the beam size, some resolution is lost by lateral diffusion of the photogenerated carriers within the wafer. To minimize this, the target is

made relatively thin (10 to 20 μ, as mentioned above). With a target thickness of 15 μ, e.g., and a center-to-center diode spacing of 15 μ, the modulation transfer function is greater than 60% at 360 TV lines (15 cycles per millimeter of target), using input light of 0.55 μ. Such resolution is comparable with that of conventional vidicons.

Unlike the vidicon and Plumbicon, the silicon target tube is very rugged, being unaffected by exposure to extremely high light levels and relatively free from target changes produced by the scanning beam. Of particular importance from the point of view of tube fabrication is the fact that the target can withstand bakeout temperatures of about 400 °C, as normally used in outgassing vacuum tubes to obtain long cathode life. Such bakeout is not possible with other photoconductive tubes.

In Fig. 5, the spectral response curves of the various photoconductive camera tubes are shown. Although the Plumbicon and silicon diode array tube both have a peak quantum yield of about 50%, the silicon target tube has an unusually broad response, extending from about 0.4 to beyond 1.0 μ. By using suitable optical filters, the response can be corrected to approximate that of the eye which peaks at about 0.55 μ. As shown, the PbO Plumbicon does not extend sufficiently into the red to cover the eye response. This deficiency is overcome, however, by the extended-red Plumbicon which has PbS added to the photoconductive target. The curve for the vidicon represents its operation at low-input levels. At high-input levels, however, the sensitivity of the vidicon drops considerably below the level shown because of the low gamma. By comparison, the sensitivity of the Plumbicon and sili-

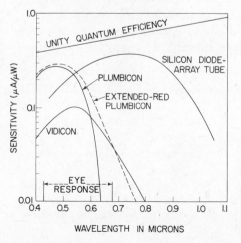

Fig. 5. Spectral sensitivity of photoconductive camera tubes.

con-target tubes remains essentially constant as the input level is increased, since their gamma is close to unity.

9.2.2 Circuit-Scanned (All-Solid-State) Camera Devices

Monolithic Semiconductor Mosaic. The diode array camera tube described above demonstrates that the important requirements of a photoconductive target such as high quantum yield, signal integration, small element size, short response time, as well as good uniformity, can be satisfied by a silicon monolithic structure. In view of this, it is not surprising that separate effort has been directed toward developing silicon sensor arrays which are scanned by solid state circuitry, instead of an electron beam.

An example of a circuit technique for sequentially sensing the signal in a linear array of photoconductive elements is shown in Fig. 6. Each photoconductive element, P, is assumed here to consist of a back-biased diode represented by a rectifier, D_e, shunted by its junction capacitance, C_e. One terminal of each photoconductive element is connected to the common bus, B, with the other terminal connected to a separate switching diode, D_s. In operation the scan generator, SG, sequentially applies a pulse, $-V$, to each of the leads $1'-5'$, while maintaining the other leads at ground potential. This pulse forward biases the switching diodes and charges the reverse-biased photodiodes to the full pulse voltage. Upon termination of the pulse, the voltage across each photodiode is retained, since the diodes, D_e and D_s, are left in their back-biased condition. (For simplicity it is assumed that the capacitance of the switching diode can be neglected.) If, during scanning, the input light, L_i, is allowed to strike any of the photodiodes, this will cause them to discharge in accordance with the integrated light falling on them. As in the case of an electron-beam-scanned diode array, as each photodiode is recharged by the switching pulse, a transient current is produced through it, resulting in a time-

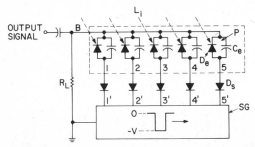

Fig. 6. Diode switching of sensor array.

varying output signal across the common load resistor, R_L .

Although no capacitance has been shown across the switching diodes, in practice some capacitance is present. It is essential for
good operation, however, that this capacitance be small compared to
that of the photodiodes. If this is not the case, a transient current will
flow through the load resistor at the termination of the switching pulse
as a result of the redistribution of charge between the photodiode and
switching diode. Similarly, when a new switching pulse is applied, a
new transient of opposite polarity will occur even if the photodiode did
not discharge between scans. Such transients produce a background
disturbance that limits the low-level signal which can be detected.

In the array shown, each picture element consists of a pair of diodes in back-to-back relationship. It is thus possible to fabricate a
pair of such diodes in the form of an n-p-n structure. This approach
leads to a further possibility, since, by suitable reduction in the width
of the p-region, a phototransistor element is created, whose collector-
base junction acts as the back-biased photodiode and whose emitter-
base junction acts as the switching diode. In this case, application of
a pulse during switching results in a charge flow through the photo-
transistor that is greater by a factor, β, than the capacitance charge required to maintain the collector-base junction at its full potential.[9]
Since the factor, β, corresponding to the current gain of a phototransistor may be of the order of 100, a substantial increase in output signal is obtained.

In the case of a two-dimensional array, the approach generally
taken is to employ a set of crossed X-Y conductors for addressing, as
shown in Fig. 7, using a pair of separate diodes (D_e and D_s) or a photo-
transistor connected to the crossover point of each pair of conductors.
In operation, a pulse voltage, $+V$ is applied sequentially to each of the
horizontal conductors by the vertical scan generator, SG_V. During this
pulse, short pulses from the horizontal scan generator, SG_H, are applied in sequence to the gate of each of the field effect transistors, FET,
in series with the vertical or x-conductors. These transistors, normally in the open circuit state, conduct when the gate is pulsed, so that
the full pulse voltage, $+V$, is applied in sequence across each element
of the selected horizontal row. Depending on the previous illumination
of these elements, varying amounts of charge will flow through them
and the common load resistor when they are recharged, producing a
time-varying output signal. By applying a pulse voltage, $+V$, to successive horizontal lines and repeating the process, the entire array can
be scanned.

The physical construction of an array of phototransistors[9] with its

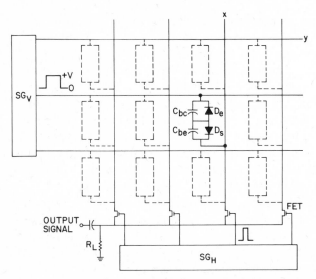

Fig. 7. Circuit for scanning photo-transistor array.

associated sets of X and Y leads is shown in Fig. 8. Initially, a p-type silicon substrate, S, is used whose surface is epitaxially coated with an n-type layer. This layer is then divided into isolated n-type strips, C, by diffusion of p-type material between them. These n-type strips act not only as the horizontal conductors for the array but also as the collector for each of the phototransistors. Following this, p-type base regions, B, are formed by diffusion into the n-type strips. Finally, n-type emitter regions, E, are formed by diffusion into the base regions. Vacuum-evaporated aluminum strips, X, are then pro-

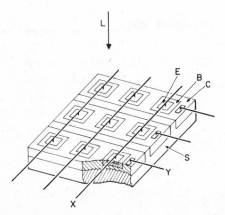

Fig. 8. Structure of photo-transistor array.

vided which contact the emitter elements of each column of the array.

Using the above construction, arrays have been fabricated with 128 columns and 100 rows on a $\frac{1}{2}$-in. square silicon chip,[10] the individual phototransistor elements being about 4×5 mils in area. For imaging purposes, the input light, L, is projected onto the top side of the wafer. Since carriers generated by visible light are produced within a few microns of the surface, the collector junction is made relatively shallow. However, to increase the minority carrier lifetime and resultant diffusion length of the base regions, the resistivity of this material is made relatively high. To minimize the loss of signal by surface recombination, an impurity gradient is created at the input surface, producing an internal field that prevents optically generated electrons from reaching the surface. To obtain an adequate response for longer wavelength photons that penetrate deeper and are absorbed in the collector region, the collector strips are made relatively thick. By coating the input surface with a silicon dioxide film about $0.7\ \mu$ thick, (not shown in Fig. 8), a spectral sensitivity is obtained that is peaked at about $0.6\ \mu$, falling to half value at about 0.45 and $0.85\ \mu$.

One of the problems with phototransistor arrays is the variation in β, or current gain, of the individual elements which results in output signal variations of more than $2:1$. In the case of elements with unusually high gain, these are seen as bright spots in the output. However, if an element is completely shorted, it acts as a low impedance, permanently shunting the emitter bus to the collector strip below it. As a result the output of all elements connected to this vertical bus is cut off and a dark vertical line is seen in the output. Another problem is image lag at low light levels, when the collector junction is only slightly discharged. Because of the high resistance of the emitter junction under these conditions, the collector junction may not be fully recharged during the switching pulse in a single scan.

Mosaics of the above type have an S-shaped transfer characteristic. When scanned at 60 frames per second the useful portion of the characteristic lies between 0.1 and 1.0 ft-c incident on the mosaic (corresponding to 0.01 to 0.1 mW/cm^2 radiation from a 2400 °K source). Using a 2000-Ω load resistor, an output signal is generated in the range of 4 to 30 mV. At the above frame rate, images containing 5 to 7 shades of gray have been produced of scenes in normal room light. By reducing the frame rate to 6/sec (allowing a longer integration time) useful images of high-contrast scenes with about $\frac{1}{10}$ of the above illumination have been obtained. With the present construction the resolution is limited not only by the number of elements in the mosaic but also by a small amount of cross-talk between elements. Also, some resolution is lost because of the combined shunt capacitance of the FET switching elements across the load resistor which attenuates the high frequencies of the output signal.

Thin-Film Photoconductor Mosaic. In all of the previous image pickup devices, the operation is based on the charge integration mode, whereby the capacitance of each picture element is discharged between scans by the integrated light falling on it. In the thin-film photosensor array discussed below, however, a different approach has been taken.[11] Here a photoconductor such as CdS or CdSe is used which has a quantum yield of the order of 10^3, compared to the quantum yield of unity or less obtained with back-biased semiconductor junctions. (Because of their high quantum yield such materials have a dark conductivity which makes their dielectric relaxation time much shorter than a TV frame time, precluding their use in the conventional charge-storage mode.)

The circuit arrangement of a thin-film photoconductive sensor array is shown in Fig. 9. As in the phototransistor array, two sets of mutually perpendicular x-y conductors are used. At each crossover of the conductors a picture element, P, consisting of a photoconductor, PC, in series with a switching diode, D, is connected. As shown in Fig. 9, a positive pulse is applied sequentially to each of the horizontal conductors by the vertical scan generator, SG_V. During the time that the pulse is applied to one of the conductors, the storage capacitors, C_s, connected to the vertical lines are charged by an amount cor-

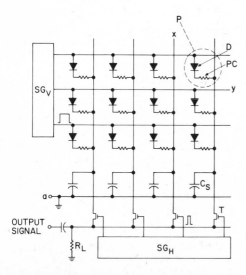

Fig. 9. Circuit for scanning thin-film photoconductor array.

responding to the conductivity of the photoelements of the selected
row (the diodes in series with the photoconductors being in the forward
direction for such charging). At the termination of the pulse the positive charge stored on the capacitors is unable to leak off through the
photoconductive elements of the array because of the reverse bias on
the switching diodes.

To produce an output signal a pulse is applied by the scan generator, SG_H, in sequence to the gate of each of the field-effect transistors, T, switching them to the conducting state and allowing the storage capacitors to discharge through the common load resistor, R_L .
Following this, the row of storage capacitors can again be charged by
applying a positive pulse to the next horizontal lead and then sequentially reading out the charges stored on the capacitors. In effect, this arrangement provides integration of the photocurrent for a line time, instead of a frame time as in the other photoconductive camera devices.

In actual operation the transistor switches are scanned at the same
time that pulses are applied to the successive horizontal leads, rather
than after the termination of the pulse applied to each horizontal lead.
A particular storage capacitor may thus be charged partly during the
time that a voltage pulse is applied to one horizontal line and partly during
the time that a pulse is applied to the next line. The readout signal
produced may thus correspond to the charging current from two vertically adjacent photoconductor elements.

The structure of a single-photoconductive element of the array is
shown in Fig. 10. For fabrication, squares of photoconductive material, either CdS or a mixture of CdS and CdSe, are first evaporated
onto the glass surface and then suitably activated to make the layer

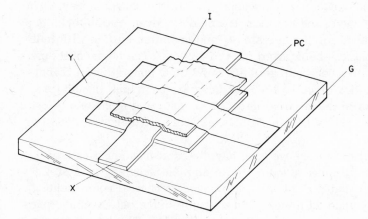

Fig. 10. Structure of single element of photoconductor array.

photoconductive. Following this, a set of narrow vertical conductors, X, is evaporated over the photoconductor. To provide ohmic contact to the photoconductor, indium is used (which is then coated with gold). Following this, thin insulating strips, I, of CaF_2, are evaporated, almost extending to the edges of the photoconductive squares. Finally, horizontal tellurium strips, Y, are evaporated onto the surface and then coated with gold.

Since the tellurium strips contact the photoconductive film only at the edges, where it extends beyond the insulating film, the flow of current through the photoconductor is parallel to the surface, rather than directly through it. By employing tellurium for the horizontal conductors, a Schottky-barrier rectifying contact is made to the photoconductor (the forward direction occurring with the tellurium positive), thus avoiding the necessity of providing a separate diode at each element. With this construction, arrays of 256×256 photoconductive elements[12] have been constructed on a 1-in. square glass plate with a center-to-center spacing of 53 μ, covering an area about $\frac{1}{2}$ in. square.

Ideally, both scan generators, also produced by thin-film techniques, would be fabricated simultaneously with the sensor array on the same substrate as the photoconductor array. In present devices, however, only the vertical scanner is fabricated on this substrate, the horizontal scanner being fabricated on a separate 1-in. square glass plate which is then subsequently joined to the sensor plate and electrically connected to it.

The vertical scan generator consists of a resistor-diode decoder arrangement having 2 sets of 16 input leads. Each set of input leads is connected to the parallel outputs of a 16-stage, shift register, whose voltage pulses result in an output voltage pulse appearing in sequence on each of the 256 decoder outputs connected to the sensor array. The resistors and diodes of the decoder are made by evaporation techniques, nichrome being used for the resistors, and cadmium sulfide with indium-tellurium contacts being used for the diodes. The horizontal scanner, also driven by two 16-stage shift registers, consists of a transistor-resistor decoder circuit which includes the switching transistors, T, of Fig. 9. As in the vertical decoder, evaporated nichrome is used for the resistors. For the thin-film transistors, evaporated CdSe is used.

Sensor arrays of the above type have been scanned at a rate of 60 frames per second, corresponding to an element scanning rate of 4.8 MHz. They are capable of picking up images in normal room light and exhibit fairly good halftones. An example of the output image produced by such a sensor array is shown in Fig. 11. The limiting resolution obtained is about 200 TV lines in both the horizontal and vertical

Fig. 11. Output image produced by thin-film photosensor array. (From Weimer *et al.* [12])

directions. It is of interest that no apparent loss of resolution occurs in the vertical direction, due to the mixing of stored charges of adjacent rows in the integrating capacitors. The vertical streaks shown are attributed to defects in the sensor array or the horizontal decoder. Because of interconnection through the decoder, spurious signals tend to be repeated 16 times each scan.

One of the problems associated with these, as well as other circuit-scanned arrays, is the pickup of transients in the output produced by the horizontal scanning pulses. With better control of such transients, it is expected that image pickup of scenes with an illumination of 1 ft-c (less than $\frac{1}{10}$ of normal room light) will be possible. Using similar thin-film techniques, it is expected that pickup devices with 500×500 elements on $25-\mu$ centers can be fabricated, providing double the resolution of the thin-film array described above.

9.3 IMAGE DISPLAY DEVICES

9.3.1 Light-Emitting Materials

Light-Emitting Diodes (Junction Luminescence). Solid state light-emitters can be divided into two classes, depending on whether their operation is based on junction luminescence or field-excited luminescence. In junction luminescence, light is generated in the neighborhood of a p-n junction of a semiconductor crystal when it is forward biased.[13] In conventional semiconductors such as silicon and germanium, the carriers injected across the junction are lost almost entirely by nonradiative recombination processes, giving up their energy as heat. In a few materials, however, such as shown in Table 1, the recombination process is accompanied by the emission of a useful amount of visible radiation. As in other types of semiconductor junctions, the forward current, J, is given by an exponential function of the applied voltage, V, in accordance with the relation:

$$J = J_0 \exp \frac{eV}{\beta kT} \tag{4}$$

where β has a value between unity and 2 and J_0 is a constant. Generally, the light emission varies linearly with the current over a large range, except for low currents, where the output may vary approximately as the square of the current.

One of the commonly used criteria for evaluating light-emitting diodes is the external quantum efficiency, N_e, defined as the number of photons leaving the crystal per charge carrier injected across the junction. For display purposes this efficiency is not an adequate measure of the usefulness of a material, since the eye is most sensitive to green light peaked at about 5500 Å and falls off markedly in the red and blue. A more useful criterion is that of lumens per watt of electrical input power, a lumen being (by definition) a measure of the light flux effective in producing a visual response. At a wavelength of 5500 Å, for example, at the peak of the eye response, 1 W of radiant power corresponds to 680 lm. At longer and shorter wavelengths, where the eye is less sensitive, 1 W corresponds to a much lower number of lu-

TABLE 1. SOME SEMICONDUCTOR MATERIALS USEFUL FOR PRODUCING JUNCTION LUMINESCENCE

Material	Color (spectral peak)		Maximum Reported External Quantum Efficiency	Approximate lm/W
GaP	Green	(5700 Å)	6×10^{-3}	3
GaP	Red	(7000 Å)	7×10^{-2}	1
$Ga_{1-x}Al_xAs$	Red	(6600 Å)	2×10^{-3}	0.1
$GaAs_{1-x}P_x$	Red	(6800 Å)	2×10^{-3}	0.1
SiC	Yellow	(5900 Å)	1×10^{-2}	0.3

mens. As indicated in the Table 1, the highest external efficiency, about 7%, is obtained for the red emission of GaP. For the green emission of GaP and for other materials, the efficiency is much lower. In terms of visible output (i. e., lumens per watt) best results are obtained, however, with green-emitting GaP, despite its limited quantum efficiency.

Although not shown in Table 1, it should be noted that visible emission can also be obtained by using a GaAs junction emitter coated with a suitable phosphor.[14] In this case, the infrared radiation from the Si-doped GaAs is converted by a two-step process into visible radiation. Possible phosphor materials for this purpose are LaF_3, $BaYF_5$, and Y_3OCl, activated with Er and Yb. With such junction-phosphor combinations it is possible to obtain green light, e.g., with an efficiency in lumens per watt comparable to GaP. The light output, however, is proportional to the square of the diode current, rather than linear with current, as in the case of conventional light-emitting diodes. Because of their low-operating voltage (e.g., 2 to 3 V), allowing them to be easily controlled by semiconductor integrated circuits and their long life (> 10,000 hr), light-emitting diodes are attractive for display devices such as alpha-numeric character generators. However, for large multielement arrays, there are factors that make their use difficult. At the present time such diodes are fabricated on small chips, either as single diodes or as a group of diodes. The problem of assembling and interconnecting an array of $> 10^4$ elements (aside from the addressing circuitry) would thus be considerable. Even at a cost of a few cents per diode, the total cost of the diodes required for such a display would not be small.

Another problem is that of power dissipation. Assuming a display device 1 ft^2 in area with a brightness of about 50 ft-L (corresponding roughly to TV screen brightness) this corresponds, by definition, to an output flux of 50 lm if the entire viewing area is emitting. With the most efficient material indicated in Table 1, this would require about 17 W of input power. However, with materials that are commercially available, such as GaAsP, whose efficiency may be less than the maximum shown in Table 1 (i. e., < 0.1 lm/W) full excitation of a 12-in. screen would require input powers of the order of a kilowatt.

Field-Excited Luminescence (*EL Phosphor*). The alternative luminescent medium for display purposes is field-excited electroluminescent phosphor.[15] A material commonly employed is ZnS powder activated with copper and chlorine. This is mixed with a plastic binder or a low-melting-temperature glass frit to form an insulating layer 1 to 2 mils thick. If an a-c voltage, e.g., 100 to 200 V rms and 1 kHz is applied across opposite surfaces of such a layer, a brightness in excess of 10 ft-L can readily be obtained.

The mechanism of light generation is believed to result from the very high fields that occur at localized regions on the surface or within the phosphor grains, where small isolated quantities of conductive Cu_2S are present. During the half-cycle of the alternating current, when this conducting phase is made negative with respect to the phosphor, a blocking action occurs and a field of the order of 10^6 V/cm or higher is created in the depletion region of the ZnS adjacent to the Cu_2S. Any free electrons present in this region (or drawn into it) are accelerated, attaining sufficient energy to ionize the copper luminescent centers by impact, producing more free electrons which are drawn deeper into the ZnS and captured in shallow traps. Upon reversal of the polarity of the applied voltage, the trapped electrons are drawn back to the luminescent centers, recombining and producing light emission. At any particular region of a phosphor grain, a pulse of light is produced once each alternate half-cycle. However, since an approximately equal number of junctions are present, oriented in opposite directions, a pulse of light is emitted from the layer as a whole during each half-cycle. In accordance with the above process, the light output should rise linearly with the a-c frequency. Although not strictly true, in actual materials, the brightness usually rises almost linearly with frequency in the range of several hundred to several thousand hertz. The impedance of electroluminescent layers is highly capacitive, the power factor being about 0.1. Over many orders of magnitude, the brightness, B, varies with the applied voltage, V, according to the following relation:

$$B = B_0 \exp \frac{-c}{\sqrt{V}} \tag{5}$$

where B_0 and c are constants determined by the material, particle size, operating frequency, and details of layer fabrication.

Although the efficiency, in terms of the ratio of power radiated to power absorbed is limited to about 1 to 2%, it is relatively easy to produce a material whose spectral emission curve peak is close to that of the eye. With such materials the luminous efficiency may be more than 10 lm/W, considerably greater than the efficiency obtained from most light-emitting diodes. However, their operating voltage (usually > 100 V) is relatively high, making difficult their control by low-voltage semiconductor circuitry. Also, their operating life is limited, the brightness falling to half value or less in about 1000 hr. Under steady a-c excitation, adequate brightness for viewing in room light can be obtained. However, attempts to obtain a peak brightness more than about 10 times this may result in voltage breakdown, severely limiting the average brightness that may be obtained by sequential excitation of a large number of elements.

Despite the above limitations, field-excited electroluminescent powder is relatively cheap and can easily be fabricated into layers 1 ft^2 or greater. Of particular importance (especially for halftone displays) is the fact that uniform characteristics can be obtained across the layer. Also, the element size is arbitrary, being determined only by the dimensions of the electrodes. For these reasons (as well as the relatively high luminous efficiency obtained), all large-area display devices to date have been based on the use of electroluminescent phosphor powder.

9.3.2 Optical-Input Electroluminescent Panels

Photoconductor-controlled Devices. From a circuit point of view, the simplest form of display device is the optical input panel in which the elements are controlled by a pattern or an image of light. Usually, each picture element consists of a series arrangement of a photoconductor and phosphor element,[16] as shown in Fig. 12, across which an a-c voltage is maintained. In the dark, it is assumed that the impedance of the photoconductor, PC, is, e.g., 5 to 10 times greater than that of the phosphor element, EL, so that only a small fraction of the supply voltage appears across the phosphor. This requires that C_{el} be considerably greater than C_{pc} and that the resistance of R_{pc} in the dark be relatively high. Upon illumination, the resistance of the photoconductor is lowered, so that the ac voltage across the phosphor is increased, producing light emission. Although the phosphor has negligible dc conductivity, it is shown shunted by a resistance, R_{el}, representing the ac losses.

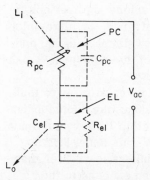

Fig. 12. Equivalent circuit of PC–EL panel.

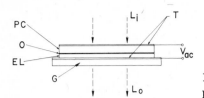

Fig. 13.   Cross section of PC–EL image panel.

For imaging purposes a structure such as shown in cross section in Fig. 13 may be used.[16] This consists of a continuous layer of photoconductor, PC, in contact with a layer of phosphor, EL, supported on a glass plate, G. In operation, an a-c voltage is maintained between the transparent conductive coatings, T, on the outer surface of the phosphor and photoconductor. If an input image of suitable radiation is projected onto the photoconductive layer, its resistance is lowered at local areas, which increases the a-c voltage across adjacent phosphor elements and produces a luminescent image corresponding to the input image. In the case in which the photoconductor is sensitive to light emitted by the phosphor layer, a thin opaque insulating layer, O, is provided between the two layers to prevent optical feedback.

The only photoconductive materials presently known which respond to visible or near-visible radiation and have a sufficiently high quantum yield and low dark conductivity are CdS and CdSe.[17] For panel application they are generally prepared as a powder, and activated with copper and chlorine. If this powder is mixed with a few percent by weight of an organic binder (such as an epoxy resin or ethyl cellulose) diluted in a volatile solvent, relatively rugged large-area layers of arbitrary size and thickness[16] can be formed which are highly photoconductive. For impedance matching purposes, the photoconductive layer is made about 10 mils thick, compared to the phosphor layer, which is usually 1 to 2 mils thick.

Using structures based on the arrangement shown in Fig. 13, a number of imaging devices have been built with various applications in mind. These include the conversion of X-ray, infrared, or ultraviolet images to visible images, the intensification of low-level images, and the storage of transient input images. For each of these purposes, panel designs have been developed, as indicated below.

For X-ray imaging, structures 12×12 in. in size, similar to that shown in Fig. 13, have been developed,[18] using CdS as the photoconductor. The outer electrode, in this case, may be a thin film of aluminum which is transparent to X-rays. With about 500 V rms and 1 kHz applied, output images can be produced whose brightness is greater than 100 times that of a conventional fluoroscope screen exposed to the same X-ray level, using 80 to 100 keV X-rays. Since the brightness of the panel is approximately proportional to the third power of the X-ray input level, the contrast of the output image is enhanced.

Because of the long response time of the photoconductor, however, seconds are required for low-level images to build up and decay, thus preventing the use of such panels for viewing moving images. Also, because of the nonlinear brightness-voltage characteristics of the phosphor and the properties of the photoconductor, the X-ray input must exceed some threshold level to produce any substantial output. Satisfactory brightness is obtained, e.g., in medical applications in which the chest or a thin portion of the body is X-rayed. In the case of more dense portions of the body, such as the abdomen, the X-ray level may be too low, however.

For the intensification of low-level input images in the visible spectrum, a structure such as shown in Fig. 13 does not function well, since most of the light is absorbed close to the input surface of the relatively thick photoconductor, leaving the deeper layers unexcited. One solution is to cut fine grooves through the photoconductor layer,[16] leaving narrow conducting lines on the surface as the top electrode. In this case, light absorbed on the groove walls provides conducting paths through the photoconductive layer.

Using this technique, panels 12×12-in. in size with 40 grooves/ in. have been constructed, also using CdS powder as the photoconductor. In operation, radiant energy gains of about 400 have been obtained using yellow input light, the quality of the output image being comparable to commercial TV images. As in the case of X-ray panels, the response time is of the order of seconds and a threshold input level exists (in this case about 10^{-3} ft-c), below which the gain falls rapidly. Using CdSe as the photoconductor, similar image panels have been developed that are sensitive to near-infrared radiation[19] with wavelengths extending to about a micron. Using other photoconductive materials such as ZnS, panels have been developed that are sensitive to ultraviolet radiation.[20]

Since the gain of the panel falls below unity at low-input levels and exceeds unity at higher input levels, if optical feedback is permitted, a bistable storage condition is possible in which an element triggered on by light will remain on indefinitely.[21] In principle, a bistable storage panel could be made by simply eliminating the opaque insulating layer normally used between the photoconductor and phosphor layers. If this is done, however, it is difficult to prevent light from excited elements from reaching adjacent off elements and triggering them on. Although experimental panels have been fabricated which can store bistable images indefinitely, they involve considerable complexity in construction to avoid image spreading.[21] Also, the erasing of an image requires that the a-c supply voltage be cut off for a sufficient time (a second or more) to allow the photoconductor to decay.

Field-Effect Storage Display Panels. For the temporary storage of halftone images, attempts have been made to use nonfeedback panels of the previous type with photoconductive layers having a very long decay time (e.g., $\frac{1}{2}$ min). This approach has had limited success because of the short storage time obtainable and the long time required for old images to decay completely. A more recent method for obtaining image storage is based on the use of the field-effect action in ZnO powder.[22]

In such material, if the surface is negatively charged by a corona discharge in air, a depletion layer is formed below the surface, decreasing the conductance in the same manner that the control electrode decreases the conductance of junction-type, field-effect triodes. This effect can be demonstrated using a thin layer of ZnO powder in a plastic binder spread across an electrode gap, as shown in Fig. 14. For test purposes the current through the ZnO is monitored, while the surface is either illuminated or exposed to a negative corona discharge produced by a fine wire or point about 1 cm from the surface. (To obtain sufficient conductivity, ZnO powder is used which is chemically reduced to provide an excess of zinc atoms that act as donors.)

A somewhat idealized description of the sample behavior is shown

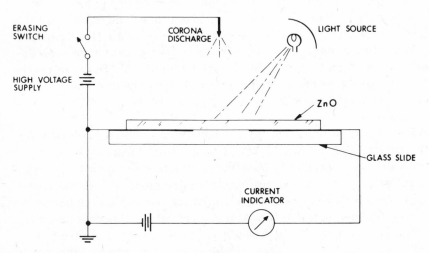

Fig. 14. Arrangement for measuring field-effect properties of ZnO. (From Kazan and Winslow.[22])

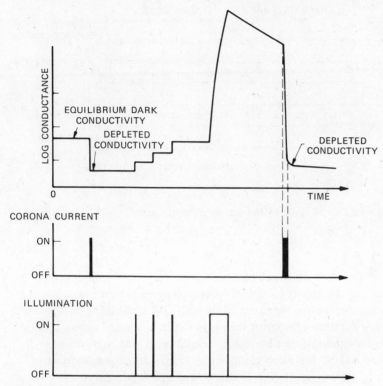

Fig. 15. Conductance changes produced in ZnO by light and corona charging. (From Kazan and Winslow. [22])

in Fig. 15. The sample is assumed to have been initially dark-adapted for a long period, having reached the equilibrium dark conductance level shown. If a negative potential is applied for a fraction of a second to the corona wire, the surface is charged, remaining at a negative potential. This causes the conductivity to drop by about an order of magnitude. Following this, if the sample is exposed to a short light pulse, the surface is partially discharged by the optically generated free holes and the conductivity is increased to a higher level. In a similar manner, successive light pulses cause the conductivity to rise further in stepwise fashion. The conductance levels shown tend to remain unchanged for long periods, e.g., varying less than about 20% over a period of an hour. If the illumination is allowed to remain on for a long period, as shown, the conductivity is raised to a much higher level, and then drops slowly. When desired, the sample can be corona-charged again, reducing its conductance to the previous low level. The sample may then be set to an arbitrary new conductance level by exposing it to a controlled amount of light.

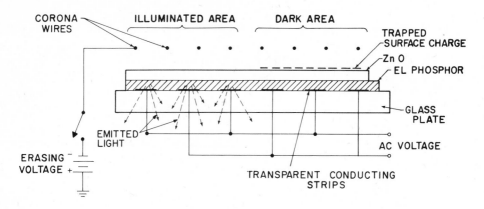

Fig. 16.   Cross section of field-effect image storage panel.

Insofar as the ZnO element can be set to different conductance levels, it may be employed to control the power delivered to a series load such as an electroluminescent element. By combining a large number of such picture elements into a two-dimensional array, an optical input display device can be made, capable of image storage. A panel structure useful for such imaging is shown in cross section in Fig. 16. Here, an array of fine transparent electrode strips is provided on the surface of a glass plate, with alternate strips being connected to opposite sides of the a-c line. Above the electrode strips is an electroluminescent layer about 2 mils thick, coated in turn with a ZnO powder layer a fraction of a mil thick. About 1 cm from the ZnO is a set of fine corona wires, about 1 mil in diameter and 1 cm apart.

Before imaging, the panel is erased by momentarily applying about – 6 kV to the corona wires. This charges the entire ZnO surface negative, leaving it in its low-conductivity state. Because of the spacing between electrodes, the flow of a-c current through the phosphor is very low and negligible light output is produced. If an area of the ZnO is illuminated, however, the surface is discharged, leaving the ZnO in a conductive state. Current can now flow through the phosphor material above adjacent electrodes and across the conductive ZnO between the electrodes. An increased a-c voltage now appears across the phosphor, causing light emission. In the same manner, if the ZnO is exposed to an optical image, a charge pattern is produced which, in turn, is seen as a luminescent image.

Since the negative charges remaining on the ZnO in unexposed or partly exposed areas are retained for hours, the luminescent image

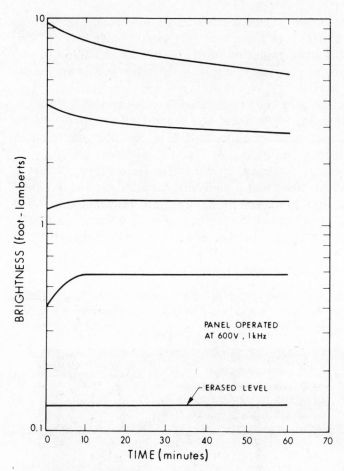

Fig. 17. Panel brightness as a function of time after optical excitation. (From Kazan and Winslow.[22])

can be viewed for long periods with little loss in brightness or half-tones. In Fig. 17 is shown a set of brightness curves produced by initially exciting the panel to different output levels by varying the input light energy. As can be seen, the brightness levels remain relatively unchanged over a period of an hour, indicating that good halftone images can be retained for this time. When desired, a stored image can be erased by applying high voltage to the corona wires to recharge the ZnO.

Since the band gap of the ZnO is about 3.2 eV, it is responsive only to input radiation whose wavelength is shorter than about 4000 Å. Using such radiation, an integrated energy of about 1 $\mu J/cm^2$ is required to produce an image. Over a large range of exposure times (between 10^{-2} and 10^3 sec) it is found that reciprocity holds between the exposure time and the input radiation level. It is thus possible to record transient images of very short duration or to integrate very weak images over a long period.

Field-effect panels have been fabricated in sizes up to 12×12 in. with 50 electrode lines/in. With about 600 V and 1 kHz applied, a picture highlight brightness of about 10 to 20 ft-L is obtained, sufficient for viewing in moderate room light. In Fig. 18, a photograph of a stored luminescent image is shown, produced by projecting a TV test pattern onto the ZnO surface from a conventional slide projector. As can be seen, a resolution in excess of 800 TV lines is obtained in the horizontal direction (parallel to the electrode lines), while a limiting resolution of about 400 TV lines is obtained in the vertical direction.

9.3.3 Electrically Addressed Display Panels

Nonstorage X-Y Addressed Panels. Earliest attempts to generate an image from electrical input signals were based on the use of a con-

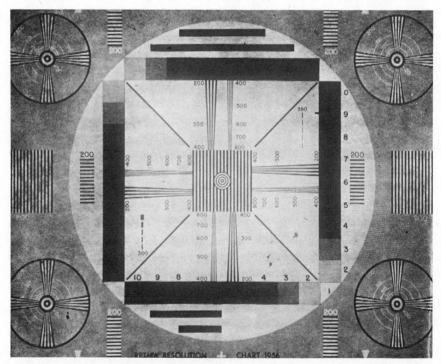

Fig. 18. Stored image of TV test pattern. (From Kazan and Winslow.[22])

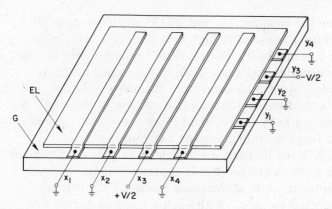

Fig. 19. X-Y addressed electroluminescent phosphor layer.

tinuous phosphor layer, as shown in Fig. 19, with two sets of electrode
strips provided on opposite surfaces, oriented perpendicular to each
other.[23] For viewing purposes it is assumed that at least one set of
electrodes is transparent. By applying a voltage $+V/2$ to a particular
X-line and a voltage $-V/2$ to a particular Y-line, a voltage V will appear across the phosphor element at the crossover of the electrodes,
causing it to emit light. (It is assumed here that the applied voltages
are a-c but of opposite phase.) In principle, by applying such voltages
to selected pairs of X-Y electrodes in rapid sequence, an output image
can be generated.

   With such schemes, however, a number of problems arise. As
can be seen, if the unselected conductors are maintained at ground potential, a voltage of magnitude, $V/2$, will appear across the remaining
elements connected to the selected X and Y conductors, producing an
illuminated cross. During a single scan of the array, each unselected
element may be excited many times at half-voltage. The resultant
image contrast may thus be very low if the array contains a large number of elements.

   To some degree the problem of low contrast is minimized by the
fact that the brightness-voltage relation of electroluminescent phosphor layers is very nonlinear, as already mentioned. Assuming a
brightness-voltage relationship as indicated by Eq. 5, we find that the
ratio, R, between the brightness of a selected element with a voltage,
V, and that of an element with a voltage, $V/2$, is

$$R = \exp \frac{0.414b}{\sqrt{V}} \qquad (6)$$

If a value of 70 (which is representative of common powder phosphor layers) is assumed for the constant, b, and 200 V applied across selected elements to obtain a reasonably high brightness, a brightness ratio of less than 10 : 1 is obtained. Such a ratio is insufficient to produce an image with acceptable contrast, even in a scanned array of relatively few elements. As can be seen from the above relation, the brightness ratio can be increased if the applied voltage is reduced. Increasing the brightness ratio by this method to an acceptable degree, however, results in a large drop in phosphor brightness.

One method for obtaining increased contrast is to coat the phosphor with a resistive layer such as SiC powder in a plastic binder.[23] This material is highly nonlinear, with the current rising as the fifth or higher power of the applied voltage. With such a layer in series with the phosphor, it is possible to obtain a brightness ratio as high as 10, 000 : 1, with a 2 : 1 ratio in applied voltage. The addition of nonlinear material, however, requires that the applied voltage be substantially increased because of the voltage drop in this material. Also, small variations in the threshold characteristics of the nonlinear material may result in large variations in the output brightness of individual elements.

Another method for obtaining a high-brightness ratio is to use evaporated films of manganese-activated ZnS[24] instead of powder material. Although such films are more difficult to fabricate, their change in light output with voltage is much greater than with powder layers. For example, with 150 V rms at 5 kHz applied across a selected element, a reduction of only 25% in the applied voltage results in a brightness reduction by a factor of 10^3.

Another problem arising when individual elements of a large array are sequentially excited is the low average brightness obtained. Unlike electron beam excited phosphors, whose peak brightness may be as high as 10^6 ft-L, the peak brightness obtained from electroluminescent phosphors generally does not exceed 10^3 ft-L. The average light obtainable from even a modest array of 10^3 elements would thus be limited to a level far below TV brightness.

One method for increasing the average brightness is to excite all the elements of each horizontal row simultaneously, rather than sequentially, thus lengthening the time they are excited during each frame. Based on this approach, an experimental flat display device has been developed,[25] capable of generating halftone TV images. The phosphor screen used, about 27. 6 × 20. 7 cm in area, is provided with 230 horizontal electrode strips on the rear surface and 230 vertical transparent strips on the front surface, the individual picture elements being 1. 0 × 0. 75 mm in size.

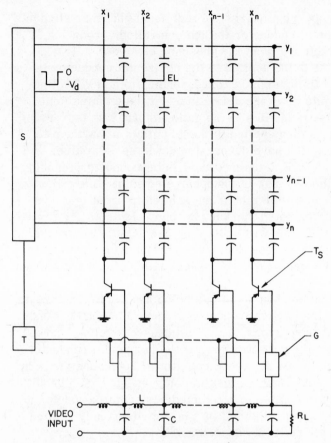

Fig. 20. Circuit arrangement for TV display panel.

The circuit arrangement for exciting the screen is shown in Fig. 20. In operation, a video signal is fed to the input of the delay line, composed of lumped L-C elements. The delay time of this line is 50.6 μsec, somewhat less than the line time of commercial TV. At the moment that the propagating signal reaches the terminating resistor, R, a pulse, $-V_d$, of -300 V is applied by the vertical scan generator, S, to one of the horizontal conductors of the array. At the same time a short sampling pulse from the synchronizing generator, T, allows the storage capacitor within each of the gating-storage units, G, to be charged to a level determined by the instantaneous potential of the corresponding tap of the delay line. This stored charge in turn sets the bias applied to the base of the transistors, T_s, determining the resistance in series with each vertical conductor. The magnitude of the pulse voltage appearing across the individual phosphor elements of the selected horizontal conductor is thus controlled, producing pulses of varying brightness from the phosphor elements.

The duration of the pulse applied to the horizontal conductor is 40 μsec. During this time, signals corresponding to a new TV line are allowed to propagate down the delay line. At the termination of this pulse a short pulse is applied to the gating-storage units to reset them. Another pulse is then applied to these units to allow them to sample the new signals that have propagated down the delay line. A pulse, $-V_d$, is then applied to the next horizontal conductor, causing light emission from the new row of phosphor elements. In similar manner, successive rows of elements are excited to produce a complete image. It should be noted that the phosphor elements are excited once each frame time ($\frac{1}{60}$ sec) with a 40 μsec d-c pulse. The resultant brightness is thus somewhat less than if the phosphor elements were excited at the same frequency (60 cycles/sec) with a sine wave of equal peak-to-peak voltage.

In operation, a fraction of the voltage pulse applied to a horizontal conductor appears on the vertical conductors, depending on the conductance of the individual transistors, T_s. This causes a voltage across the phosphor elements of unselected rows if the horizontal conductors of these rows are maintained at ground potential. To reduce the light output from these elements and the loss in contrast that would result, each time that a pulse is applied to a selected horizontal conductor a voltage pulse about $\frac{1}{3}$ the magnitude of this is simultaneously applied to each of the unselected horizontal conductors.

Display devices of the above type have been successfully employed to produce halftone moving images, using the video input obtained from TV broadcast signals. An example of such an image is shown in Fig. 21. Although the resolution is limited by the size of the elements, the uniformity appears good. Also, as expected, the contrast is limited (less than 3:1). A more serious limitation, however, is the low brightness, requiring a darkened room for viewing the image.

Bistable Storage Display Devices. In X-Y addressed phosphor layers, aside from the limited brightness and low contrast, the generation of even a stationary image requires repetitive excitation of the picture elements at a sufficiently high rate to avoid flicker. In the case of static images or images that are updated at a low rate, there are advantages to be gained by providing some type of local storage at each picture element, so that it will remain on after being excited.

One method of accomplishing this is to employ, in series with each phosphor element, an element that will switch from a high-impedance

Fig. 21. Output image produced on X-Y addressed EL panel. (Reprinted from Proc. of 1970 IEEE International Computer Group Conference, p. 269.)

to a low-impedance state above some threshold voltage and remain in this state until the applied voltage (or current) is reduced below some extinguishing level. Since a-c voltage is required for phosphor excitation, it is necessary that the bistable element be electrically symmetrical. Although electrical characteristics of this type are exhibited, e.g., by gas cells or triac semiconductor control elements, these are not easily compatible with phosphor layers.

A material that has been explored for bistable image storage is CdSe powder,[26] doped with copper and chlorine and held together with a plastic binder. If the voltage (either d-c or a-c) across such a sample is gradually increased, its conductance remains low, until a threshold voltage, V_t, is reached. At this point the current rises suddenly to a level that may be several orders of magnitude higher. The conductance then remains high, until the voltage is reduced below some level, V_e (where $V_e < V_t$), at which point it drops suddenly back to its initial low-conductance state. By maintaining an intermediate voltage, V_s, across the sample, between V_t and V_e, the sample can remain in either a high- or low-conductance state.

From a device fabrication point of view, CdSe powder material has the advantage that it can be easily produced in large-area layers of arbitrary thickness. Also, it can be prepared so that its impedance per unit area in the conductive state is comparable to that of the phosphor layer and several orders of magnitude higher in its nonconductive state, allowing good control action to be obtained with equal areas of CdSe and phosphor.

Using a double-layer structure of phosphor and CdSe powder, with outer X-Y electrodes, small experimental devices have been built with 10×10 elements on 50-mil centers.[27] With a sustaining voltage of 200 V rms and 400 Hz across the two sets of electrodes, individual elements could be triggered on by superimposing a 300-V pulse of 20 μsec duration across selected pairs of X-Y conductors. These elements would then remain on, producing an image whose brightness was about 20 ft-L, sufficient for viewing in moderate room illumination. Erasing could be accomplished by reducing the sustaining voltage to zero for about 20 msec.

Despite their structural simplicity, problems arise with such display devices if many elements are used, since the electrical characteristics of the CdSe may vary from element to element and may change with time, ambient conditions, or continued operation. More recently, consideration has been given to the use of amorphous semiconductor films of chalcogenide glasses[28] (referred to in some cases as Ovonic threshold switches), composed, e.g., of tellurium, arsenic, silicon, and germanium in various ratios. However, at the present time, limited information is available on the uniformity and stability of such materials, especially when designed to operate at switching potentials of the order of 100 V.

Halftone-Storage Display Devices. Although bistable control elements allow images to be obtained with high average brightness, they do not permit the generation of halftone images. For this purpose a control element is necessary that has storage and can also be set at intermediate levels. The most successful solution to this problem to date has involved the use of ferroelectric elements.

Before discussing the operation of ferroelectric-controlled displays, it is useful to consider some of the basic electrical characteristics of ferroelectric materials. Referring to Fig. 22, if an a-c voltage, V_{ac}, is applied across such a material and the polarization charge, P, passing through it is measured, a curve such as A-B-C-D-E-F is obtained, whose shape is similar to the hysteresis curve of ferromagnetic materials. In accordance with the excursion of the applied voltage from $-V_p$ to $+V_p$, the charge flow through the sample varies from $-P_s$ to $+P_s$. If an a-c voltage, V'_{ac}, of the same magnitude as V_{ac} with

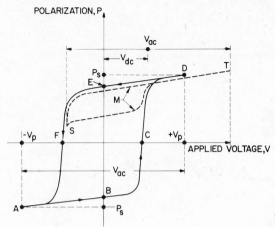

POLARIZATION, P

APPLIED VOLTAGE,V

Fig. 22. Effect of d-c bias on hysteresis curve of ferro-electric material.

a d-c bias component, V_{dc}, is applied, a hysteresis loop such as shown by the dashed curve, M, will result. The polarization will now vary between points, S and T, limiting the charge flow through the sample. By placing an electroluminescent phosphor element in series with the ferroelectric element, the current flow through the phosphor can thus be modulated by varying the d-c bias voltage on the ferroeletric element. However, since the a-c current cannot be completely cut off in such a series arrangement, other circuits are generally used.

A ferroelectric bridge circuit[29] that produces good control action is shown in Fig. 23. One loop, i_1, includes a ferroelectric element, FE_1, and the phosphor element, EL; while the second loop, i_2, includes the ferroelectric elements, FE_2 and FE_3, and the phosphor element. It is assumed that, with no d-c bias voltage across any of the ferroelectric elements, the hysteresis curve of the series combination of FE_2 and FE_3 is identical to that of FE_1. If an a-c voltage is applied from the center-tapped secondary of the transformer, T, to points 1, 2, and 3, the net current flow through the phosphor element will be zero and no light will be emitted. (It is assumed that the impedance of the resistor, R, and the back-biased diode, D, are sufficiently high and can be neglected.)

Assuming the a-c voltage to be cut off, if repetitive pulses, $-V_b$, equal in magnitude to the d-c supply voltage are produced by generator, G_r, in series with line, H, point 5 will be charged to a potential, $-V_b$, leaving no potential across the ferroelectric elements. However, if a positive pulse of amplitude, V_s, is produced simultaneously by the generator, G_c, in series with line, V, this will result in forward con-

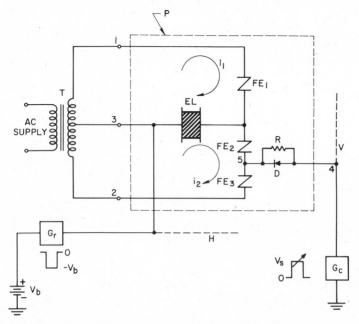

Fig. 23.   Circuit for ferroelectric control of single-picture element.

duction through the diode, leaving a potential, V_s, at point 5, biasing
the ferroelectric elements. FE_2 and FE_2. in mutually opposite directions.   Following the termination of the pulses, the diode will be left
in the cutoff condition (neglecting leakage through the shunt resistor,
R), preventing the polarization charge from leaking off.   If the a-c
voltage is now switched on, the charge transfer through loop i_2 will be
reduced and an a-c current will flow through the phosphor, producing
light emission.

Since the ferroelectric elements can be biased to varying degrees,
depending on the magnitude of the signal pulse, V_s, halftone control of
the phosphor brightness is possible.   If the shunt resistor, R, were
omitted, the brightness would remain fixed for a relatively long period,
depending only on the leakage of the diode and remaining circuit components.   For TV applications, however, where it is desired to reset
the brightness periodically 30 times/sec, it is convenient to shunt the
diode with a resistance, R, as shown which allows the bias charge
trapped at point 5 to leak off in about $\frac{1}{30}$ sec, thus avoiding the need for
a separate erasing procedure.

For imaging purposes, an array of picture elements, each of which
consists of the circuit shown in the dashed box, P, has been used.[29]

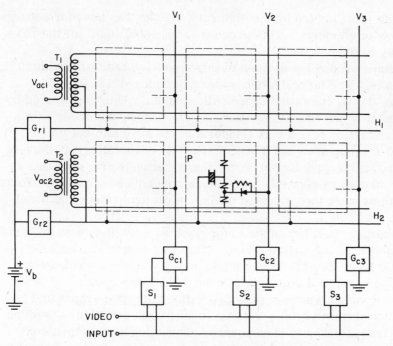

Fig. 24. Circuit arrangement for addressing ferroelectric display panel.

The circuit arrangement employed for controlling such an array is shown in Fig. 24. For addressing purposes, a set of horizontal lines, H, and a set of vertical lines, V, are used (only a few of which are shown). To generate an image, video signals from a suitably scanned camera tube are fed in parallel to the storage elements, S_1, S_2, and S_3. During a line time of the input signal, the storage elements (normally cut off) are sequentially gated on by pulses from a horizontal scanning generator (not shown). As each element is gated on, it capacitively stores a charge corresponding to the instantaneous video signal. At the end of a line time (during the retrace time of the electron beam in the camera tube), the a-c voltage applied to the rows is cut off. During this retrace time, a negative pulse, $-V_b$, is produced by the generator, G_{rl}, in series with the line, H_1, and positive pulses are simultaneously produced by the generators, G_c, in accordance with the charge stored in the elements, S. The coincidence of these pulses "sets" the ferroelectric circuits of the first row of picture elements. Following this, the a-c voltage is switched on again for about a line time, producing light emission from the elements corresponding to their electrical setting. During this time, the video signals for the

next TV line are sampled by the elements, S, and then used for setting
the next row of elements. This process is repeated, until all the rows
of the array have been set.

Experimental display devices of this type have been built with 40
×30 elements. The ferroelectric material employed consists of a
Pb (Zn, Sn, Ti) O_3 compound doped with niobium. This is prepared by
doctor-blading a mixture of the oxides of the various materials onto a
substrate and firing it at a high temperature to form thin ceramic
strips 4×0.5 in. in area and 3 mils thick. By evaporating a pattern of
gold electrodes on opposite surfaces, the 60 ferroelectric elements required for 20 picture elements were fabricated on a single strip. Forty
of these strips were then mounted behind an electroluminescent panel,
6×8 in. in size.

The phosphor panel consisted of a glass support plate with 30 horizontal transparent conductive strips. This was coated with an electroluminescent layer about 0.7 mils thick, each conductive strip serving
as one of the horizontal electrodes of the addressing system. The surface of the phosphor was then provided with an array of evaporated
aluminum squares in registry with the transparent conductors, with an
individual lead attached to each square. These leads were then connected to the corresponding ferroelectric circuits, mounted behind the
phosphor plate. For each picture element a diode and a 22-MΩ resistor were also attached.

In operation, the array was scanned at 30 frames/sec, the line
repetition rate being 1080 Hz. During most of the 926-μsec line time
the input signals from the camera tube were sequentially sampled by
the 40 storage elements, S, at 20.6-μsec intervals. At the end of the
line time, a pulse of -45V and 40-μsec duration was applied to the
selected horizontal conductor and voltage pulses, varying in amplitude
between 0 and $+45$ V (also of 40 μsec duration) were simultaneously
produced by the column generators in accordance with the sampled input signals. The a-c voltage for exciting the phosphor elements consisted of 900-V peak-to-peak square waves (instead of sine waves),
whose period was made equal to a line time.

The images obtained had a highlight brightness of about 16 ft-L,
adequate for viewing in moderate room light. There was essentially
no flicker and, in the case of rapidly moving images, only a small
amount of smear. Although halftones could be produced, the halftone
quality was limited because of nonuniformity in the characteristics of
the picture elements. For example, in the case of picture elements
excited with a given input signal to a relatively high level, brightness
variations of about ±15% occurred. These nonuniformities are caused

primarily by variations in the coercive force of the ferroelectric elements resulting from differences in thickness of the wafers.

9.4 CONCLUSION

In the case of image pickup the use of silicon crystal wafers for photosensor arrays has resulted in significant new devices. One of these is the silicon diode array camera tube which is not only extremely rugged but has characteristics superior to previous photoconductive camera tubes. Another is the all solid state, circuit-scanned image sensor. Although such image pickup devices are still inferior to camera tubes in terms of cost and performance, they are being steadily improved and have important potential advantages of greater scanning accuracy, reduced size, weight and power requirements, and improved life.

In the case of display devices, although there is presently considerable activity on new approaches, the emergency of devices with performance close to, or comparable with, cathode ray tubes does not seem to be a likelihood in the near future. This situation is partly attributable to the fact that no cheap, large-area, solid state medium with satisfactory electrical and optical characteristics is available for generating or modulating light. Also, because of their large size display devices do not necessarily profit from the availability of small semiconductor chips with high-element density. However, considerable effort is presently being directed toward the development of alphanumeric display devices based on light-emitting diodes. Advances made in such materials, as well as new ferroelectric light modulators,[30] may well extend into large-area display devices, despite the many existing limitations of such materials.

REFERENCES

1. B. Kazan and M. Knoll, *Electronic Image Storage*, Academic Press, New York, 1968.
2. P. K. Weimer, S. V. Forgue, and R. R. Goodrich, "The Vidicon Photoconductive Camera Tube," *Electronics*, 23, 70–73 (May 1950).
3. A. Rose, "An Outline of Some Photoconductive Processes," *RCA Rev.*, 12, 362–414 (1951).
4. A. Rose, "Space-Charge Limited Currents in Solids," *Phys. Rev.*, 97, 1538–1544 (1955).
5. A. Rose, "Maximum Performance of Photoconductors," *Helv. Phys. Acta.*, 30, 242–244 (1957).

6. E. F. de Haan, A. van der Drift, and P. P. M. Schampers, "The Plumbicon, A New Television Camera Tube," *Philips Tech. Rev.*, 25, 133–151 (1963/1964).

7. E. F. de Haan, F. M. Klaassen, and P. P. M. Schampers, "An Experimental Plumbicon Camera Tube with Increased Sensitivity to Red Light," *Philips Tech. Rev.*, 26, 49–51 (1965).

8. M. H. Crowell and E. F. Labuda, "The Silicon Diode Array Camera Tube," *Bell System Tech. J.*, 48, 1481–1528 (1969).

9. I. Tepper, R. A. Anders, and D. H. McCann, "Transfer Functions of Imaging Mosaics Utilizing the Charge Storage Phenomena of Transistor Structures," *IEEE Trans. Electron Devices*, ED-15, 226–237 (1968).

10. R. A. Anders, D. E. Callahan, W. F. List, D. H. McCann, and M. A. Schuster, "Developmental Solid-State Imaging System," *IEEE Trans. Electron Devices*, ED-17, 191–196 (1968).

11. P. K. Weimer, G. Sadisiv, J. E. Meyer, Jr., L. Meray-Horvath, and W. S. Pike, "A Self-Scanned Solid-State Image Sensor," *Proc. IEEE*, 55, 1591–1602 (1967).

12. P. K. Weimer, W. S. Pike, G. Sadisiv, F. V. Shallcross, and L. Meray-Hervath, "Multielement Self-Scanned Sensor Arrays," *IEEE Spectrum*, 6, 52–65 (1969).

13. M. R. Lorenz, "The Generation of Visible Light from p-n Junctions in Semiconductors," *Trans. Met. Soc.* AIME, 245, 539–549 (1969).

14. H. J. Guggenheim and L. F. Johnson, "New Fluoride Compounds for Efficient Infrared-to-Visible Conversion," *Appl. Phys. Lett.* 15, 51–52 (1969).

15. H. F. Ivey, *Electroluminescence and Related Effects*, Academic Press, New York, 1963.

16. B. Kazan and F. H. Nicoll, "An Electroluminescent Light-Amplifying Picture Panel," *Proc. IRE*, 43, 1888–1897 (1955).

17. R. H. Bube, "Mechanism of Photoconductivity in Microcrystalline Powders," *J. Appl. Phys.*, 31, 2239–2254 (1960).

18. B. Kazan, "Description and Properties of the Panel X-Ray Amplifier," *Nondestructive Testing*, 16, 438–447 (1958).

19. F. H. Nicoll and A. Sussman, "A Two-Color Input, Two-Color Output Image Intensifier Panel," *Proc. IRE*, 48, 1842–1846 (1960).

20. H. Graff and R. Martel, "A Display Screen with Controlled Electroluminescence," *Inform. Display*, 2, 53–57 (September/October 1965).

21. B. Kazan, "A Feedback Light-Amplifier Panel for Picture Storage," *Proc. IRE*, 46, 12–19 (1959).

22. B. Kazan and J. S. Winslow, "Image-Storage Panels Based on Field-Effect Control of Conductivity," *Proc. IEEE*, 56, 285–295 (1968).

23. H. F. Ivey and W. A. Thornton, "Preparation and Properties of Electroluminescent Phosphors for Display Devices," *IRE Trans. Electron Devices*, ED-8, 265–279 (1961).

24. E. J. Soxman, "Ultrahigh Contrast, Solid-State Teletype Display," Office of Naval Research, Technical Note ELTN-1 (AD 687-740), March 1969.

25. M. Yoshiyama, "Lighting the Way to Flat-Screen TV," *Electronics*, 42, 114–118 (Mar. 17, 1969).

26. F. H. Nicoll, "A Hysteresis Effect in Cadmium Selenide and Its Use in a Solid-Stage Image Storage Device," *RCA Rev.*, 19, 77–85 (1958).
27. G. K. Zin and L. J. Krolak, "A Solid-State Matrix-Addressed Display," *IEEE Trans. Electron Devices*, ED-12, 632–637 (1965).
28. S. R. Ovshinsky, "Reversible Electrical Switching Phenomena in Disordered Structures," *Phys. Rev. Lett.*, 21, 1450–1453 (1968).
29. G. W. Taylor, "The Design and Operating Characteristics of a 1200-Element Ferroelectric-Electroluminescent Display," *IEEE Trans. Electron Devices*, ED-16, 565–575 (1969).
30. C. E. Land and P. D. Thacher, "Ferroelectric Ceramic Electrooptic Materials and Devices," *Proc. IEEE*, 57, 751–768 (1969).

Chapter 10

ADVANCES IN LSI TECHNOLOGY

A. S. Grove

INTEL Corporation
Mountain View, California

10.1 INTRODUCTION

Integrated circuits consist of the combination of active electronic devices such as transistors and diodes with passive components such as resistors and capacitors within and upon a single-semiconductor crystal.[1,2] The construction of these elements within the semiconductor is achieved through the introduction of electrically active impurities into well-defined regions of the semiconductor. The fabrication of integrated circuits thus involves such processes as vapor-phase deposition of semiconductors and insulators, oxidation, solid state diffusion, and vacuum deposition, among others.[3,4]

Generally, integrated circuits are not a straightforward replacement of electronic circuits made of discrete elements. They represent an extension of the technology by which planar silicon transistors are made. Because of this, integrated circuit design is based on two important facts:

1. Transistors of good quality can be made easily in large numbers. Thus they are the principal elements of integrated circuits.

2. Passive components are difficult to obtain. Resistors and capacitors can be made only in a limited range of values, but inductors not at all.

The easiest form of circuitry to adapt to these changes is the logic circuitry employed in digital computers. Thus it is in this area that integrated circuits are employed predominantly.

Integrated circuits can be classified into two groups on the basis

404

TABLE 1. DESIGN TECHNOLOGIES

Discretionary wiring
Cell-library approach
Conventional design

of the type of transistors that they employ: (a) *bipolar integrated circuits*, in which the principal element is the bipolar junction transistor; and (b) *MOS integrated circuits*,[5] in which the principal element is the MOS transistor. Both depend upon the construction of a desired pattern of electrically active impurities within the semiconductor body and upon the formation of an interconnection pattern of metal films on the surface of the semiconductor.

The economic factors that have influenced the path of technological developments in the integrated circuit field in the past are such that those technologies which have permitted an increase in the functional density of electronic circuitry have invariably emerged as the dominant ones. This accounts for the intense current activity associated with *large-scale integration* (LSI).[6]

10.2 DESIGN TECHNOLOGIES

The three important approaches to the design of LSI circuits are listed in Table 1.

The discretionary wiring technique[7] makes very extensive use of computers in identifying and mapping the functionally good logic gates on a given silicon wafer, then in generating a mask automatically for that particular silicon wafer which interconnects the good circuits. A somewhat less heavily computer-oriented approach is the cell-library approach, in which a library of masks for commonly used functions is stored in a computer.[8] The computer accepts a logic diagram as its input, rearranges this library of cells, and generates a mask to interconnect them.

Notwithstanding the large amount of attention given to these techniques, the overwhelming majority of LSI circuits to date have been generated by conventional human design.

10.3 BIPOLAR TECHNOLOGIES

Table 2 lists the possible choices available in bipolar integrated circuit technology. The millions of circuits that are produced each

TABLE 2.   BIPOLAR TECHNOLOGIES

Diffused isolation, buried layer, gold-doped process
   with single-layer metallization
   with multiple-layer metallization
Dielectric isolation
Schottky diode process
Collector-diffusion isolation
Base-diffusion isolation
Three-mask process

week by the industry are made by the conventional technology,[1,2] which employs diffused isolation between components, a buried n^+ layer to reduce the collector resistance, and gold doping to speed up the circuits by reducing the lifetime of minority carriers.   Most of these circuits are made with a single layer of metallization.

Several technologies have been proposed to replace conventional diffused isolation.   One of them uses a dielectric for isolation in the place of a diffused p-n junction.[2]   This technique was originally introduced for circuits that were meant to operate in a radiation environment, but now at least one manufacturer attempts to use them in the fabrication of complex integrated circuits.   Likewise, the last three technologies—the collector-diffusion isolation,[9] the base-diffusion isolation,[10] and the three-mask process[11]—all attempt to replace the conventional isolation by different techniques.

The Schottky diode process[12] attacks another aspect of the conventional process, gold doping, while leaving the other features of the conventional process unchanged.

10.4 MOS TECHNOLOGIES

Table 3 lists the technologies that have been proposed for MOS integrated circuit fabrication.   One of the important advantages of MOS devices in integrated circuit applications is that they do not, in principle, require isolation between components.   However, parasitic conduction paths between neighboring devices can easily destroy the isolation inherent in MOS circuits.   As a result, in the early stages of development, diffused isolation was a necessary part of MOS circuits. Then, a family of technologies called the thick-oxide technology was developed[13] which combatted this isolation problem by using a thick oxide layer over all nonactive parts of the device.   This technology can be used in conjunction with (111) or (100) oriented silicon, resulting in

TABLE 3. MOS TECHNOLOGIES

Diffused isolation technique
Thick-oxide technology (111) Si orientations
(100)
Thick-oxide technology with aluminum oxide
Thick-oxide technology with silicon nitride
Silicon gate technology
Molybdenum gate technology
Ion-implanted MOS
n-channel technology with substrate bias
Complementary MOS technology
Silicon-on-sapphire technology

circuits that operate with different signal and power supply voltages. More recently, it has been used in conjunction with different dielectrics like aluminum oxide[14] and silicon nitride in the gate region of the MOS transistors, in order to lower further the operating voltage range of the circuits.

There have been attempts to eliminate the need for alignment between the gate region, and the source and drain regions, through technologies like the silicon gate technology,[15] the molybdenum gate technology,[16] and the use of ion implantation with conventional aluminum technology.[17]

There is n-channel technology,[18] with certain desirable features but additional technological difficulties; complementary MOS technology[19] which uses n- and p-channel devices on the same chip; and finally, silicon-on-sapphire technology,[20] in which devices are built in a thin layer of silicon on a sapphire substrate.

10.5 PACKAGING TECHNOLOGY

Table 4 lists some of the choices available in packaging technology. Again, the millions of circuits that are being manufactured today are

TABLE 4. PACKAGING TECHNOLOGIES

Eutectic die-attach, wire bond
dual-in-line package
flat-pack
Flip-chip bonding
Flip-chip bonding with solder bumps
Beam-lead bonding
Spider bonding

being produced with a conventional eutectic gold-silicon die attach and wire bonds, mostly in dual-in-line package.[3] Hundreds of people all over the world have been working on replacing this packaging scheme with different ones,[21] such as flip-chip bonding, a face-down bonding technique that may employ ultrasonic energy or thermal compression, or a combination of these to create a direct bond between the bonding pads of the device and the package. The same technique can be modified to employ solder bonds, in which a ball of solder joins the pads to the package. Beam lead bonding and spider bonding are other simultaneous lead bonding techniques.

It should be noted that there is a total of 225 possible combinations between the various technological choices—design, chip technology, and packaging technology—and it is evident that empirically optimizing these variations would be difficult if not impossible. Likewise, to review these choices in any detail would unduly tax both the writer and the reader. Accordingly, in the remainder of this paper we shall merely discuss the two silicon fabrication techniques, the Schottky diode bipolar technology and the silicon gate MOS technology that INTEL Corporation chose to use in the fabrication of LSI circuits.

10.6 SCHOTTKY DIODE BIPOLAR TECHNOLOGY[12]

Figure 1 shows the current voltage characteristics of a Schottky diode and a p-n junction diode. Note that the Schottky diode begins to conduct current at a lower voltage than the p-n junction. As a result, if the two of them are connected in parallel, and a forward voltage is applied to them, almost all of the current will flow through the Schottky diode. A Schottky diode is a majority carrier device: it conducts forward current without injecting minority carriers. By contrast, a p-n junction in forward bias injects minority carriers; it is these minority carriers that are stored in the collector region of the transistor and lead to an increase of propagation delay.

Figure 2 shows two ways in which such a parallel combination of a Schottky diode and the collector-base junction diode can be incorporated into the structure of an integrated circuit. Both of these are simple and both accomplish the same aims: the Schottky diodes are constructed in such a way that high-field effects at the edge of the diodes are reduced.

The use of such structures essentially eliminates the injection of minority carriers in the base-collector junction, as illustrated in Fig. 3. This figure shows no observable storage time in the case of the

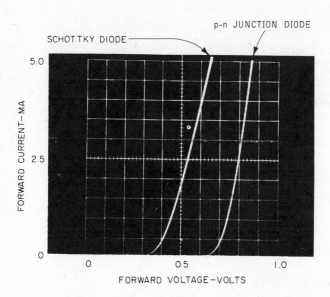

Fig. 1.   Current–voltage characteristics of Schottky diode and *p–n* junction diode.   (From Noyce, Bohn, and Chua. [12])

Schottky diode, clamped *p–n* junction.   Because the Schottky diode remains a majority carrier device, even as the temperature is increased, the storage effect does not become evident even at high temperatures.

    These results are achieved without the need for gold in the circuit. Since gold is one of the most difficult to control impurities, by eliminating it from the process higher yields result.   The elimination of gold also leads to another desirable feature: increased flexibility for the designer.   He can now have fast transistors and slow transistors right next to each other; he can use components like substrate and lateral *pnp*'s, whose characteristics would be completely destroyed if there were gold in the system.   The availability of these additional components leads to better designs and, therefore, to higher functional densities.

    Figure 4 shows a 64-bit, high-speed, bipolar memory circuit. This circuit contains 64 bits of storage, all decoding and reading and writing circuitry.   It fits into a 16-pin, standard, dual-in-line package. It makes extensive use of Schottky diodes; for instance, the memory cell of the circuit consists of two cross-coupled, Schottky diode clamped transistors.

    Another example of a complex Schottky integrated circuit is shown

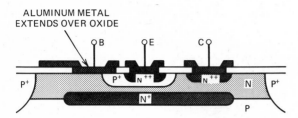

SCHOTTKY EXTENDED–METAL TRANSISTOR

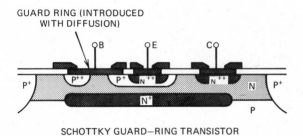

SCHOTTKY GUARD–RING TRANSISTOR

Fig. 2. Schottky diodes incorporated into integrated circuits. (From Noyce, Bohn, and Chua.[12])

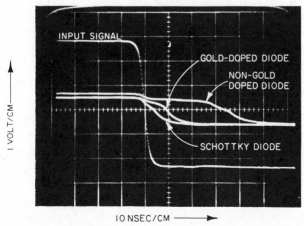

Fig. 3. Switching wave forms of various diodes. (From Noyce, Bohn, and Chua.[12])

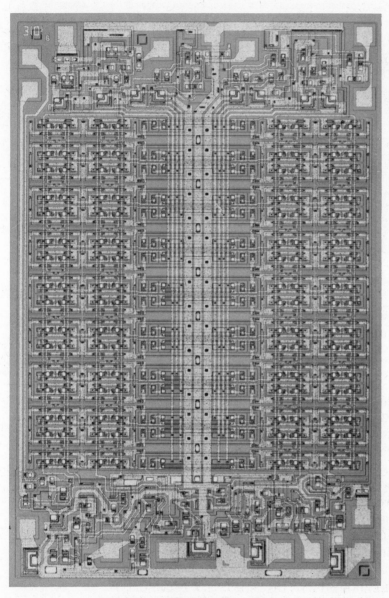

Fig. 4. A 64-bit, bipolar, read/write memory.

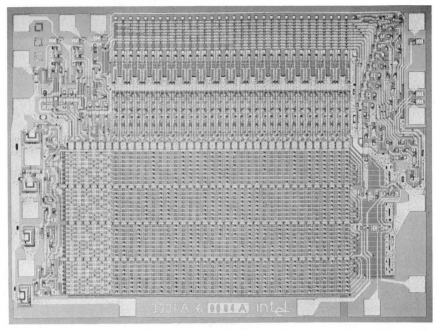

Fig. 5. A 1024-bit, bipolar, read only memory.

in Fig. 5. This is a 1024-bit, high-speed, bipolar read-only memory
circuit which also contains all the necessary peripheral circuitry. It
uses Schottky diodes even more extensively than the previous circuit;
the storage cell of the circuit consists of a single substrate *pnp* transistor, a component not available in gold-doped circuits.

Because they are a new integrated circuit component, these Schottky diodes were subjected to very extensive reliability testing, at high
temperatures, under forward and reverse bias conditions. Some results obtained on the 64-bit memory circuit are presented in Table 5.
These results, in agreement with results obtained on discrete components, indicate no reliability problems whatsoever.

TABLE 5. RELIABILITY DATA: SCHOTTKY TECHNOLOGY

Test conditions:
  full-rated power
  forward-, zero-, and reverse-biased Schottky junctions
  junction temperature: ~160 °C
Length of lifetest to last readout: 3800 hr
Number of bit hours: 50 million
Number of failures: 0

10.7 SILICON-GATE MOS TECHNOLOGY

Silicon-gate technology[15] departs from all other MOS technologies in major respects. As in conventional MOS technology, the starting material for silicon-gate technology is n-type silicon. The wafer is first placed into an oxidizing atmosphere at high temperature and a relatively thick (about 1-μm) layer of silicon dioxide is formed on its surface. Next, regions for the source and drain of the final device and the eventual channel regions are defined by photomasking. The oxide is then etched from this area, as shown in Fig. 6a. The slice is placed into an oxidizing ambient again, and a layer of silicon dioxide about 0.1 μm thick is formed in the window (Figs. 6b and 6c).

Subsequently, a thin layer of silicon nitride (Si_3N_4), another insulator, is deposited onto the entire surface of the wafer. This layer affects both the electrical characteristics and the reliability in a desirable manner.

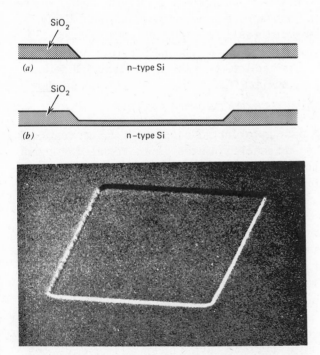

Fig. 6. (a) First oxide cut after initial oxidation for subsequent source, drain, and gate region. (b) Second oxidation to define gate oxide thickness. (c) Scanning electron microscope (SEM) photograph of a transistor fabrication at stage B. (From Vadasz *et al.*, Ref. 15, p. 28.)

Next, a thin layer of amorphous silicon is deposited on the wafer. The device structure at this point is as illustrated in Fig. 7a.

The structure now is returned to photomasking for removal of the silicon and silicon nitride layers, except where the gate area is to be or where the silicon film is intended for use as a circuit interconnection layer. The thin oxide is also removed by exposing it to an oxide etch, as shown in Fig. 7b.

Finally, the wafer is placed into a diffusion furnace and boron is diffused into the surface of the structure. Consequently, after exposure, boron impurities will convert both the n-type silicon wafer, where exposed, and the deposited silicon layer into p-type silicon; but the n-type regions under the oxide-covered areas are unaffected (Fig. 7b, 7c, and 7d).

The device structure, except for the necessary interconnections, is now complete. A layer of silicon dioxide is deposited onto the entire surface: the exposed regions of the silicon wafer, the thick oxide layer covering most of the surface, and the deposited gate silicon. Openings are photoetched in this deposited silicon dioxide layer, whereever a contact between the subsequent metallization and the underlying silicon wafer or deposited silicon is desired, as shown in Fig. 8a. Aluminum is evaporated onto the surface, so that it enters into these contact openings, and the desired interconnection patterns are defined by another photomasking operation. The device now appears as shown in Figs. 8b and 8c.

The principal advantage of the silicon gate technology is that, by using deposited silicon for the gate electrode instead of the conventional aluminum, we achieve low-threshold voltages. As a result, the signal voltages with which silicon gate devices can operate are lower, and direct compatibility with bipolar circuits becomes possible. Furthermore, because the source and drain regions were defined by the use of the deposited gate electrode as the mask, the overlap capacitance between the diffused regions and the gate is lower. It turns out, in addition, that other parasitic capacitances in a silicon gate circuit will be lower also. Because the speed of MOS circuits is generally RC time-constant-limited, the C being parasitic capacitances, the reduction of these capacitances leads to higher-speed circuits.

The elimination of the need for alignment tolerances between the gate and source and drain, as well as the crossover possibilities that come about because the silicon gate is buried under a layer of dielectric, result in less area needed for each electronic function, and hence to higher yields. Since the gate is buried under a layer of dielectric, we also achieve a better reliability.

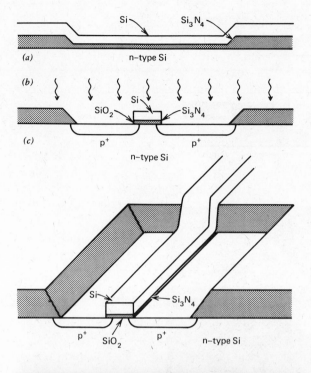

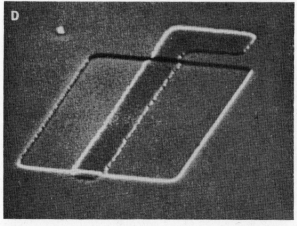

Fig. 7. (a) Device of Fig. 6b after deposition of nitride and amorphous sili-con sandwich. (b) Definition of gate insulator and gate electrode dimensions and subsequent boron diffusion. (c) Perspective view of structure of (b). (d) SEM photograph of the structure at the end of this fabrication stage. (From Vadasz *et al.*, Ref. 15, p. 29.)

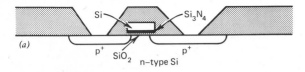

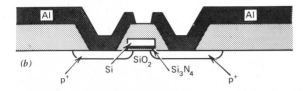

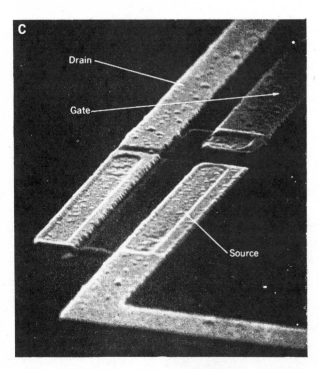

Fig. 8. (a) Device of Fig. 7b after deposition of silicon dioxide and definition of the contact regions. (b) Final structure after deposition and definition of metal interconnections. (c) SEM photograph of the completed transistor structure. (From Vadasz et al., Ref. 15, p. 29.)

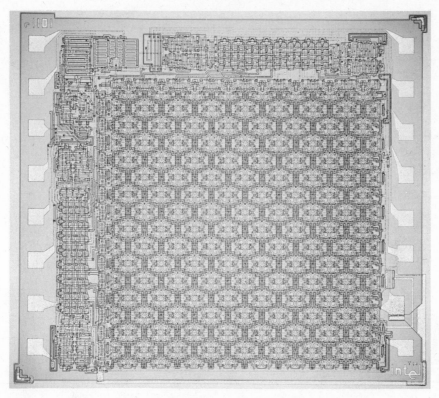

Fig. 9. A 256-bit, MOS read/write memory.

Figures 9 and 10 show some examples of LSI circuits fabricated with this technology, a 256-bit, static, random access memory, containing about 2000 transistors; and a 1024-bit, dynamic shift register, containing about 6000 transistors.

Table 6 summarizes some reliability data obtained with the 256-bit

TABLE 6. RELIABILITY DATA:
Si-GATE TECHNOLOGY

Test conditions:
  full-rated power
  reverse-biased inputs
  junction temperature: ~160 °C
Length of life test to last readout: 2200 hr
Number of bit hours: 30 million
Number of failures: 0

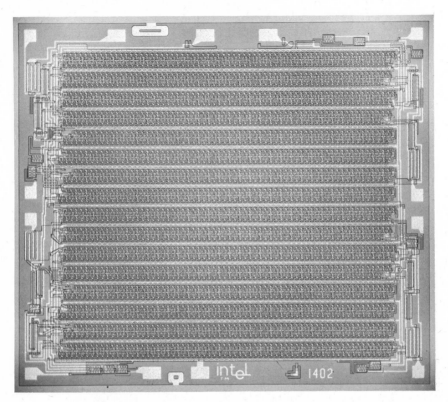

Fig. 10. A 1024-bit, dynamic shift register.

memory circuit. In agreement with other data obtained on discrete components, the silicon gate circuit shows no reliability problem.

10.8 CONCLUSIONS

We have seen that a formidable battery of technologies is available for use in the fabrication of LSI circuits. Some of these technologies are undoubtedly superior to others. The tradeoff between their advantages and shortcomings, of course, strongly depends on the particular requirements placed on the manufacturer.

The two silicon device fabrication technologies discussed in detail here, the Schottky bipolar technology and the silicon gate MOS technology, were found to possess a particularly useful combination of characteristics in the framework of mass production of LSI semiconductor memory circuits: they lead to high manufacturing yields and large functional density, permitting low cost, and they are very reliable.

REFERENCES

1. G. E. Moore, *Semiconductor Integrated Circuits*, in *Microelectronics*, E. Keonjian, Ed., McGraw-Hill, New York, 1963, Chap. 5.
2. *Integrated Circuits: Design Principles and Applications*, R. M. Warner and J. N. Fordenwalt, Eds., McGraw-Hill, New York, 1965.
3. *Integrated Circuit Engineering*, Boston Technical Publishers, Boston, 1966.
4. A. S. Grove, *Physics and Technology of Semiconductor Devices*, John Wiley & Sons, New York, 1967.
5. R. H. Crawford, *MOSFET in Circuit Design*, McGraw-Hill, New York, 1967.
6. A. J. Khambata, *Introduction to Large Scale Integration*, John Wiley & Sons, New York, 1969.
7. J. Kilby and J. W. Lathrop, "Discretionary Wiring Approach to Large-Scale Integration," WESCON, 1966.
8. L. Vadasz, "The Micromatrix Approach to MOS Complex Arrays," WESCON, 1966.
9. V. J. Glinski and B. T. Murphy, "Bipolar Integrated Circuits Formed in p-type Epitaxial Layers," International Electron Devices Meeting, Washington, D. C., 1968.
10. L. S. Senhouse, D. L. Kushler, and B. T. Murphy, "Base Diffusion Isolation for Transistors," International Electron Devices Meeting, Washington, D. C., 1969.
11. V. J. Glinski, "A Three-Mask Bipolar Integrated Circuit Structure," International Electron Devices Meeting, Washington, D. C., 1968.
12. R. N. Noyce, R. E. Bohn, and H. T. Chua, *Electronics*, (July 21, 1969).
13. J. L. Seely, "Advances in the State-of-the-Art of MOS Device Technology," *SCP Solid-State Technology*, 59 (1967).
14. H. E. Nigh, J. Stach, and R. M. Jacobs, Solid-State Device Research Conference, 1967.
15. L. L. Vadasz, A. S. Grove, T. A. Rowe, and G. E. Moore, "Silicon-Gate Technology," *IEEE Spectrum*, 6, 28, 29 (1969).
16. D. M. Brown, W. E. Engeler, M. Garfinkel, and P. V. Gray, "Refractory Metal Silicon Device Technology," *Solid-State Electronics*, 11, 1105 (1968).
17. R. W. Bower, H. G. Dill, K. G. Aubuchon, and S. A. Thompson, "MOS Field-Effect Transistors Formed by Gate Masked in Implantation," *IEEE Trans. Electron Devices*, ED-15, 757 (1968).
18. H. Yamamoto, M. Shiroishi, and T. Kurosawo, "A 40ns, 144-bit, n-channel MOS-IC Memory," International Solid-State Circuits Conference, Philadelphia, 1969.
19. T. Klein, "Technology and Performance of Integrated Complementary MOS Circuits," IEEE J. Solid-State Circuits, SC-4, 122 (1969).
20. "Silicon and Sapphire Getting Together for a Comeback," *Electronics*, (June 8, 1970).
21. G. K. Fehr, "Microcircuit Packaging and Assembly—State of the Art," *Solid-State Technology*, 41 (August 1970).

Chapter 11

MICROWAVE INTEGRATED CIRCUITS

Martin Caulton

David Sarnoff Research Center
RCA Corporation
Princeton, New Jersey

11.1 INTRODUCTION

Microwave integrated circuits or MICs have only existed since 1965. All prior equipment used wave guide, coaxial, or stripline circuitry. Significant progress has been made during the past few years in introducing integration into the microwave frequency range. Much of this progress is a result of advances in microwave solid state devices; the achievements could not have been accomplished, however, without the development of a sound technology base for microwave integrated circuits. The feasibility of integration of other types of microwave circuits with solid state devices has been demonstrated and most companies in the microwave industry have groups working on microwave integrated circuits. For these reasons a chapter on microwave integrated circuits will aid in bringing solid state electronics up to date.

This chapter will present a survey of microwave integrated circuits. The types of integration, the circuits forms, and the material technology (substrates, conductors, dielectrics, and resistors) will be described. A discussion of circuit design and performance will be followed by some fabrication techniques. Some circuit techniques particular to microwave integrated circuits will be developed. Examples of circuits will be presented but there will be no discussion of active devices. Throughout we shall attempt to point out the areas in which MIC technology differs from that pertaining to lower-frequency circuits.

11.2 CIRCUIT TYPES

11.2.1 Circuit Size Reduction

In order to use solid state devices with microwave circuits, it is necessary to transform the world of the miniature into the real world. On the other hand, by reducing the real world of large-size microwave circuits to the small one of the solid state device, it is often advantageous to integrate the device and the circuit.

The size reduction of microwave circuits was accomplished in two general ways. One way was to use lumped components, keeping them smaller than a wavelength. This was accomplished using conventional elements at frequencies as high as 1 GHz, barely approaching the microwave region. Photolithographic techniques now allow lumped elements to operate into the microwave region.

The other and, to date, more common reduction technique is to use sections of transmission line (classified as distributed circuitry). Distributed circuits use sections of the line less than a wavelength to provide the needed function. By shortening the wavelength in the line through appropriate dielectric material, size reduction can be achieved.

The semi-infinite, parallel plate transmission line (Fig. 1a) is a simple line in which a transverse electromagnetic (TEM) wave is slowed down by using a material with a high dielectric constant. In 1952 the microstrip line was introduced.[1] This is a practical version of the parallel plate, in which the top plate is narrowed to a width W. The fields are no longer perpendicular to the plates. Another version, in which an additional dielectric and ground plane is above the center conductor, was also established and is known as stripline. A commercial version is also known as Triplate. All of these distributed lines

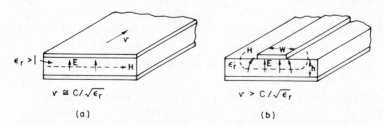

Fig. 1. Development of microstrip line from parallel plate transmission line. (a) Parallel plate transmission line (E and H perpendicular to v); (b) microstrip line.

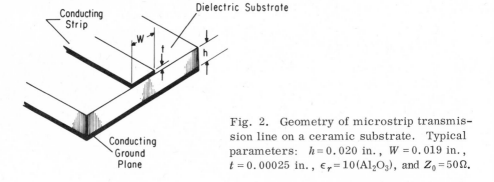

Fig. 2. Geometry of microstrip transmission line on a ceramic substrate. Typical parameters: $h = 0.020$ in., $W = 0.019$ in., $t = 0.00025$ in., $\epsilon_r = 10 (Al_2O_3)$, and $Z_0 = 50\Omega$.

can slow the velocity of propagation by a factor $1/\sqrt{\epsilon_r'}$, where ϵ_r' is the effective dielectric constant of the propagation medium. Thus the guide wavelength is shortened

$$\lambda_g = \frac{v}{f} = \frac{c}{f\sqrt{\epsilon_r'}}$$

Figure 1 illustrated the evolution of parallel plate transmission lines to microstrip.[1] Figure 2 demonstrates the microstrip line with its most widely used material, a metal line adhering to an alumina substrate. Since the propagation is only "quasi-TEM" (the fringe fields of Fig. 1 extend beyond the conductors), the effective dielectric constant ϵ_r' of the line as shown, is 6.5, instead of the alumina ϵ_r of 10, although the major part of the propagation field is confined to the region of the dielectric below the strip conductor. The ratio of the width of the center conductor to the thickness of the substrate is about 1 for a 50-Ω transmission line. A metal thickness of several skin depths is required. Matching sections and filters are made using side stubs and lines of varying widths and lengths (impedances and phase). Figure 3 is a photograph of a microwave integrated circuit that illustrates microstrip lines on alumina. This module (so-called MERA[2]) is the result of a large program initiated by Texas Instruments and the Air Force that started active investigation of microwave integrated circuits for radar systems and showed their feasibility. We shall consider later some of the functions of circuits fabricated by microstrip distributed systems.

A second form of distributed circuitry, the suspended substrate,[3] is shown in Fig. 4. Here the ground metal surrounds the center conductor deposited on one side of the alumina. The wave propagation is again quasi-TEM, but most of the energy in the propagating fields is concentrated in the air space above and below the center conductor, and

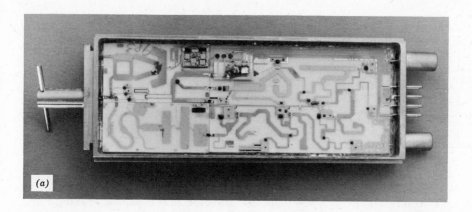

Fig. 3. Photograph of MERA rf module. (Developed by Texas Instruments under Contract No. AF33(615)-2525, sponsored by the U. S. Air Force Avionics Laboratory, Wright-Patterson Air Force Base, Dayton, Ohio.) Microstrip circuits on alumina, composed of different-width impedance lines and stubs, as well as capacitors, resistors, and inductors. The module is 2.5 in. long. Of these 604 are used in a phased array radar. (a) X4 frequency multiplier, lower left; pulsed amplifier and module, right. Upper left has a TR switch, mixer, and 500-MHz i. f. preamp, fabricated on a separate substrate and mounted on a circuit. (b) Left, X4 local oscillator multiplier, logic and phase shifter circuits, and S-band preamplifier and duplexer.

the effective ϵ_r' of the propagation is about 1.6. For this reason the resulting package is larger than one of microstrip. The transmission characteristics are determined by the substrate thickness, air space height, and the width of the strip conductor. The system has been used by Bell Telephone Laboratories and has shown high reliability. Figure 5 illustrates an L-band (1 to 2 GHz) amplifier from Bell Telephone

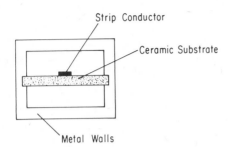

Strip Conductor

Ceramic Substrate

Metal Walls

Fig. 4. Suspended substrate transmission line.

Laboratories, incorporating a suspended-substrate system.

Two more recent types of lines, slotline and coplanar wave guide, are shown in Figure 6. The slotline[4] of finite width is a geometrical and electrical dual of the coplanar guide.[5] The propagation is non-TEM and there are longitudinal, in addition to transverse, rf magnetic fields. The polarization properties (magnetic fields on the surface of

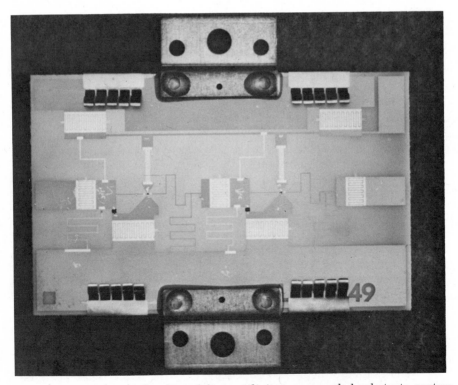

Fig. 5. Two-stage, L-band amplifier, utilizing a suspended substrate system. Lumped element and distributed lines on 24-mil glazed alumina. (Courtesy of K. F. Sodomsky, Bell Telephone Laboratories.)

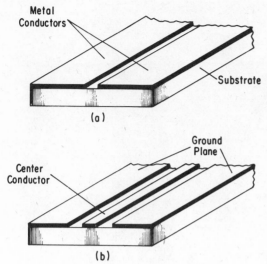

Fig. 6. (a) Slot line and (b) coplanar wave guide (CPW).

the substrate) allow the placement of magnetic material on the substrate surface, as well as the use of magnetic substrate material, and the simple fabrication of magnetic nonreciprocal devices. The metal is generally confined to the top surface of the substrate and permits shunt mounting of devices without requiring holes through the substrate, as in the case of microstrip lines.

A photograph of a coplanar 3-dB directional coupler fabricated by C. P. Wen[6] is shown in Fig. 7. This coupler has better directivity than equivalent microstrip versions, as will be described later. A combination of slotline and microstrip on opposite sides of a substrate can be used to provide directional couplers with very wide band properties.[7]

Lumped elements were not normally used above 1 GHz because they are difficult to fabricate by conventional means. When the wavelength decreases to 2 or 3 times the element size, the phase varies across the element, and the reactance no longer varies directly with frequency. The element also can radiate. Wave guide, coaxial, and microstrip circuits confine the fields and use the transmission line aspects at high frequencies to accomplish the required tuning. By using photolithographic techniques developed with transistor and integrated circuit technology, lumped elements are now practical up to 10 GHz. The elements can now be fabricated in sizes much smaller than a wavelength.

Typical lumped element components suitable for microwave frequencies are shown in Fig. 8. An inductor, a strip of ribbon that suf-

Fig. 7. 3-dB directional coupler, utilizing CPW line. (1.0× 1.0-in. Trans
Tech D-16 substrate, centerline separation less than 1 mil, wires to maintain
ground at line bends.)

fices for the few nanohenrys required at microwave frequencies is
shown in Fig. 8a, and Fig. 8b shows the metal-dielectric sandwich
that serves as a capacitor. Interdigital capacitors are also used. It
will be recalled that the term, lumped element (LE), implies: (a) constant inductance, capacitance, and resistance for the element under
frequency variation, and (b) no variation of phase over the element,
such as occurs in the lumped analog of a distributed circuit.

Lumped element circuits can be much smaller than equivalent distributed circuits at microwave frequencies. Figure 9 is a photograph
of a lumped element, 2-GHz, 1-W amplifier (150×120 mils) setting on
an equivalent microstrip amplifier.[8] The size of the microstrip tuning
elements is of the order of a quarter-wavelength, but in this amplifier
the layout was not compressed, as in Fig. 3, so that the size is larger

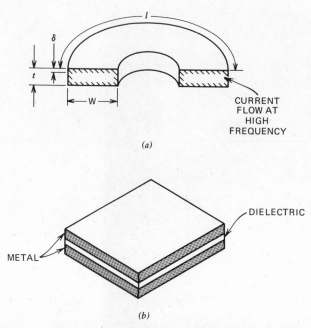

Fig. 8. Typical lumped element components. (a) Ribbon inductor; (b) metal-oxide-metal thin-film capacitor.

than it need be. The cost of fabricating integrated circuits varies inversely as the circuit size, since batch processing is possible with small circuits. The cost of LE circuits can thus be lower than that of distributed ones. Batch processing of LE circuits is demonstrated in Fig. 10, where many amplifier circuits are shown processed on a single $\frac{3}{4} \times 1$-in. sapphire substrate which is 5 mils thick.

The various circuit techniques are ranked in Table 1 with respect to size and weight, cost, reproducibility, reliability, and performance.[9] The ranking varies from the best (1) to worst (5). The conventional circuits, coax and wave guides, provide the best performance with the lowest-circuit loss, but they are the highest in cost, size, and weight. Microwave integrated circuits show the best reproducibility and reliability. The circuit losses of all microwave integrated circuits are intermediate in rank; lumped elements are the lowest in cost, size, and weight.

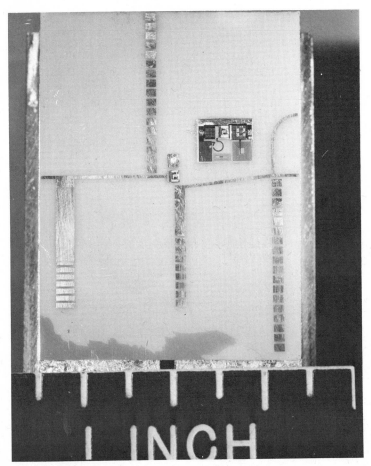

Fig. 9. Lumped element, 2-GHz, 1-W amplifier (150×120 mils of sapphire) on an equivalent tunable microstrip amplifier (0.7×1.0-in. alumina).

11.3 MATERIALS AND TECHNOLOGY

11.3.1 Substrates

Some of the desirable characteristics of substrates for microwave integrated circuits are low dielectric loss, good adherence to metal conductors, polished surfaces for the attainment of good line definition and yield in thin-metal oxide capacitors, and maintenance of mechanical integrity during processing. Heat conductivity is important, as well, in certain applications.

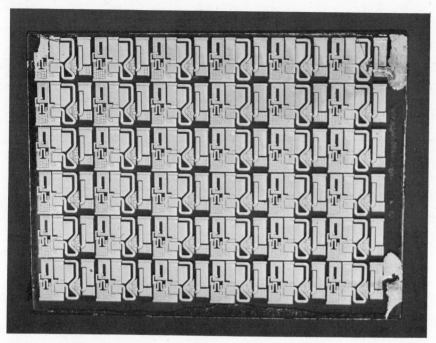

Fig. 10. Batch fabrication of lumped element, microwave amplifier circuits. (72 circuits on a 1.0×0.75×0.0045-in. sapphire substrate.)

TABLE 1. RANKING OF CIRCUIT TECHNIQUES*

	Micro Strip MIC	Slot Line; Coplanar MIC	Lumped Element MIC	Conventional Strip Line	Coax and Wave Guide
Size and weight	2	3	1	4	5
Cost	2	2	1	2–3	4
Reproducibility	1	1	2	2–3	4
Reliability	1	1	1	2	3
Circuit losses	2–3	3	2–3 to C Band 4–5 Higher	2	1

*1 is best, 5 is worst.

429

TABLE 2.   PROPERTIES OF SUBSTRATES

Material	Tan $\delta \times 10^4$ at 10 GHz	ϵ_r	K (W/ cm-°C)	Application
Alumina	2	10	0.3	Microstrip, suspended substrate
Sapphire	< 1	9.3–11.7	0.4	Microstrip, lumped element
Glass	> 20	5	0.01	Lumped element, quasi-monolithic microwave integrated circuit
Beryllia	1	6	2.5	Compound substrates
Rutile	4	100	0.02	Microstrip
Ferrite/ garnet	2	13–16	0.03	Microstrip, slotline, coplanar, compound substrates. Nonreciprocal components
GaAs	16	13	0.3	High-frequency microstrip, monolithic microwave integrated circuit
Silicon	10–100	12	0.9	(Limited) monolithic microwave integrated circuits

Table 2 describes the properties of some popular substrates that have been used for microwave integrated circuits. The table lists rf loss or tan δ, dielectric constant ϵ_r, thermal conductivity K, and also applications of each material. Glass is the only lossy material, and it has poor thermal properties. Beryllia has a good thermal conductivity but is difficult to handle. For this reason it is used in conjunction with other substrates, and it is listed as a compound substrate. Rutile is also used for slot- and coplanar lines. Ferrite and garnet are used for nonreciprocal components, and semiinsulating GaAs is included because it can also be used for monolithic integration. High-resistivity silicon was also considered for use as a substrate, but the insulating properties could not be maintained during processing.

11.3.2 Conductors

Some considerations for conductors of microwave integrated circuits are the rf resistance and skin depth (determining the thickness required), deposition techniques, substrate adherence, and the thermal

TABLE 3.   CHARACTERISTICS OF THREE TYPES OF CONDUCTORS

Material Type	Surface Resistance at 2 GHz, $\Omega/\square$	Skin Depth at 2 GHz, μm	α_T Thermal Expansion, $/°C \times 10^6$	Adherence to Dielectric
(1) (Ag, Cu, Au, Al)	0.011–0.015	1.4–1.9	15–26	Poor
(2) (Cr, Ta, Ti)	0.03–0.07	4.0–10.5	8.5–9.0	Good
(3) (M_0, W)	0.12	2.6	6.0, 4.6	Fair

expansion during processing. The ability to define the conductor circuit must play a role and will be discussed separately.

Conductors can be divided into three categories as indicated in
Table 3: (1) good conductors, low resistivity, but with poor adhesion
to dielectrics. Aluminum is an anomaly, since it has fair adhesive
properties. The second category (2) is that of poorer conductors but
with good adherence to dielectrics. We have listed the more popular
ones: chromium, tantalum, and titanium. Thin flashes of these provide adhesive layers for category (1) conductors. The metals of the
first two categories are generally deposited by vacuum evaporation,
resistance-boat or electron beam heating. The third category (3) contains fair conductors with fair adherence to dielectrics. Tungsten and
molybdenum are refractory materials, and vacuum evaporation using
electron beam heating is required for deposition. Sputtering, however,
works with all of these conductors, and is particularly useful for category 3 materials. Looking at category 1 metals in Table 3, we see that
several microns of metal are required to obtain the maximum conductivity.

Metals adhere to ceramics by either a mechanical or chemical
bond. The mechanical bond requires a rough surface, defeating the
precise definition of metals. The chemical bond is therefore required
for surface finishes better than 5 μin. Figure 11 indicates the mechanism for a chemical bond. A reducing material such as chrome or
titanium is oxidized by the heated substrate; the oxidized material is
then firmly bonded to most dielectric substrates. Metal will adhere to
other unoxidized metal. Thus pure chrome without oxide will adhere
to a pure metal of category 1 if they are all deposited in the same vacuum run. The required thickness of good conductor metal can be built
up by either evaporating in the same vacuum run, or later by plating
over the seed metal.

Two techniques to define patterns (metal layers several skin depth
thick) are illustrated in Fig. 12. By using only a thin seed of evapo-

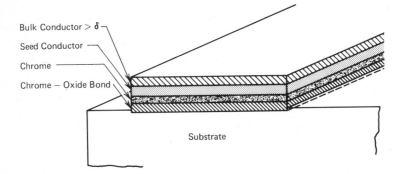

Fig. 11. The chemical bond of metal to dielectric (chrome–oxide layer can be ~50 Å or greater). The deposition shown must be carried out in one vacuum run.

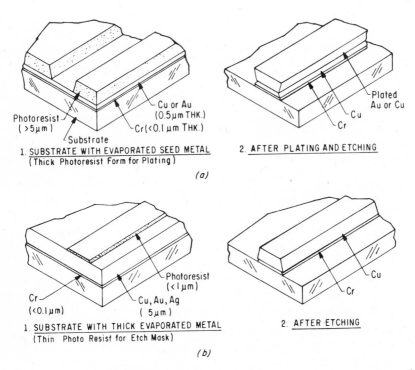

Fig. 12. Techniques for defining patterns in several μm of metal. (a) Circuit fabrication by plating and etching; (b) circuit fabrication by etching thick metal.

rated metal, one can plate through a thick photoresist form as shown in
(a) and obtain a 5-μ-thick line or greater with rather precise definition.
This is because only a thin seed metal less than 1 μ thick is etched
and the undercut is reduced. On the other hand, straight etching of
thick metal as shown in (b) leaves one with an undercut that is of the
order of twice the line thickness, or in this case about a half a mil.
The plating-etching technique allows more precise definition, although
the evaporated metal provides smoother material with bulk properties.

Metal films may also be deposited and defined by so-called "thick-
film" techniques involving silk screening through a mask. The term
"thick film" refers to the process used and not the film thickness (our
"thin film" metals are several microns thick, thicker than those of
low-frequency integrated circuits). The thick-film process usually
involves the printing and screening of silver or gold in a glass frit that
is applied on the ceramic and fired at 850 °C. Silk screening about $\frac{1}{2}$
mil of material through a metal mask gives reasonable definition which
may be improved later by etching. After firing, the initial layer may
be covered with plated gold. The difficulty with thick films is that the
material is not pure—it is interdispersed with glass. Because of this,
a well-defined etch pattern such as that obtained with thin-film processing is not possible, although etching of the screened pattern improves the definition. The dc resistance of the fired conductor is much
higher than that of the pure evaporated films, and the rf loss is increased by a factor of 2 or more.[10]

The use of a thin layer of high-resistivity adhesive films such as
chromium, tantalum, or titanium has very little influence on the rf
loss. Figure 13 shows the computed[9] relative loss of chrome and a
category 1 metal forming the microstrip line. Normally, about 200 to
400 Å of chrome is deposited. This produces a negligible increase of
rf resistivity, even at higher frequencies.

In Fig. 14 we see an X-band receiver front end from Microwave
Associates, fabricated using thick-film technology. This distributed
circuit is on a 1×2-in. $\times 20$-mil alumina substrate. The IF circuit uses
NiCr and gold on 25-mil alumina.

11.3.3 Dielectrics and Resistors

Dielectrics for isolation and for capacitors in microwave integrated
circuits require the properties of reproducibility; the capability of
withstanding high voltages; the ability to undergo processing without
developing pin-holes; and, for many applications, low rf dielectric
loss. Some of these properties are difficult to achieve in thin-films.

Some of the dielectrics used with microwave integrated circuits

TABLE 5. PROPERTIES OF RESISTIVE FILMS

Material	Resistivity, $\Omega/\square$	TCR, %/°C	Stability
Cr (evap.)	10–1000	−0.1 to +0.1	P
NiCr (evap.)	40–400	+0.002 to +0.1	G
Ta (sputtered in A-N)	5–100 +	−0.01 to +0.01	E
Cr–SiO (evap.) (Cermet)	Up to 600	−0.005 to −0.02	F
Ti (evap.)	5–2000	−0.1 to +0.1	F

the design of lumped elements of known value and loss.

11.4.1 Microstrip and Distributed Circuits

As indicated earlier, the fundamental propagation mode of a microstrip line is TEM. Most of the energy propagates in the dielectric below the center conductor strip. A fringe field, indicated in Fig. 1b, extends out of the dielectric in the air space. Conformal mapping procedures are used to determine the properties of propagation in microstrip. Many workers[1,12,13] have published numerous analyses (taking into account the fringing electric fields) that were accurate for low-characteristic impedances ($w \gg h$), when the line was closer to the parallel plate line of Fig. 1a.

The proper mapping procedure must also account for the dielectric discontinuity. This was proposed by Wheeler[14,15] in 1964–1965. In 1966 Caulton, Hughes, and Sobol[16] used Wheeler's results to obtain design data that have been very accurate through 3 GHz. These data hold for high-impedance lines ($w < h$), as well as low-impedance lines, which the earlier analyses did not achieve.

Figure 15 summarizes much of the data for microstrip on alumina ($\epsilon_r = 10$) at 2 GHz, as predicted by the TEM analysis of Wheeler[15] and verified by experiment.[16,17] A common ordinate for wavelength, attenuation, and w/h is used. Note that the wavelength varies only slightly with Z_0 and correspondingly with the width-to-height ratio. Also note the narrow lines required for higher impedances and the correspondingly high attenuation associated with these lines. A 50-Ω line has a w/h of close to 1. A general expression for the characteristic impedance of the unshielded microstrip line is given by Sobol.[18]

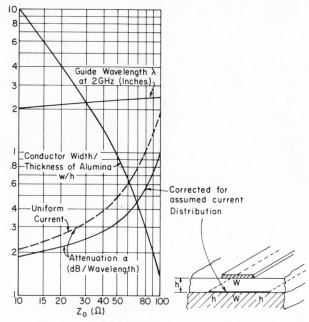

Fig. 15. Wavelength λ, width-to-height w/h, and attenuation characteristics of microstrip on alumina at 2 GHz, all versus characteristic impedance. The attenuation is normalized for $h = 10$ mil and varies as $10/h$ (mils).

$$Z_0 = \frac{377}{\sqrt{\epsilon_r}} \frac{h}{w} \frac{1}{1 + 1.735(\epsilon_r)^{-0.0724}(w/h)^{-0.836}} \tag{1}$$

This expression has been found to be accurate to within 1% for $w/h > 0.4$, and $\epsilon_r > 1$, and within 3% for $w/h > 0.4$ and $\epsilon_r > 1$.

The wavelength for the line of Fig. 5 is given by

$$\frac{\lambda_g}{\lambda_{TEM}} = \left[\frac{\epsilon_r}{1 + 0.63(\epsilon_r - 1)(w/h)^{0.1255}} \right]^{1/2} \quad w/h > 0.6 \tag{2}$$

$$\frac{\lambda_g}{\lambda_{TEM}} = \left[\frac{\epsilon_r}{1 + 0.6(\epsilon_r - 1)(w/h)^{0.0297}} \right]^{1/2} \quad w/h \leq 0.6 \tag{3}$$

where λ_g is the guide wavelength and $\lambda_{TEM} = \lambda_0/\sqrt{\epsilon_r}$, where λ_0 is the free-space wavelength.

A design nomagraph, derived from Wheeler's curves, that is very helpful, is given by Presser.[19]

The loss per unit length of a microstrip line is inversely proportional to the substrate thickness h. The unloaded Q_0 of a line $\lambda/4$ long is:

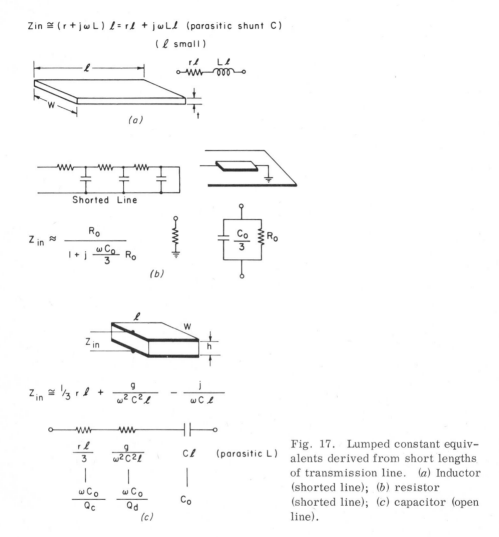

$$Z_{in} \cong (r + j\omega L)\, \ell = r\ell + j\omega L\ell \quad (\text{parasitic shunt } C)$$

$$(\ell \text{ small})$$

$$(a)$$

Shorted Line

$$Z_{in} \approx \frac{R_0}{1 + j\frac{\omega C_0}{3} R_0}$$

$$(b)$$

$$Z_{in} \cong \frac{1}{3} r\ell + \frac{g}{\omega^2 C^2 \ell} - \frac{j}{\omega C \ell}$$

| $\dfrac{r\ell}{3}$ | $\dfrac{g}{\omega^2 C^2 \ell}$ | $C\ell$ | (parasitic L) |

$$\frac{\omega C_0}{Q_c} \qquad \frac{\omega C_0}{Q_d} \qquad C_0$$

$$(c)$$

Fig. 17. Lumped constant equivalents derived from short lengths of transmission line. (a) Inductor (shorted line); (b) resistor (shorted line); (c) capacitor (open line).

smaller inductances are generally required. At S-band the Qs can be predicted from Eqs. 5 and 6.

The previous discussion is for free-space inductors with ground shields far removed. In practice the integrated inductors are usually in close proximity to metal surfaces underneath and to the side, as shown in Fig. 20, lowering both the inductance and Q. By treating the resulting configurations as transmission lines and computing Z_0, we find that inductances and Qs will be less affected if the side metal is at least 4 stripwidths away, and ground planes forming microstrip are 20 widths away.

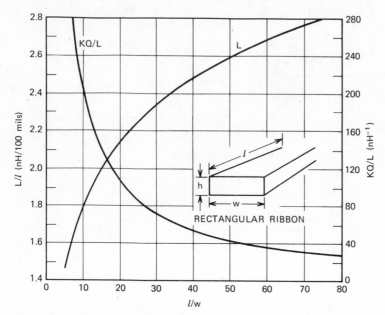

Fig. 18. Inductance and Q versus geometry ratio (length-to-width) of a ribbon inductor (Eqs. 9 and 10). $KQ/L_R = 2.153 \times 10^{12}\, w/l\; (w \gg h)$; $f = 2$ GHz; and $\rho = \rho_{cu}$.

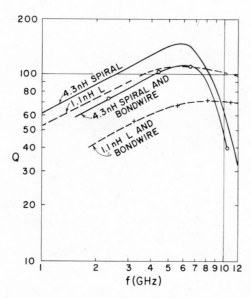

Fig. 19. Q versus frequency for a 4.3-nH spiral- and a 1.1-nH single-turn inductor (correction for bond-wires shown).

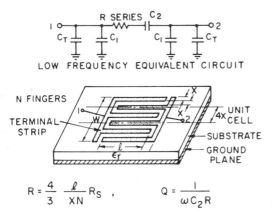

$$R = \frac{4}{3} \frac{\ell}{XN} R_S \ , \qquad Q = \frac{1}{\omega C_2 R}$$

Fig. 22. Interdigitated capacitor. (From Alley.[24])

about an order of magnitude improvement over those reported in earlier papers. The Q_c (metal) expression derived from these curves has an experimental number in Eq. 9 greater than 5400, to be compared with the 29,000 theoretically expected.[8,23] This is at present unexplained. Capacitor Q's greater than 100 at 12 GHz have been achieved and they behave in a lumped fashion to X-band. The thin-film capacitors are approximately 2.5 pF per 100 sq mils.

There is another type of capacitor that gives Q's close to those reported that is being widely used at Bell Laboratories.[24] The so-called interdigital capacitor makes use of the fringe fields of adjacent metals, an example of which is shown in Fig. 22. These capacitors show distributed effects at frequencies above 2 GHz and are much larger than simple metal-oxide-metal capacitors. However, for low frequencies, they are advantageous in their relative ease of manufacture, as the use of an additional dielectric can be avoided. Some of these capacitors can be seen in Fig. 4.

11.5 FABRICATION

The fabrication procedures for microwave integrated circuits will now be considered. Figure 23 illustrates the basic three layers that suffice for both lumped and distributed circuits. (If only chip capacitors are used with microstrip, there will be only one layer.) We have shown each metal layer consisting of an adhesive layer such as any of the category 2 metals (Table 3), a good conductor metal, and often a combination of all three categories. For instance, systems such as Ti-Au, Ta-Ti-Au, Ti-Mo-Au, Cr-Au, and Cr-Cu-Au have been used.

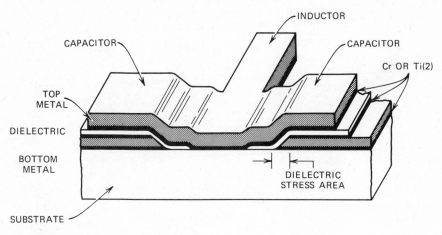

Fig. 23. Three-layer, metal-dielectric-metal sandwich for microwave integrated circuits.

Chrome-gold introduces problems because diffusions of the two metals at high temperatures causes the formation of an intermetallic state with very high resistance. However, there is no one metal technology that has been accepted to the exclusion of all others.

A fabrication procedure is illustrated in Fig. 24. The starting substrate in Fig. 24a has the layers of chrome-copper-chrome deposited in one vacuum evaporation. A photoresist etch defines bottom layers of capacitors, leaving chrome on top, and grounding connections without chrome. Then SiO_2 is deposited and defined for the capacitors, using a thick photoresist to protect the metal sides from being etched. (By redefining the top metal layout, and keeping all grounds on top, it is possible to avoid etching the SiO_2.) After densifying the oxide, a thin top layer of metal is deposited, a thick photoresist form is prepared, and the gold is selectively plated, using the top metal layer as a continuous electrode. The thin layer of copper uncovered is then etched to form the circuits. Note that this leaves gold as the exposed metal.

It is now appropriate to examine the question of monolithic versus hybrid integration. Monolithic techniques usually involve a semiinsulating, semiconducting substrate to provide resistive isolation. For microwave ICs the resistivity of the substrate should be much greater than 1000 Ω-cm for good circuit performance. Active devices are fabricated in pockets by growing them epitaxially in the insulating material. Silicon substrates have not been successful; it was not possible to maintain resistivity during processing. On the other hand, GaAs does have the required characteristics and should be successful, once

$$S_{11} = \frac{b_1}{a_1}\bigg|_{a_2=0} \tag{17}$$

$$S_{22} = \frac{b_2}{a_2}\bigg|_{a_1=0} \tag{18}$$

Similarly, it may be shown that $|S_{21}|^2$ is then the power gain or loss of the device (power out of port 2, $|b_2|^2$, divided by the power incident into port 1, $|a_1|^2$). The S_{12} is a feedback term, important for stability.

The matrix representations derived from Fig. 25 depend on definitions involving open or short circuits on the device terminals. Scattering parameter representation involves matched loads and is particularly attractive for microwave use, since these parameters, reflection and transmission coefficients, are the quantities measured at microwave frequencies. By avoiding the use of open- and short-circuits references that can cause device instability and are difficult to achieve at microwave frequencies, S-parameters with 50-Ω references are a better representation. These parameters work very well for small-signal circuits and are amenable for computer circuit design. The measurement of the four or six (if a three-port system) S-parameters allows the automatic computation of the possible gain conditions for maximum stability. A circuit design for these devices using the S-parameters and the computer has become standard.

Figure 26 is a photograph of a wideband 40-dB gain, small-signal amplifier (20 mW), available from Hewlett-Packard, which incorporates thin-film circuits. It is designed to provide stable gain over the large frequency range by careful measurement of the device S-parameters as a function of frequency. This circuit uses both lumped and distributed elements.

The measurement of small-signal quantities is presently useless to provide the characteristics of high power nonlinear devices operating class B and C. In these cases it is necessary to measure the device parameters (usually impedances) under dynamic conditions at the desired power level, as well as at the frequencies of interest. Figure 27 is a typical system. The tuners are used to optimize the impedances the device sees under the desired operating conditions. The device jig is removed and the impedances looking into the input and output circuits are then measured. The impedance values are then transformed back through the measured electrical lengths of line (l_1, l_2, Fig. 27) to the device planes. A circuit can then be synthesized to realize (in distributed or lumped form) the conjugate of these impedances over the desired frequency band and operating power level. This empirical technique is both time-consuming and does not readily lend itself to computer analysis or circuit design.

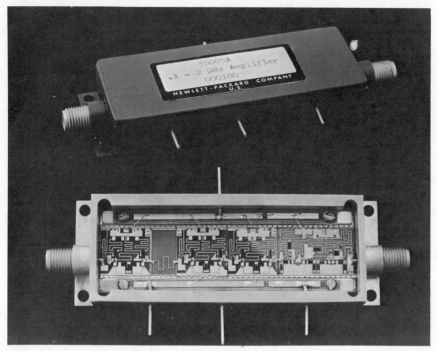

Fig. 26. Photograph of wide-band amplifier. Circuitry is 0.5×2.3 in. The amplifier provides linear phase, 40-dB gain, and over 20-mW output over the band. Both lumped and distributed networks were constructed on sapphire using a tantalum, gold technology. (Courtesy of George Bodway, Hewlett-Packard, Inc.)

In both techniques the desired device plane from which the design circuit takes over must be clearly defined. The reference plane for the measurement of the device parameters must be carefully determined, so that parasitic reactance of the characterization jig and the final circuit can be accounted for.

Figure 28 illustrates the empirical design of a common-base transistor which is characterized with the measured input impedance Z_{in}

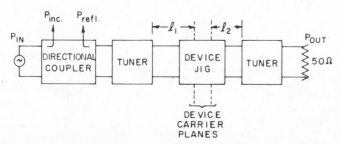

Fig. 27. Empirical characterization of devices under operating conditions.

Fig. 29. Single-stage lumped element 2.25 GHz 1-W amplifier. The circuit is 115 × 115 mils and uses TA7487 (2N5470) transistors. Microstrip lines are bonded to the input and output.

block, also contributed to the good performance. This amplifier and the 1-W amplifier shown previously are examples of microwave integrated circuits and devices in which design curves do not exist. One cannot characterize these transistor chips by linear S-parameters because the desired source and load impedances change with power. Only an empirical technique has so far been successful. The proper design of microwave integrated circuits would be facilitated by the characterization of high-power devices. At the higher frequencies and higher powers, any characterization is further muddled by the mounting and bonding techniques. Here the minimization of parasitics becomes very important.

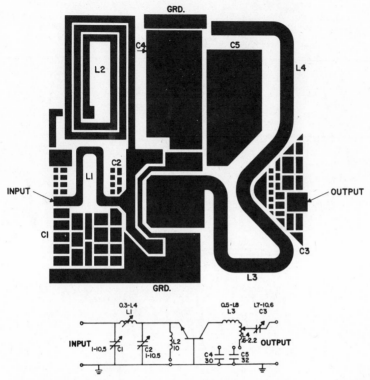

Fig. 30. Circuit diagram and metallization pattern for the amplifier of Fig. 22. Ground plane or metal extends under capacitors. The input is shunt C–series, L–shunt C network with spiral rf choke.

11.6.3 Directional Couplers

Other important circuits in the microwave region are directional couplers, filters, and power splitters. In the microstrip and coplanar version the directional coupler and band pass filter make use of edge coupled lines, as shown in Fig. 33. In the past, design data based on the theory of line properties often led to incorrect results. Recently, some definitive measurement techniques have been developed, and the problems are now solvable.[28] For this reason, we shall discuss the properties of edge-coupled lines and the problems of forming a directional coupler.

Figure 34 shows a cross section of the coupled lines with their associated fields. There are two modes of propagation (as shown), (a) an even mode in which the electric fields are the same direction under both lines, and (b) an odd mode in which the electric fields are in

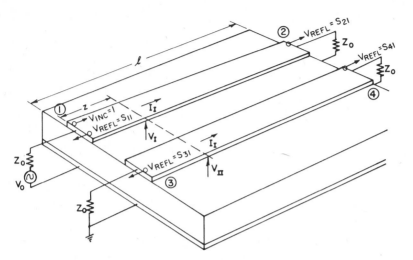

Fig. 33. Edge-coupled microstrip lines and S-parameters.

$$S_{41} = 0 \qquad (22)$$

The reflection coefficient ρ is

$$\rho = \left| \frac{Z_0 - Z}{Z - Z_0} \right| \qquad (23)$$

where Z is either Z_{0e} or Z_{00}, and Z_0 is the characteristic impedance of an uncoupled microstrip. The coupling coefficient in dB from port 1 to port 3 is

$$\mathrm{dB} = 20 \log_{10} \left| S_{31} \right| = 20 \log_{10} \frac{2\rho}{1 + \rho^2} = 20 \log_{10} \frac{Z_{0e} - Z_{00}}{Z_{0e} + Z_{00}} \qquad (24)$$

As an example, if we want a 10-dB coupler, $\mathrm{dB} = -10$, $Z_0 = 50\ \Omega$, $Z_{0e} = 69.5\ \Omega$ and $Z_{00} = 36\ \Omega$, since for a matched system $Z_0^2 = Z_{00} Z_{0e}$. Thus it is important to be able to choose the correct values of Z_{00} and Z_{0e} to produce a device employing coupled lines. We have illustrated a directional coupler, but these choices hold for in-line resonators and filters. In the in-line resonator one line (open at both ends) acts as an absorption cavity coupled to the other lines.

If the odd and even mode phase velocities differ, S_{41} is not zero. It is shown[28] that the relative power out of port 4 is

$$S_{41} = \frac{\pi}{4} \left(\frac{1 - \rho^2}{1 + \rho^2} \right) \Delta \qquad (25)$$

where

$$\Delta = \frac{v_0 - v_e}{v_e} \qquad (26)$$

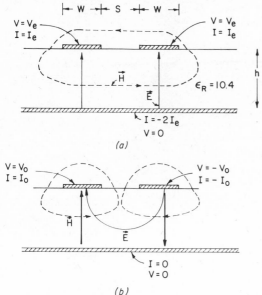

Fig. 34. Electric and magnetic fields of coupled microstrip. (a) Even mode; (b) odd mode.

is the relative difference in phase velocity.

The directivity (power out of desired port 3 relative to that of un-desired port 4) is

$$\left|\frac{S_{31}}{S_{41}}\right|^2 = \left(\frac{4}{\pi(1 - |S_{31}|^2)} \frac{|S_{31}|}{\Delta}\right)^2 \tag{27}$$

The difference in phase velocities causes a decrease in directivity.

Thus, for the design of circuit functions using coupled lines, one needs to know the even and odd mode impedances, since these determine the coupling (Eq. 24). On the other hand, the differences in even- and odd-mode phase velocities destroy the directivity (Eq. 27). Napoli and Hughes[28] have measured these parameters directly by separately exciting the even and odd modes. The data indicate that there is of the order of 15% difference in the even- and odd-mode wavelengths for both lightly coupled (10-dB) and tightly coupled (5-dB) lines. This limits the directivity possible, especially with lightly coupled lines. Figure 35 (from Napoli and Hughes[28]) shows how the phase velocity difference varies with coupler directivity and coupling coefficient.

There are several ways to reduce the difference in even- and odd-mode phase velocities. One way[29] is to place a slot in the ground plane, located symmetrically with respect to the pair of coupled microstrip-lines. The slot interrupts the transverse (odd-mode) current in the ground plane, increases the odd-mode inductance slightly, and thus re-

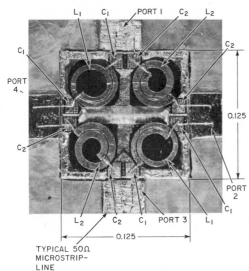

Fig. 37. Photograph of quadrature hybrid. $L_1 = 3.60$ nH, $L_2 = 2.57$ nH, $C = C_1 + C_2 = 3.48$ pF, $C_1 = 1.44$ pF, $C_2 = 2.04$ pF, and $f = 2.2$ GHz. (Courtesy of S. P. Knight, RCA Corporation).

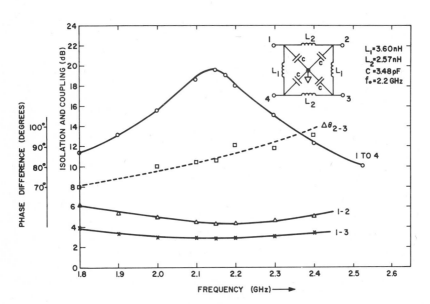

Fig. 38. Performance of quadrature hybrid.

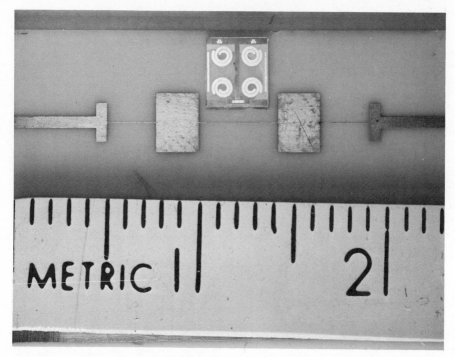

Fig. 39. Photograph of LE low-pass filter on microstrip one. (Courtesy of B. Hershenov and L. S. Napoli, RCA Laboratories.)

gold. The transistors are connected to the circuit by beam leads.

An example of a lumped element hf tuning network for power devices is shown in Fig. 40. This is an impedance transformer (designed using tables of Matthaei[30]) to transform 50 Ω to about 3 Ω. Figure 41 shows the top metalization mask layout and circuit schematic. The transformers were designed to be back-to-back, with two pairs of L and C elements, matched with input and output 50-Ω loads. The low-impedance plane is the dotted line in Fig. 41. This combination forms a wideband Chebyshev filter, centered at 6 GHz. The tuning elements of the photograph indicate close to a 20:1 transformation.

The transformer, tuned as shown in the photograph, has a 1.5-GHz, 3-dB bandwidth, centered at 6 GHz. This is not quite as wide as the theoretical 3-GHz bandwidth expected. (The element values were imperfectly adjusted.) The transmission loss at midrange was predicted to be of the order of 0.6 dB, using element Q values, as described earlier. This compared well with the measured loss of 0.7 dB, showing that the Q of the circuit, tuned as shown, agrees with

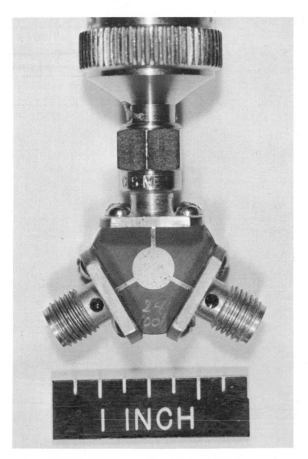

Fig. 42. Microstrip circulator. (Courtesy of B. Hershenov, RCA Laboratories.)

tors have Q's that are lower. The high-Q LE capacitors and inductors have a performance comparable to that of distributed circuits up to C-band. Capacitors and short-ribbon inductors are lumped up through S-band. At 10 GHz the Q's of small (less than 1-pF) capacitors are greater than 100. Small inductors are all that are needed here; their performance holds up better at higher frequencies, since they remain lumped.

As described earlier, microstrip circuits are dispersive at higher frequencies; therefore, thinner substrates, which cause higher losses, are needed, so that the lower end of the shaded distributed Qs apply at the higher frequencies. Lumped elements appear to be practical to 10 GHz.

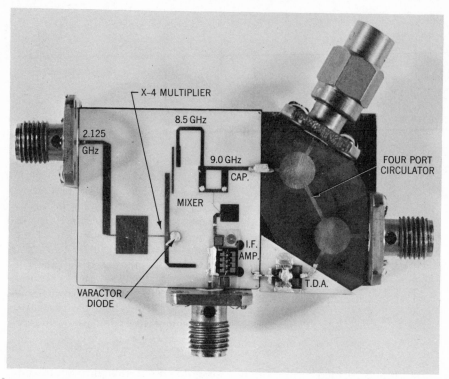

Fig. 43. A 9.0-GHz receiver. (1.5 × 1.0-in. alumina and garnet) Minimum noise figure 6.4 dB. The IF amplifier is a lumped element R and C 0.160 × 0.140 in. on sapphire. (Courtesy of L. S. Napoli, RCA Laboratories).

There are still problems in microwave integrated circuits. At the higher frequencies many active devices are available in the laboratory but the technology to integrate is not. Monolithic technology is almost a necessity at 50 GHz and above; GaAs is a strong contender. In addition, lumped element circuits await further exploitation of their possibilities. For instance, computer-aided programs for microstrip circuits are available. While most of the work on microwave computer-aided design has used lumped element models, a complete accurate characterization of thin-film lumped elements at microwaves was not available until recently. Thus it is now possible to develop a comprehensive characterization and layout of thin-film passive components, including parasitics, and place the design on a basis more comparable with that of microstrip.[35] As noted earlier, characterization of devices for microwave integrated circuits is not yet solved, especially for higher-power devices.

From a technological point of view, one sees many paths to follow. As we become more experienced, we see that the metals used are be-

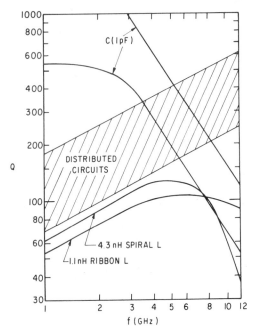

Fig. 44. Performance of microwave integrated circuit elements versus frequency. The capacitor Q_S are measured limits for 1-pF, densified-SiO_2 capacitors. The inductor Qs can be improved according to theory; capacitor Q_S vary as $1/C$. The microstrip Q_S vary with substrate thickness and Z_0. Capacitors made from distributed lines have Q_S higher than that shaded; inductors so made have Q_S lower.

coming more refined. Simple chrome-gold is not enough. Sputtered Ta-Ti-Au or Cr-Cu-Au, Ta-Ti, Ta-Mo, Pt-Au, and Ti-W-Au metalization will become prevalent. Tantalum technology is being increasingly used for resistors. No one apparently has a technique that stands out above the others in performance and economics. There are strong indications however, that the thick-film techniques described earlier are not suitable for microwave circuits.[36]

We have attempted to survey the field of microwave integrated circuits, showing a lot of methods that work, but without any strong conclusions on what is the one best way to proceed. It will be interesting to see what improvements develop in the 70s which should be the decade of MIC systems.

ACKNOWLEDGMENTS

The author is indebted to the representatives of various industrial laboratories who have made photographs and drawings of their work available to him. Credit is given either in the figure captions or text to all work outside of RCA Laboratories. In general, all photographs without credit captions represent work performed at RCA. In addition some of this material has appeared in an article by the author and H. Sobol, published in the *Journal of Solid-State Circuits*.[37]

REFERENCES

1. F. Assadourian and E. Rimai, "Simplified Theory of Microstrip Transmission Systems," *Proc. IRE*, 40, 1651 (December 1952).
2. G. C. Bandy, L. J. Hardeman, and W. F. Hayes, "MERA Modules—How Good in an Array?" *Microwaves*, 8, 39–49 (August 1969).
3. R. S. Englebrecht and J. W. West, "Microwave Integrated Circuits," *Bell Laboratories Record*, 44, 328–333 (October/November 1966).
4. S. G. Cohn, "Slot-Line—An Alternative Transmission Medium for Integrated Circuits," IEEE G-MTT International Microwave Symposium Digest, 104–109 (May 1968).
5. C. P. Wen, "Coplanar Waveguide, A Surface Strip Transmission Line Suitable for Nonreciprocal Gyromagnetic Device Applications," *IEEE Trans. Microwave Theory Techniques*, MTT-17, 1087–1096 (December 1969).
6. C. P. Wen, "Coplanar-Waveguide Directional Couplers," *IEEE Trans. Microwave Theory Techniques*, MTT-18, 318–322 (June 1970).
7. F. C. DeRonde, "A New Class of Microstrip Directional Couplers," 1970 *G-MTT Int. Symposium Digest*, 184–189 (May 12, 1970).
8. M. Caulton, "The Lumped Element Approach to Microwave Integrated Circuits," *Microwave J.*, 13, 51–58 (May 1970).
9. H. Sobol, "Technology and Design of Hybrid Microwave Integrated Circuits," *Solid-State Technology*, 13, 49–57 (February 1970).
10. R. N. Patel, *"Microwave Conductivity* of Thick-Film Conductors," *Electronics Lett.* 6, No. 15, 455 (July 23, 1970).
11. N. Goldsmith and W. Kern, "The Deposition of Vitreous Silicon-Dioxide Films from Silane," *RCA Rev.*, 28, 153–165 (March 1967).
12. J. M. C. Dukes, "An Investigation into Some Fundamental Properties of Strip Transmission Lines with the Aid of an Electrolytic Tank," *Proc. IEE (London)*, 103B, 319 (1956).
13. T. T. Wu, "Theory of the Microstrip," *J. Appl. Phys.*, 28, 299 (1957).
14. H. A. Wheeler, "Transmission-Line Properties of Parallel Wide Strips by a Conformal-Mapping Approximation," *IEEE Trans. Microwave Theory Techniques*, MTT-12, 280 (May 1964).
15. H. A. Wheeler, "Transmission-Line Properties of Parallel Strips Separated by a Dielectric Sheet," *IEEE Trans. Microwave Theory Techniques*, MTT MTT-13, 172 (March 1965).
16. M. Caulton, J. J. Hughes, and H. Sobol, "Measurements on the Properties of Microstrip Transmission Lines for Microwave Integrated Circuits," *RCA Rev.*, 27, 377–391 (September 1966).
17. R. A. Pucel, D. J. Masse, and C. P. Hartwig, *IEEE Trans. Microwave Theory Techniques*, MTT-16, 342–350 (June 1968).
18. H. Sobol, "Extending IC Technology to Microwave Equipment," *Electronics*, 112–124 (March 20, 1967).
19. A. Presser, "RF Properties of Microstrip Line," *Microwaves*, 53 (March 1968).
20. G. I. Zysman and D. Varon, "Wave Propagation in Microstrip Transmission Lines," *IEEE-G-MTT Int. Microwave Symposium Digest* (3-9) (May 3–7, 1969).

21. L. S. Napoli and J. J. Hughes, "High Frequency Behavior of Microstrip Transmission Lines," *RCA Rev.*, 30, 268—276 (June 1969).

22. M. Caulton, S. P. Knight, and D. A. Daly, "Hybrid Integrated Lumped–Element Microwave Amplifier," *IEEE Trans. Electronic Devices*, ED-15, 459—466 (July 1968).

23. R. DeBrecht and M. Caulton, "Lumped Elements in Microwave Integrated Circuits in the 1 to 12 GHz Range," 1970 *IEEE G-MTT Int. Symposium Digest*, 14—18 (May 11, 1970).

24. G. D. Alley, "Interdigital Capacitors for Use in Lumped-Element Microwave Integrated Circuits," 1970 *IEEE G-MTT Int. Symposium Digest*, 7—13 (May 11, 1970).

25. S. Ramo, J. R. Whinnery, and T. V. VanDuzer, *Fields and Waves in Communication Electronics*, John Wiley & Sons, New York, 1965, pp. 592, 603.

26. Hewlett-Packard Application Note 95, "S-Parameters—Circuit Analysis and Design," September 1968.

27. K. Kurokawa, "Power Waves and the Scattering Matrix," *IEEE Trans. Microwave Theory Techniques*, 194—202 (March 1965).

28. L. S. Napoli and J. J. Hughes," Characteristics of Coupled Microstrip Lines," *RCA Rev.* (September 1970).

29. L. S. Napoli, private communication.

30. G. L. Matthaei, "Tables of Chebyshev Impedance-Transforming Networks of Low-Pass Filter Form," *Proc. IEEE*, 52, 939 (1964).

31. D. R. Taft and G. H. Robinson, "MIC Ferrite Devices," Symposium on Microwave Integrated Circuits (Digest), Goddard Space-Flight Center (NASA), Greenbelt, Md., March 10, 1970.

32. B. Hershenov, "Microstrip Junction Circulator for Microwave Integrated Circuits," *IEEE Trans. Microwave Theory Techniques*, MTT-15, No. 12, 748—750 (December 1967).

33. L. Napoli and J. Hughes, "Low-Noise Integrated X-Band Receiver," *Microwave J.*, 11, 37—42 (July 1968).

34. A. Botka, J. Bunker, and M. Gilden, "Integrated X-Band Radar Receiver Front End," *Microwave J.*, 11, 65—71 (July 1968).

35. M. Caulton, B. Hershenov, S. P. Knight, and R. E. De Brecht, "Status of Lumped Elements in Microwave Integrated Circuits—Present and Future," *IEEE Trans. Microwave Theory Techniques*, MTT-19, 588—599 (July 1971).

36. M. Caulton, "Film Technology in Microwave Integrated Circuits," *Proc. of IEEE*, Vol. 59, 1481-1489 (October 1971).

37. M. Caulton and H. Sobol, "Microwave Integrated-Circuit Technology—A Survey," *IEEE J. Solid-State Circuits*, SC-5, 292—303 (December 1970).

Chapter 12

MAGNETIC MATERIALS

Peter J. Wojtowicz

RCA Laboratories
Princeton, New Jersey

The subject of magnetic materials is very extensive, and many new and exciting developments have appeared on a wide front. Since it would be impossible to include all new advances, a narrow selection of topics had to be chosen. The selection of these topics was based on several criteria: high current interest and activity, the likelihood of eventual use in engineering applications, and the author's personal interests. The emphasis of this chapter will be on *materials*, their magnetic and other physical properties, and the phenomena on which new devices and systems are likely to be based. The discussion will center on those materials that offer the possibility of optimizing the performance of devices and systems utilizing magnetic phenomena. The chapter is divided into three independent sections, describing magnetooptic materials, materials for use in domain-wall devices, and future trends in magnetic material development. The last section emphasizes materials in which strong interactions between the magnetism and other physical properties exist; the magnetic semiconductors and materials for the photomagnetic effect are featured.

The reader who is more seriously interested in the current status of magnetic material research and development is encouraged to examine the spring issues of the *Journal of Applied Physics* (where the proceedings of the annual Conference on Magnetism and Magnetic Materials are published) and the fall issues of the *IEEE Transactions of Magnetics* (where the proceedings of the annual Intermag Conference are published). Many excellent review articles appear each year, summarizing the most recent advances in both magnetics research and technology. Several of these fine articles have, in fact, formed the basis of many of the discussions to follow.

12.1 MAGNETOOPTIC MATERIALS

The invention and development of the laser has prompted considerable activity in the application of optical phenomena and techniques in computer and communications technologies. Magnetooptic materials and devices will play central roles in the realization of optical computers and optical communications. In computer applications magnetooptic materials will be used to store information which will be read and/or written by interaction with light. In optical communication devices magnetooptic materials will manipulate light at high frequencies by means of interactions with rapidly varying magnetizations. The kinds of devices envisioned include memory systems, visual displays and magnetic holograms; light modulators, isolators, rotators, switches, and beam deflectors.

12.1.1 Magnetooptic Phenomena

A brief review of the most important magnetooptic effects will be presented before beginning our considerations of the relevant materials. The discussion will closely follow that given in recent survey articles by Freiser[1] and Dillon.[2]

The most common of the magnetooptic phenomena is the Faraday effect, the rotation of the plane of polarization of linearly polarized light upon propagation through a magnetized material. The effect results from the existence of circular magnetic birefringence[2]; for light propagating parallel to the magnetization, the two normal modes of propagation are right and left circularly polarized waves, each with its own different index of refraction, n_+ and n_-, respectively. Now, when linearly polarized light travels along the direction of magnetization in a reasonably transparent crystal, it is resolved into the two normal modes. These move through the medium with different velocities, so that when they emerge from the crystal they will recombine with altered relative phase and produce a rotated linear polarization. The phase difference corresponds to the rotation per unit thickness which is

$$\varphi_F = \left(\frac{\omega}{2c}\right)(n_+ - n_-)$$

where ω is the optical angular frequency, and c the velocity of light. In most cases the total rotation, R, varies as the projection of the magnetization on the direction of propagation:

$$R = t\varphi_F(\omega)\frac{(\vec{M}\cdot\vec{1})}{|\vec{M}|}$$

where t is the thickness of the sample, and φ_F the rotation per unit thickness. The remainder gives the relative projection of the magnetization $\vec{M}$ on $\vec{1}$, a unit vector in the propagation direction. This important relation includes the antireciprocity property of the Faraday rotation: if the direction of $\vec{M}$ is reversed, the sign of R changes.

The rotation per unit length may be further related to the elements of the dielectric tensor $\vec{\epsilon}$.[1,2] The form of $\vec{\epsilon}$ for a medium that is optically isotropic, except for a distortion along the magnetization direction (here the z axis), is given by

$$\vec{\epsilon} = \begin{pmatrix} \epsilon_\perp & -i\epsilon_2 & 0 \\ i\epsilon_2 & \epsilon_\perp & 0 \\ 0 & 0 & \epsilon_\parallel \end{pmatrix}$$

The ordinary real and imaginary optical constants come from the diagonal elements, while the rotation depends on the off-diagonal elements. The imaginary parts of the elements give rise to absorption. The complex rotation can be written in terms of the complex refractive indexes and matrix elements:

$$\varphi_F{}^c = \varphi_F' = i\varphi_F''$$

$$= \left(\frac{\omega}{2c}\right)(n_+^c - n_-^c) = \left(\frac{\omega}{2c}\right)\left(\frac{\epsilon_2}{n^c}\right)$$

where n^c is the average refractive index, $(n_c^+ + n_c^-)/2$. If, in addition, there exists a difference in the attenuation (absorption) for the two senses of circular polarization (the circular magnetic dichroism), an originally linear polarized wave will emerge elliptically polarized. Thus the real part of ϵ_2 provides the rotation, while the imaginary part is responsible for the ellipticity.

The rotation of the plane of polarization of linearly polarized light upon reflection from the surface of a magnetic medium is called the Kerr effect. Three configurations may be distinguished.[1] In the polar Kerr effect the magnetization $\vec{M}$ is normal to the reflecting surface. In the longitudinal Kerr effect, $\vec{M}$ is parallel to the surface and is in the plane of incidence; in the equatorial (or transverse) Kerr effect, $\vec{M}$ is parallel to the surface and is perpendicular to the plane of incidence. For normal incidence the two latter effects are indistinguishable. The Kerr effect has been a useful tool in the study of the ferromagnetic metals but has not been widely considered in technological applications. Thus, other than recounting these definitions, we shall continue to concentrate on the magnetooptic effects in transmission.

12.2.2 Magnetooptic Materials

The principal requirements on materials to be used in magneto-optic applications and devices are high-specific Faraday rotation, accompanied by low absorption. In addition, the Curie temperature (the temperature below which spontaneous magnetization appears) should be above room temperature, unless the operation of the devices at low temperatures is permitted. Other requirements must also be met, depending on the specific applications being considered. Memory applications, for example, require square loop properties; low magnetizations, so that demagnetizing fields do not interfere with the operation of the material in open structures; coercive fields applicable to switching; etc. High-frequency light modulation, on the other hand, requires desirable microwave properties such as low resonance line width, and so on. In most all applications, however, the common figure of merit will be Faraday rotation per unit attenuation (usually expressed in units of degrees per decibel).

The temperature and wavelength dependence of the Faraday rotation has been measured for a large number of magnetic materials. Many of these results are available not only because of the search for technically interesting magneto-optic materials but because the Faraday effect is an important tool in sorting out the spectroscopy of the electronic states of magnetic substances. Table 1 gives a short compilation of Faraday rotation data on several different classes of magnetic materials; the absorption coefficients α and the figures of merit are included. Figures 1 and 2 display the wavelength dependence of the Faraday rotation for a number of different kinds of materials.[2,3] Figures 3 through 5 depict the wavelength dependence of the figures of merit, again for a wide selection of different material categories.[2-4] The ferromagnetic metals, iron, cobalt, nickel, and MnBi, have the

TABLE 1. FARADAY ROTATION DATA FOR MAGNETIC MATERIALS[a]

Material	$\lambda(m\mu)$	T(°K)	ϕ_F (degree/cm)	$\alpha\,(cm^{-1})$	$0.23\ \phi_F/\alpha$ (degree/dB)
Fe	546	300	3.5×10^5	7.6×10^5	0.11
Co	546	300	3.6×10^5	8.5×10^5	0.09
Ni	400	300	7.2×10^5	2.1×10^5	0.79
YIG	1200	300	240	0.069	800
GdIG	600	300	2500		
CrBr$_3$	493	1.5	2×10^5	2×10^3	23
EuSe	755	4.2	1.4×10^5	45	720

[a]After Freiser.[1]

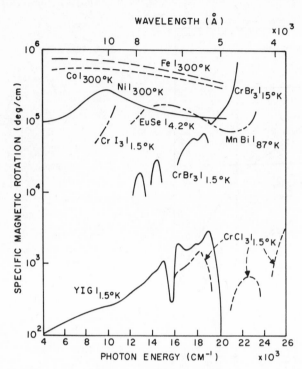

Fig. 1. Wavelength dependence of the Faraday rotation of several magnetic materials. (After Dillon.[2])

highest specific rotations on the order of 10^5 to 10^6 degree/cm. The europium chalcogenides and the chromium trihalides also have high specific rotations (to 10^5 degree/cm) but suffer from having Curie points well below room temperature. The oxidic ferrites such as yttrium and gadolinium iron garnets (YIG and GdIG) and lithium ferrite have lower specific rotations of 10^2 to 10^3 degree/cm. The same is true of other compounds having light anions such as the fluorides $RbFeF_3$ and $RbNiF_3$ (10^2 to 10^3 degree/cm). The comparison of the figures of merit, however, completely changes the order of desirability of these materials. The metallic ferromagnets, iron, cobalt, nickel, and MnBi, being highly absorptive, give figures of merit that are quite low, on the order of 10^{-2} to 10^{-1} degree/dB. The oxidic ferrites provide figures of merit of about 10^{-1} to 10 degree/dB, with YIG giving up to 10^3 degree/dB in the infrared. The chromium trihalides are quite transparent and have figures of merit reaching 50 degree/dB. Extremely high values of the figure of merit are achieved in the europium chalcogenides (to 10^4 degree/dB) but, again, the Curie points are very

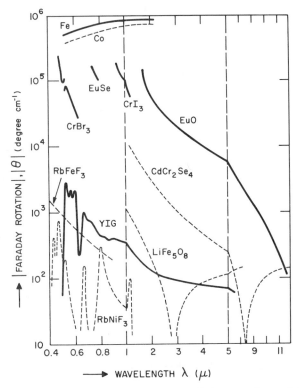

Fig. 2. Wavelength dependence of the Faraday rotation of several magnetic materials. (After Bongers.[3])

low. In the infrared $CdCr_2Se_4$ has an appreciable figure of merit (10 to 10^2 degree/dB) but has a Curie point of only 130 °K.

Figure 5 contains the figures of merit for three materials only recently examined for magnetooptic applications: FeF_3, $FeBO_3$, and $LuFeO_3$. While the figures of merit are quite good in the visible (1 to 10 degree/dB), these materials all suffer a common disadvantage: their low symmetry crystal structures admit a large birefringence. In the presence of birefringence the Faraday rotation is no longer proportional to the sample thickness but becomes a periodic function of optical path length.[4,5] The maximum Faraday rotation attainable is then severely limited; in $FeBO_3$ and FeF_3 the maximum achievable rotation for green light is only 0.7 and 0.3 degree, respectively.[4]

12.2.3 Guidelines for New Materials

To be technologically acceptable in device applications, a magneto-optic material must possess at least the following properties:

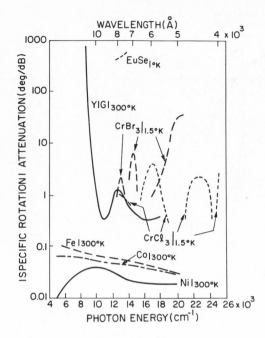

Fig. 3. Wavelength dependence of the Faraday figure of merit (rotation per unit attenuation) of several magnetic materials. (After Dillon.[2])

1. Curie temperature above room temperature.

2. Large specific Faraday rotation.

3. As low absorption as possible in the frequency range of operation.

4. Absence of birefringence.

5. Low values of the spontaneous magnetization in order to permit operation in planar or other open geometries without suffering from demagnetizing effects.

The various mechanisms that contribute to both the Faraday rotation and the absorption have been extensively studied by many investigators. The correlation of the properties of many known materials, coupled to the theoretical elucidation of the mechanisms of rotation and absorption, have provided several important suggestions to aid in the search for new and improved magnetooptic materials. The current understanding of these materials has been summarized by Bongers[3] and by Wolfe, Kurtzig, and Le Craw.[4] Using this information, these authors have derived a list of guidelines for optimizing the performance of magnetooptic materials. These follow in the same order as the material requirements listed above.

1. High Curie temperatures are easily achieved in the metallic ferromagnets; these are generally unsuitable, however, because of

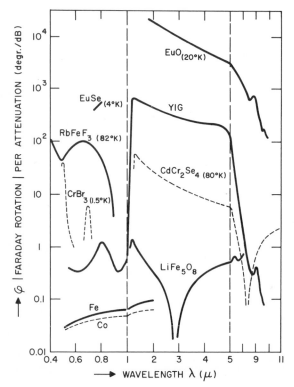

Fig. 4. Wavelength dependence of the Faraday figure of merit (rotation per unit attenuation) of several magnetic materials. (After Bongers.[3])

their low transparency and high values of magnetization. Experience with insulating materials shows, on the other hand, that high Curie points can be obtained in compounds of the $3d$ transition elements, those containing Fe^{3+} in particular. As examples we have the rare earth iron garnets with Curie points in the vicinity of 550 °K, the rare earth ortho-ferrites with Curie temperatures in excess of 600 °K, and iron-based halides such as FeF_3 with a Curie point of 363 °K.

2. The major large contributions to the Faraday effect in insulating magnetic materials come from optical absorptions arising from electronic transitions having large orbital splittings. These can be optical transitions between an orbital singlet ground state and an excited state that is split by spin-orbit coupling and the exchange field; or transitions from a ground state that already has an orbital moment. In either case the Faraday rotation (at frequencies far from the frequency of the relevant optical transition) can be shown to be approximately proportional to the transition probability of the electronic ab-

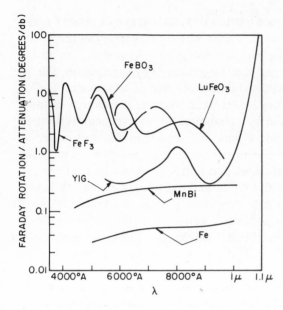

Fig. 5. Wavelength dependence of the Faraday figure of merit (rotation per unit attenuation) of several magnetic materials. (After Wolfe, Kurtzig, and Le-Craw.[4])

sorption, to the magnetization, and (for orbital singlet ground states) to the strength of the spin-orbit coupling in the excited state.

In most of the magnetic compounds of interest (containing Fe^{3+}, Cr^{3+}, and Eu^{2+} ions), the magnetic ion has a singlet ground state and the large Faraday rotations arise from strongly allowed transitions with large spin-orbit splitting in the excited state. The first strongly allowed absorption in these cases is usually a charge transfer transition of an electron from the anion to the magnetic cation. It is then the spin-orbit coupling on the anion that causes the orbital splitting of the excited state. Heavy anions with their large spin-orbit coupling will therefore give rise to large Faraday rotations. Reference to Table 1 and to Figs. 1 and 2 shows that compounds with heavy anions such as Br^-, I^-, and Se^{-2} do indeed have large specific Faraday rotation, while compounds containing the lighter anions such as F^- and O^{-2} have smaller rotations. In the case of the europium compounds the rotation results from transitions within the Eu^{2+} ion (electronic transitions from the $4f$ to $5d$ shell with large spin coupling in the excited state). Rare earth cations with their large spin-orbit interactions would therefore appear promising if incorporated into compounds with high Curie point (such as the garnets or the orthoferrites; GdIG in Table 1 is an example).

3. The metallic magnets are generally too absorbing to be useful, except possibly as very thin films. The insulating magnetic compounds,

on the other hand, usually have a region of transparency between optical absorptions, due to lattice vibrations at long wavelengths and charge transfer transitions at short wavelengths. Transparency in the infrared (and higher frequencies) can be achieved by depressing the frequencies of the optical phonons through the use of compounds containing heavy ions of low valency. The frequency of the charge transfer absorption can be increased by enhancing the electronegativity difference between the cations and anions. Charge transfer will occur at lower frequencies, the larger and more highly charged the anions and the smaller and more highly charged the cations. For use in the visible and near-infrared regions fluorides and oxides are certainly superior, as seen in Figs. 3 through 5. Relatively weak d-d transitions within the transition metal cations generally cause absorptions of highly variable intensity within the visible and infrared. These transitions are not associated with the Faraday effect mechanism and can be minimized by the proper choice of the transition metal cation. It is these kinds of absorption, for instance, which are responsible for the oscillations in the figures of merit of FeF_3, $FeBO_3$, and $LuFeO_3$, shown in Fig. 5.

4. Since birefringence limits the maximum attainable Faraday rotation, as mentioned above, cubic crystals or materials whose optic axis corresponds to the direction of magnetization are desirable.

5. Low values of the magnetization are realizable in ferrimagnets having compensation points, in composition-compensated ferrimagnets and in weak ferromagnets. In ferrimagnets with compensation points, there exists a narrow temperature range in which the magnetizations of the opposed nonequivalent sublattices essentially cancel each other, providing low total magnetizations. If the Faraday effect originates predominantly within one sublattice (and this sublattice moment is high near compensation), then the rotation can be quite large even at vanishing total magnetization; GdIG is one such example. In a composition-compensated ferrimagnet, the composition of the opposed nonequivalent sublattices is chosen such that the net magnetization at $0°K$ vanishes. Then, at higher temperatures, as the two sublattice magnetizations decrease unequally, a small and slowly varying net moment will develop over a wide temperature range. A hypothetical example would be a garnet with equimolar amounts of Fe^{3+} in both the tetrahedral and octahedral sublattices. The weak ferromagnets derive their low magnetizations from a basically antiferromagnetic spin arrangement, accompanied by a canting of the opposed sublattices into a common direction. The rare earth orthoferrites, FeF_3 and $FeBO_3$, are examples of this type.

12.2 MATERIALS FOR DOMAIN-WALL DEVICES

12.2.1 Cylindrical Domains in Rare Earth Orthoferrites

The rare earth orthoferrites ($RFeO_3$, where R is any rare earth element or yttrium) are a class of magnetic materials that have received wide scientific attention over the past decade or so. The many investigations of these materials have been reviewed recently by White.[6] All the orthoferrites crystallize with the orthorhombic structure with $a < b < c$; the Curie temperature are all in excess of 600 °K. The magnetic structures are generally quite complex and undergo one or more structure changes with varying temperature, especially at low temperatures. At room temperature the magnetic structure is basically antiferromagnetic with a slight canting of the two antiparallel sublattices, giving a small (weak ferromagnetic) moment. The uniaxial anisotropy axis at room temperature is along the (001) direction in all the orthoferrites, except $SmFeO_3$, where it is the (100) direction.

Within the past several years, however, the orthoferrites have been studied extensively for their application in memory devices of various kinds. Technological interest in these materials began when it was found that the magnetic properties were appropriate to the generation and manipulation of magnetic domains of particularly simple and useful geometry in thin plates.[7,8] Workers at the Bell Telephone Laboratories have recently demonstrated that small cylindrical domains of diameter comparable to the plate thickness can be generated in thin plates if a stabilizing bias field is applied perpendicular to the plate.[7,8] The basic geometry is illustrated in Fig. 6. A thin plate of orthoferrite is cut so that the easy axis of magnetization is perpendicular to the plane of the plate. If a large saturating bias field were applied along the easy axis (in the down direction, say), the entire sample would be uniformly magnetized along the direction of the applied field.

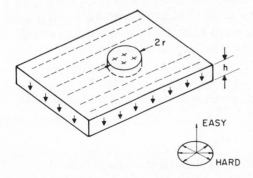

Fig. 6. Basic geometry of cylindrical domains in thin plates. (After Bobeck et al.[8])

For applied bias fields of lower magnitude, however, stable domains of reverse magnetization with the geometry indicated can be easily introduced. The magnetization within these cylindrical domains is also uniform but is directed opposite to that of the bulk of the specimen (in the up direction in this case). The very narrow region between the domains of oppositely directed magnetization is called the domain wall. The magnetization within the domain wall is strongly nonuniform, varying rapidly from all up on one side to all down on the other; the domain wall is therefore a region of relatively high energy density. As will be discussed later, $2r \approx h$ (the diameter is roughly equal to the thickness).

The cylindrical domains, popularly called "bubble" domains, can be injected into the plate or generated by the sectioning of an existing domain into halves. The domains can be further propagated about the plate by the use of local control fields, and thus permit memory and logic functions. The presence, or passage, of domains can be detected by means of the Faraday effect or by Hall and/or induced voltage readout. Using the orthoferrites, cylindrical domains can be supported to densities approaching $10^6/\text{in.}^2$ The nucleation fields in orthoferrites are quite high, so that domains do not appear spontaneously, and low-domain-wall coercivities and high-domain-wall mobilities are observed. Domains can, in fact, be moved 1 diam in less than 100 nsec, indicating that data rates in excess of 10^6 bits/sec may be achieved.

12.2.2 Properties of Cylindrical Domains

A complete understanding of the relationship between intrinsic material parameters such as the domain wall energy σ_w and the saturation magnetization $4\pi M_s$, and the properties of cylindrical domains has been obtained by Thiele.[9] A summary of the main results of this theory has been given by Bobeck et al.[8]; our presentation closely follows this excellent review.

The basic geometry of the cylindrical domains is shown in Fig. 6. Such cylindrical domains are stable under two counteracting forces. One is the combined force of the applied bias field and the domain wall energies that act to reduce the domain volume and the domain wall area, respectively. The second force derives from the magnetostatic energy which acts to increase the surface area (and hence the volume) of the domains. The stability of a cylindrical domain is achieved when the following condition is satisfied:

$$H_A + H_w - H_D = 0$$

where H_A is the applied bias field, H_w is the effective field contributed by the domain wall energy, and H_D can be interpreted as the average vertical component of magnetostatic field (demagnetizing field) present at the domain wall surface (the shell of the cylinder). The wall field H_w is given by $\sigma_w/2rM_s$ and is the controlling field at small r. The magnetostatic field H_D (quite complicated and not reproduced here) varies less rapidly with r and has a maximum of $4\pi M_s$.

Figure 7 illustrates the interplay between these two quantities and the applied bias field. When H_A is zero only a single intersection of H_w and H_D (intersection a at $r = r_a$) exists. Although this intersection satisfies the stability condition, it represents an unstable solution. In the presence of a finite bias field H_A, however, a second intersection b is present; the corresponding domain size r_b is stable. With increasing bias field the two intersections approach each other. The stable and unstable solutions run together at a critical value of H_A, giving a minimum radius r_c. Increasing the bias field beyond this value causes the domains to collapse inward and disappear. For small values of H_A, there is a critical value of r_b ($r_b \approx 3r_c$), beyond which the cylindral domains become unstable with respect to elliptical perturbations. Long serpentine strip domains become the stable configuration.

A material parameter l_d has been introduced which proves extremely helpful in comparing and optimizing the pertinent properties of materials to be used in domain wall devices. A material length is defined as

$$l_d = \frac{\sigma_w}{4M_s^2}$$

The theory of cylindral domain stability provides the following con-

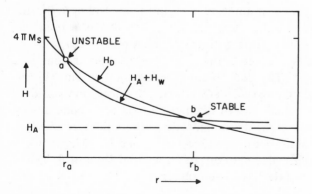

Fig. 7. Graphical solution of the stability condition for cylindrical domains. (After Bobeck et al.[8])

TABLE 2.   MATERIAL PROPERTIES AND DOMAIN CHARACTERISTICS OF RARE EARTH ORTHOFERRITES, $RFeO_3$ [a]

R	$4\pi M_s$, G	$2r$, mils	H_A, Oe	Thickness, mils	l_d, mils	σ_w, erg/cm^2
Y	105	3.0	33	3.0	2.15	1.8
Nd	62	7.5	3.2	2.0	4.4	1.1
Sm	84	6.0	3.0	1.1	2.9	1.3
Eu	83	5.5	10.5	2.0	3.7	1.6
Gd	94	3.7	16	2.4	2.9	1.7
Tb	137	1.7	51	2.2	1.4	1.7
Dy	128	2.0	32	1.6	1.7	1.8
Ho	91	4.5	12	2.1	3.3	1.7
Er	81	6.0	8	2.0	3.9	1.6
Tm	140	2.3	37	2.3	1.9	2.4
Yb	143	3.8	41	3.0	3.0	3.9
Lu	119	7.5	10.5	2.0	4.3	3.9
$Sm_{0.6}Er_{0.4}$	83	1.0	33	1.8	0.80	0.35
$Sm_{0.55}Tb_{0.45}$	108	0.75	61	2.0	0.40	0.30

[a]After Bobeck et al.[8]

clusions: (a) the diameter of the smallest stable domain realizable is 1.2 l_d; (b) the plate thickness h which yields the smallest domain diameter is l_d; and (c) the bias field required to maintain cylindral domains at optimum thickness is $H_A \approx 0.3 (4\pi M_s)$.

A comparison of the material properties and the domain characteristics of the various rare earth orthoferrites is presented in Table 2. The values of $4\pi M_s$ runs from about 60 to 140 G. The calculated wall energies σ_w run between about 1 and 4 erg/cm^2. The values of l_d vary between 1.4 and 4.4 mils. By utilizing plate thicknesses of 1 to 3 mils, cylindrical domains of diameters ranging from 1.7 to 7.5 mils have been observed (in bias fields, ranging from 3 to 50 Oe). In agreement with the theory, the materials with the highest magnetizations show the smallest domains. Table 2 also lists two mixed rare earth orthoferrite compositions. The Sm-Tb composition is particularly interesting: by mixing $TbFeO_3$ and $SmFeO_3$ (having different crystallographic directions for the easy axis of magnetization, and hence anisotropy constants of opposite sign) the magnitude of the uniaxial anisotropy can be reduced, providing lower values of σ_w and l_d. These in turn give domains of much reduced size; in an applied field of 61 Oe, domain diameters of 0.75 mils have been observed.

12.2.3 Materials Requirements

The material requirements of cylindrical magnetic domain devices have been considered in detail by Gianola et al.[10] To produce usable domain devices, a material and its domains must possess at least the following desirable properties: (a) the domains are stable in a plate of uniform thickness; (b) the magnetization is constrained to lie in a direction perpendicular to the plate by uniaxial anisotropy; (c) the domain wall width is small in comparison to the domain diameter; (d) the wall motion coercivity is sufficiently small so that the domain size and shape are independent of coercivity; (e) materials with low magnetostriction coefficients are desired to reduce the coupling between the domains and the host lattice (thereby increasing the domain mobility); and (f) the magnetic ordering temperature should be above room temperature and the material properties should be reasonably temperature-insensitive at room temperature. In addition, if optical detection of the domains is contemplated, the Faraday rotation per unit absorption should be reasonably high and the birefringence low.

Many of these requirements can be summarized in terms of the anisotropy field H_k and the saturation magnetization $4\pi M_s$. In order for domains not to self-nucleate, the nucleation field must be greater than the combined demagnetizing field ($4\pi M_s$) and the applied bias field (some fraction of $4\pi M_s$). In uniaxial crystals (and to a rough approximation in more complex structures) this condition will be satisfied if $H_k > 4\pi M_s$. This relation also satisfies conditions a to c above. It can further be shown that the domain wall mobility will be highest when H_k is just slightly larger than $4\pi M_s$, and decreases as H_k becomes very much larger.

There are a number of different classes of magnetic materials that will support isolated cylindrical domains covering a wide range of diameters. The most desirable domain size from the current device point of view is 1 to 10 μ diameter. The combined material requirements indicate the need for $4\pi M_s$ values in the low hundreds of gauss; the anisotropy field H_k should then be just slightly larger than this. The rare earth orthoferrites described above fulfill most of the requirements (including $H_k > 4\pi M_s$) at room temperature. The lowest possible magnitude of H_k can be achieved in that temperature range in which the uniaxial anisotropy is changing from c-axis-oriented to a-axis-oriented. This reorientation temperature can be adjusted to be near room temperature by altering the composition of the rare earth orthoferrites. Mixtures of any of the rare earth orthoferrites with $SmFeO_3$ (which alone is a-axis-oriented at room temperature) can provide suitable materials. As already shown in Table 2, $Sm_{0.55}Tb_{0.45}FeO_3$ has the smallest domain size seen in the orthoferrites; other mixed orthoferrites of this type giving similar performance include $Sm_{0.5}Gd_{0.5}FeO_3$, $Sm_{0.5}Y_{0.5}FeO_3$, $Sm_{0.6}Er_{0.4}FeO_3$, and $Sm_{0.6}Dy_{0.4}FeO_3$.[10-13] The adjustment of the reorientation temperature to be near room temperature introduces other problems, however; the material properties become temperature-sensitive at room temperature.

Many hexagonal ferrimagnetic materials exist, in which the anisotropy is strongly uniaxial with the hexagonal axis the easy axis of magnetization; the condition $H_k > 4\pi M_s$ is easily satisfied. Most of these materials, moreover, have room temperature values of $4\pi M_s$ that are quite large, giving cylindrical domains of diameters as small as 0.3 μ. Materials such as the magnetoplumbites $MFe_{12}O_{19}$ (M is Ba, Pb, or Sr), $BaFe_{18}O_{27}$, and the intermallic MnBi have $H_k \approx 10^4$ Oe with $4\pi M_s \gtrsim 4000$ G. In several of the magnetoplumbites the values of $4\pi M_s$ have been reduced by substituting aluminum for the iron. The aluminum substitution also reduces the anisotropy field, but the relative change is not as great. These adjustments increase the domain size to approximately 10 μ.[10] The substitution of $Ti_{0.5}Co_{0.5}$ for iron in the magnetoplumbites reduces the anisotropy without appreciably altering the magnetization; domain diameters of about 1 μ are seen.[10] The aluminum and the $Ti_{0.5}Co_{0.5}$ substitutions also have the effect of reducing the Curie point which makes the properties of these compositions somewhat temperature-dependent.

Cubic materials are generally inapplicable to cylindrical domain devices, since they are not uniaxial. In addition the anisotropy of the more well known cubic materials such as the spinels and garnets is usually not large enough to give $H_k > 4\pi M_s$. Cylindrical domains have been observed in gadolinium iron garnet,[14] but such observations are attributed to induced (rather than intrinsic) uniaxial anisotropy.

In a very recent development, however, workers at the Bell Telephone Laboratories have discovered[15,16] that many nominally cubic flux grown garnet crystals possess uniaxial magnetic anisotropy. A large number of mixed, rare earth iron garnets have been grown and examined and found to be satisfactory for cylindrical domain device applications. The most promising materials are those in which the magnetostriction has been deliberately minimized by carefully choosing the rare earth ion composition. Examples of mixed rare earth iron garnets that have uniaxial magnetic anisotropy and near-zero magnetostriction are $Er_2TbAl_{1.1}Fe_{3.9}O_{12}$, $Gd_{0.94}Tb_{0.75}Er_{1.31}Al_{0.5}Fe_{4.5}O_{12}$, and $Gd_{2.31}Tb_{0.60}Eu_{0.09}Fe_5O_{12}$. Platelets of flux-grown garnets, having the compositions $Gd_{2.34}Tb_{0.66}Fe_5O_{12}$ and $Er_2TbAl_{1.1}Fe_{3.9}O_{12}$ cut to 0.6 mils thick,[15] have been operated as shift registers at storage densities in excess of 10^6 bits/in.^{2} These compositions have saturation magnetizations of about 100 to 200 G, uniaxial anisotropy fields of about 10^3 Oe, and display domains of 0.25 mils diameter.

The origin of the uniaxial anisotropy in the otherwise cubic garnets has not yet been firmly established. The uniaxial anisotropy is believed to result from chemical inhomogeneities introduced in the crystals during the growth from flux. It has been suggested that,[16] in mixed rare earth garnets, the growth mechanism proceeds in such a way as to provide alternate layers of material that are first rich in one rare earth element and then rich in another, and so on. Such a layered chemical inhomogeneity would then necessarily exhibit a uniaxial magnetic anisotropy whose axis is perpendicular to the layering and thus parallel to the growth direction, as is indeed observed.[15, 16] In any case it is quite clear at this time that the mixed rare earth garnet materials will be prominently featured in future cylindrical domain devices.

12.3 FUTURE TRENDS IN MAGNETIC MATERIALS

At the present time the major trend in basic research on magnetic materials emphasizes the investigation of the interconnection of the magnetism with other solid state properties.[17] A parallel trend is also evident in the development of magnetic materials for applications: the emphasis lies in the utilization of materials that display strong interactions between the magnetic behavior and other solid state phenomena. The prime example of this trend is the investigation and application of magnetic materials that display significant magnetooptic effects. Other examples derive their interest from interactions between the magnetism, on the one hand, and microwaves, elastic properties, electrical transport properties, dielectric properties, on the other hand. In this section we shall briefly consider two classes of materials in which strong interactions with other properties may prove to be of future technological importance. These materials are (a) the magnetic semiconductors, and (b) materials exhibiting photomagnetic effects.

12.3.1 Magnetic Semiconductors

Several groups of materials have recently been discovered to be simultaneously ferromagnetic and semiconducting. The interest in these compounds increased substantially when numerous experiments revealed that the electronic states of these materials were strongly coupled to the degree of magnetic order. Now, while donor and acceptor concentrations, traps, and scattering are all fixed in a given semi-

conductor and cannot be altered, the magnetic state can be changed by
the application of magnetic fields of moderate magnitude. Thus, just
as ordinary semiconductors possess sensitive and nonlinear reactions
to applied electric forces, the magnetic semiconductors will, in addition, possess sensitive and nonlinear reactions to applied magnetic
forces.[18] The existence of the interaction between the electronic states
and the magnetism offers the possibility of constructing electronic devices having magnetic control.

The first group of magnetic semiconductors to receive extensive
scientific attention were the europium chalcogenides. The properties
of these interesting materials have been comprehensively reviewed by
Methfessel and Mattis.[18] The parent compounds, EuO, EuS, and EuSe,
have the simple NaCl crystal structure and are insulating ferromagnets with Curie points of approximately 70 °K, 17 °K, and 7 °K, respectively. The substitution of trivalent rare earth elements for the europium has been found to decrease the resistivity by as much as 12 orders
of magnitude. The magnetic properties are then observed to be strongly influenced by the increases in the carrier concentration. Most notable are the increases in Curie temperature brought about by the enhancement of the magnetic exchange interactions via the increased electrical conductivity. As examples, a 20% substitution of gadolinium for
europium in EuS raises the Curie point from 17 to 34 °K, while a 4%
substitution of gadolinium for europium in EuO raises the Curie point
from 70 to 135 °K.

The most dramatic manifestations of the strong interaction between
the magnetism and the electronic properties are the existence of large
changes in the resistivity at temperatures near the Curie point and the
existence of giant magnetoresistance effects. Examples of these phenomena are illustrated in Figs. 8 and 9. Figure 8 depicts the resistivity-temperature characteristic for the composition $Eu_{0.95}La_{0.05}S$, as
measured in zero field and in several large applied magnetic fields.
In the absence of an applied field the resistivity is seen to increase by
seven orders of magnitude in the vicinity of the Curie point (where rapid changes in the degree of magnetic ordering are occurring). Upon
the application of magnetic fields, giant negative magnetoresistances
are observed; at temperatures near the Curie point the application of
a 14 kOe field reduces the resistance by three orders of magnitude.
Figure 9 displays the same behavior for the composition $Eu_{0.99}Gd_{0.01}Se$,
where the effects are somewhat larger; in the vicinity of the Curie
point the resistance decreases by about five orders of magnitude upon
the application of a 13.5 kOe field. A description of the physical
mechanisms responsible for these phenomena is beyond the scope of
this article; the various possibilities are discussed in the reviews of

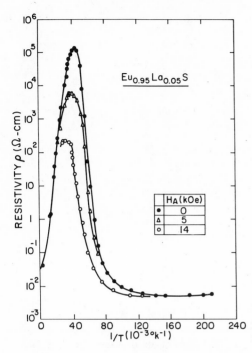

Fig. 8. Resistivity versus reciprocal temperature for the composition $Eu_{0.95}La_{0.05}S$ in several different applied magnetic fields. (After Methfessel and Mattis[18] and von Molnar.[19])

Methfessel and Mattis[18] and von Molnar.[19]

Another group of magnetic semiconductors that has been extensively studied is the chromium chalcogenide spinels. The properties of these materials have recently been reviewed by Wojtowicz.[20] The most important of these materials, $CdCr_2S_4$, $CdCr_2Se_4$, and $HgCr_2Se_4$ have the spinel crystal structure and are insulating ferromagnets with Curie points of 84.5 °K, 129.5 °K, and 106 °K, respectively. The related spinels $ZnCr_2S_4$ and $ZnCr_2Se_4$ are antiferromagnetic below 20 °K, while $HgCr_2S_4$ is a metamagnet below 36 °K. The magnetizations of the three ferromagnetic compounds are clse to the six Bohrmagnetons/formula unit $(4\pi M_s \approx 4$ to 5 kG) expected from a true ferromagnetic alignment of the Cr^{3+} moments. The studies of the semiconductor properties have largely concentrated on $CdCr_2Se_4$. Undoped $CdCr_2Se_4$ is usually observed to be a p-type semiconductor. Small substitutions of monovalent elements such as silver, copper, or gold for the cadmium increase the p-type conductivity, while substitutions of trivalent elements such as gallium or indium for the cadmium render the crystals n-type. The temperature and magnetic field dependencies of the electrical transport properties are found to be strikingly different in n-type and p-type materials. In n-type crystals the conduction electrons interact strongly with the magnetism; the mobilities tend to be low (~1 to

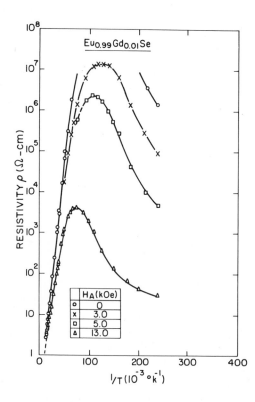

Fig. 9. Resistivity versus reciprocal temperature for the composition $Eu_{0.99}Gd_{0.01}Se$ in several different applied magnetic fields. (After Methfessel and Mattis[18] and von Molnar.[19])

20 cm²/V-sec) and most of the electrical transport properties display anomalous behavior in those temperature ranges in which the magnetic order is undergoing significant changes. In p-type crystals, however, the holes apparently interact only weakly with the magnetism; the mobilities tend to be higher (~ 20 to 700 cm²/V-sec), and the transport properties resemble those of conventional semiconductor materials.

An example of the anomalies observed in the transport properties of n-type materials is shown in Fig. 10. Displayed here is the resistivity-temperature characteristic of n-type $CdCr_2Se_4$: In. The prominent feature is the reduction of the resistivity by four orders of magnitude as the sample becomes ferromagnetic below its Curie point of about 130 °K. Large anomalies are also observed in the Seebeck effect, the Hall effect, and the magnetoresistance in the temperature range immediately near the Curie point. Just as in the europium salts, the magnetoresistance of n-type $CdCr_2Se_4$ is negative and peaks at the Curie point: $\Delta\rho/\rho_0 \approx -0.9$ for a field of 10 kOe at 130 °K.

The most dramatic evidence for the influence of the magnetic ordering on the electronic states of the chromium chalcogenide spinels is the temperature and magnetic field dependence of the optical absorption

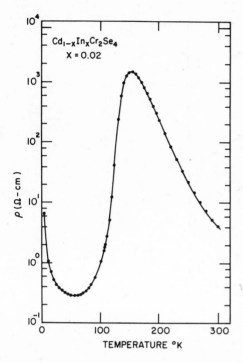

Fig. 10. Resistivity versus temperature for the n-type material $Cd_{1-x} In_x Cr_2 Se_4$ with $x \approx 0.02$. (After Amith and Gunsalus.[21])

edge or bandgap. All the chalcogenide spinels investigated have displayed significant large shifts in the optical absorption edge in those temperature ranges in which the magnetic ordering is setting in. Figure 11 shows the temperature dependence of the optical absorption edge of $CdCr_2Se_4$. The absorption edge at room temperature is 1.32 eV. On lowering the temperature, the edge first shifts to higher energies (as in a conventional semiconductor), reaching 1.35 eV at 190 °K. Upon further reducing the temperature, the absorption edge begins to shift to lower energies. The shift is gradual at first, while the short-range magnetic order is developing, then becomes very pronounced at and below the Curie point, where the long-range magnetic order develops. A quadratic extrapolation of the high-temperature data (as expected for a conventional semiconductor and shown as the dotted line in Fig. 11) would put the low temperature edge at 1.43 eV if the magnetism had no influence. The effect of the magnetic ordering on the band edge is thus estimated to be 0.27 eV. Further shifts to lower energies are observed on applying external magnetic fields: at the Curie point a field of 8.5 kOe lowers the band edge an additional 0.007 eV. Such observations provide sufficient evidence that strong interactions between the electronic states and the localized magnetic moments do

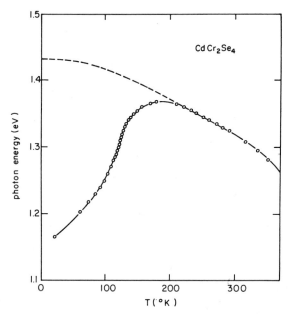

Fig. 11.   Temperature dependence of the optical absorption edge of CdCr$_2$Se$_4$. (After Harbeke and Pinch.[22])

indeed exist.   The possibilities for modulating and controlling the electrical properties through the alteration of the magnetic state therefore appear very real.

12.3.2 Photomagnetic Effect

In the magnetooptic phenomena described in Section 12.1.1, the magnetism acts to alter the quality of optical radiation passing through or reflecting from a magnetic material.   In the photomagnetic effects, on the other hand, optical radiation is observed to influence the magnetic properties of a material.   The several photomagnetic effects and the materials that display them have recently been reviewed by Enz *et al.*[23]

Photomagnetic effects have been found at low temperatures in both Si-doped, yttrium iron garnet ($Y_3Fe_{5-x}Si_xO_{12}$) and in Ga-doped, cadmium chromium selenide ($Cd_{1-y}Ga_yCr_2Se_4$).   At high levels of substitution ($0.1 < x < 0.3$) the photomagnetic effect consists of the significant alternation of the magnetocrystalline anisotropy of a crystal upon optical illumination.   At lower levels of doping ($x < 0.05$; $y \approx 0.015$), the chief manifestations of the photomagnetic effect are the reduction of the permeability and the enhancement of the coercive force upon irradia-

tion.   Light of wavelengths shorter than 1.3 μ is required; the effects
depend on the product of the light intensity and irradiation time.   The
mechanism of the photomagnetic effect involves light-induced electron
transfers, resulting in the redistribution of the Fe^{2+} (or Cr^{2+}) ions
among the background of Fe^{3+} (or Cr^{3+}) ions.

Experiments on the photomagnetic effects involving the magnetocrystalline anisotropy have been performed on Si-doped, yttrium iron
garnet having the nominal composition $Y_3Fe^{3+}_{4.8}$ $(Si^{4+}Fe^{2+})_{0.1}O_{12}$.   The
Fe^{2+} ions are assumed to reside in the octahedral sites of the garnet
lattice.   There are four inequivalent octahedral sites in the structure,
each distinguished by the particular (111) direction along which its local trigonal axis lies.   The Fe^{2+} ions (or equivalently the extra electrons
that distinguish Fe^{2+} ions from the background of Fe^{3+} ions), residing
in those sites whose local trigonal axes lie nearest to the magnetization
direction, have a somewhat lower energy than those in the other three
sets of sites.   At temperatures at which electron migration is possible,
these sites are preferentially occupied by the Fe^{2+} ions (the extra electrons).   The magnetization is thus stabilized in this initial direction;
this stabilization may be viewed in terms of providing an additional contribution to the anisotropy of the material.

In the photomagnetic experiments the single-crystal sample is
cooled in the dark to 20 °K, with the magnetization constrained to the
(111) direction by a large applied field.   The magnetization is stabilized
in this direction by the migration of the Fe^{2+} ions (extra electrons) into
the appropriate subset of octahedral sites.   The resulting distribution
of the Fe^{2+} ions is, at low temperatures, almost entirely frozen.   The
contribution to the magnetocrystalline anisotropy provided by this
"magnetic annealing" process can be measured by the ferromagnetic
resonance technique; alterations in the anisotropy caused by the photomagnetic effect can likewise be monitored by ferromagnetic resonance.
Figure 12 illustrates the ferromagnetic resonance results for one such
photomagnetic experiment.   After cooling to 20 °K in a large field applied along the (111) direction, ferromagnetic resonance (at 9.4 GHz)
is observed at a field value of about 2.5 kOe as indicated by H_1 in the
figure.   The applied field (and the magnetization with it) is then rotated
into the $(11\bar{1})$ direction (which is no longer equivalent to the original
(111) direction because of the "magnetic anneal").   The field for resonance is now found to have increased to about 2.7 kOe.   The contribution to the anisotropy field from the "magnetic anneal" is thus about
200 Oe.   The field for resonance (and hence the induced anisotropy) is
then monitored in time, first in the dark (curve A) and then under IR
irradiation (curve B).   Even in the dark there is a slight decrease in the

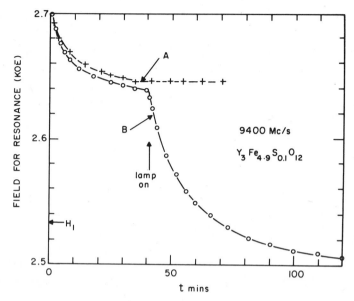

Fig. 12. Time dependence of the field for resonance in photomagnetic experiments on a $Y_3Fe_{4.9}Si_{0.1}O_{12}$ single crystal. (After Enz *et al.*[23])

field for resonance with time corresponding to a relaxation of the "magnetic anneal" by the purely thermal redistribution of the Fe^{2+} ions. Upon IR irradiation, however, the Fe^{2+} ions (extra electrons) are provided with sufficient energy to migrate completely into those octahedral sites whose local trigonal axes are in the $(11\bar{1})$ direction; the field for resonance slowly returns to the value H_1, indicating that the irradiation has effectively reoriented the original induced anisotropy from the (111) into the $(11\bar{1})$ direction.

The photomagnetic effects involving the permeability and the coercive force are observed in samples with much lower concentrations of dopant elements: $Y_3Fe_{5-x}Si_xO_{12}$ with $x < 0.05$ and $Cd_{1-y}Ga_yCr_2Se_4$ with $y \approx 0.015$. The effects are seen in both single-crystal window frames and polycrystalline toroids. The influence of optical irradiation on the permeability is most pronounced in the polycrystalline toroids. Figure 13 illustrates the results of experiments on a polycrystalline toroid of composition $Y_3Fe^{3+}_{4.99}(Se^{4+}Fe^{2+})_{0.005}O_{12}$. The figure shows the time dependence of the permeability in response to irradiation at $77°K$. After cooling to $77°K$ in the dark, followed by a-c demagnetization, the permeability is found to be quite high at 120. Upon illumination with white light the permeability is observed to decrease rapidly to about 10 in a few seconds. For temperatures up to $100°K$ the permeability

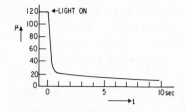

Fig. 13. Time dependence of the permeability in photomagnetic experiments on a polycrystalline toroid of composition $Y_3Fe_{4.995}Si_{0.005}O_{12}$. (After Enz et al. [23])

remains low after the removal of the light. It cannot, moreover, be returned to its original value, either by demagnetization or by a combination of strong transverse fields plus illumination. At higher temperatures thermal recovery takes place.

Light-induced changes in the coercive force are also observed. Measurements of the hysteresis loop of a polycrystalline toroid of composition $Y_3Fe^{3+}_{4.988}(Si^{4+}Fe^{2+})_{0.006}O_{12}$ at $77\,°K$ reveal that the coercive force can be increased from a dark value of 0.6 to 2.0 Oe after illumination. The shape of the hysteresis loop is also modified; the loop becomes considerably more square after illumination. In high-speed switching experiments, similar increases in the threshold field and switching speed were found after irradiation. These latter effects may be technologically important; in cycling the hysteresis loop in the dark state with a field just slightly larger than the coercive force, the switching between the two senses of magnetization can be suppressed by means of a light pulse.

The mechanism behind the permeability and coercive force variety of photomagnetic effects must be different from that controlling the induced anisotropy behavior. Not only is there a considerably reduced amount of Fe^{2+} present but the permeability and coercive force effect are found to be irreversible and independent of the magnetization distribution of the sample during illumination. A model that can explain, in principle, the effects at low-dopant concentrations is the following: in a slightly doped yttrium iron garnet the Si^{4+} ions are few and far apart. An equivalent amount of Fe^{2+} ions occupy the octahedral sites, along with the predominant Fe^{3+} ions. In addition to distinguishing between octahedral sites having different trigonal axis directions, it is now also necessary to differentiate between octahedral sites next to, or far away from, sites occupied by Si^{4+} ions, type I and type II sites, respectively. Because of the electrostatic interaction between Fe^{2+} and Si^{4+} ions (excess negative and positive charges, respectively, within the background of the Fe^{3+} lattice) Fe^{2+} ions in type I sites have a lower energy than those in type II sites. Thus, after cooling, type I sites should be preferentially occupied by Fe^{2+} ions. Upon illumination, light-induced electron transitions from type I Fe^{2+} ions to remote type II Fe^{3+} ions occur creating type II Fe^{2+} ions. If one now assumes that the contribution of type I Fe^{2+} ions to the magnetic properties is in some way different from that of type II Fe^{2+} ions, then the observed behavior is readily explained.

The observations of the photomagnetic effects in $CdCr_2Se_4$: Ga are similar to those described above. Here Ga^{3+} plays the role of the Si^{4+}, while Cr^{2+} ions among the dominant Cr^{3+} behave in much the same way as the Fe^{2+} ions in the garnets. Possibilities for the application of photomagnetic effects exist in the areas of radiation detection, recording with light, and information storage.

REFERENCES

1. M. J. Freiser, "A Survey of Magneto-optic Effects," *IEEE Trans. Magnetics*, MAG-4, 152 (1968).
2. J. F. Dillon, "Origin and Uses of the Faraday Rotation in Magnetic Crystals," *J. Appl. Phys.*, 39, 922 (1968).
3. P. F. Bongers, "Faraday Rotation of Infrared and Visible Light by Magnetic Materials," *IEEE Trans. Magnetics*, MAG-5, 472 (1969), and private communication.
4. R. Wolfe, A. J. Kurtzig, and R. C. LeCraw, "Room Temperature Ferromagnetic Materials Transparent in the Visible," *J. Appl. Phys.*, 41, 1218 (1970).
5. W. J. Tabor and F. S. Chen, "Electromagnetic Propagation through Materials Possessing Both Faraday Rotation and Birefringence: Experiments with Ytterbium Orthoferrite," *J. Appl. Phys.*, 40, 2760 (1969).
6. R. L. White, "Review of Recent Work on the Magnetic and Spectroscopic Properties of Rare-earth Orthoferrites," *J. Appl. Phys.*, 40, 1061 (1969).
7. A. H. Bobeck, "Properties and Device Applications of Magnetic Domains in Orthoferrites," *Bell System Tech. J.*, 46, 1901 (1967).
8. A. H. Bobeck, R. F. Fischer, A. J. Perneski, J. P. Remeika, and L. G. Van Uitert, "Application of Orthoferrites to Domain-Wall Devices," *IEEE Trans. Magnetics*, MAG-5, 544 (1969).
9. A. A. Thiele, "The Theory of Cylindrical Magnetic Domains," *Bell System Tech. J.*, 48, 3287 (1969).
10. U. F. Gianola, D. H. Smith, A. A. Thiele, and L. G. Van Uitert, "Material Requirements for Circular Magnetic Domain Devices," *IEEE Trans. Magnetics*, MAG-5, 558 (1969).
11. F. C. Rossol, "Temperature Dependence of Rare-earth Orthoferrite Properties Relevant to Propagating Domain Device Applications," *IEEE Trans. Magnetics*, MAG-5, 562 (1969).
12. R. C. Sherwood, L. G. Van Uitert, R. Wolfe, and R. C. LeCraw, "Variation of the Reorientation Temperature and Magnetic Crystal Anisotropy of the Rare-earth Orthoferrite," *Phys. Lett.*, 25A, 297 (1967).
13. R. D. Pierce, R. Wolfe, and L. G. Van Uitert, "Spin Reorientation in Mixed Samarium-Dysprosium Orthoferrites," *J. Appl. Phys.*, 40, 1241 (1969).

14. C. D. Mee, "The Magnetization Mechanism in Single Crystal Garnet Slabs near the Compensation Temperature," *IBM J. Res. Develop.*, <u>23</u>, 468 (1967).

15. A. H. Bobeck, E. G. Spencer, L. G. Van Uitert, S. Abrahams, R. Barns, W. H. Grodkiewicz, R. C. Sherwood, P. Schmidt, D. H. Smith, and E. M. Walters, "Uniaxial Magnetic Garnets for Domain Wall 'Bubble' Devices," *Appl. Phys. Lett.*, <u>17</u>, 131 (1970).

16. L. G. Van Uitert, W. A. Bonner, W. H. Grodkiewicz, L. Pietroski, and G. J. Zydzik, "Garnets for Bubble Domain Devices," *Materials Res. Bull.*, <u>5</u>, 825 (1970).

17. H. S. Jarrett, "Potentially Useful Properties of Magnetic Materials," *J. Appl. Phys.*, <u>40</u>, 938 (1969).

18. S. Methfessel and D. C. Mattis, *Magnetic Semiconductors, Encyclopedia of Physics*, Vol. 18/1, Ed. H. P. J. Wijn, Springer-Verlag, Berlin, 1968, p. 389.

19. S. von Molnar, "Transport Properties of the Europium Chalcogenides," *IBM J. Res. Develop.*, <u>14</u>, 269 (1970).

20. P. J. Wojtowicz, "Semiconducting Ferromagnetic Spinels," *IEEE Trans. Magnetics*, <u>MAG-5</u>, 840 (1969).

21. A. Amith and G. Gunsalus, "Unique Behavior of Seebeck Coefficient in N-type $CdCr_2Se_4$," *J. Appl. Phys.*, <u>40</u>, 1020 (1969).

22. G. Harbeke and H. L. Pinch, "Magnetoabsorption in Single Crystal Semiconducting Ferromagnetic Spinels," *Phys. Rev. Lett.* <u>17</u>, 1090 (1966).

23. U. Enz, W. Lems, R. Metselaar, P. J. Rijnierse, and R. W. Teale, "Photomagnetic Effects," *IEEE Trans. Magnetics*, <u>MAG-5</u>, 467 (1969).

AUTHOR INDEX

SUBJECT INDEX